TEXT-BOOKS OF SCIENCE

ADAPTED FOR THE USE OF

ARTISANS AND STUDENTS IN PUBLIC AND SCIENCE SCHOOLS.

ELECTRICITY AND MAGNETISM.

LONDON: PRINTED BY
SPOTTISWOODE AND CO., NEW-STREET SQUARE
AND PARLIAMENT STREET

ELECTRICITY AND MAGNETISM.

BY

FLEEMING JENKIN, F.R.SS. L. & E., M.I.C.E.

PROFESSOR OF ENGINEERING
IN
THE UNIVERSITY OF EDINBURGH.

THIRD EDITION.

D. APPLETON AND CO.
NEW YORK.
1876.

INTRODUCTION.

WHEN the author was asked to write the following little treatise he acceded to the request with much pleasure, because he had long known that an elementary treatise on Electricity and Magnetism of a somewhat novel character was much needed. In England at the present time it may almost be said that there are two sciences of Electricity—one that is taught in ordinary text-books, and the other a sort of floating science known more or less perfectly to practical electricians, and expressed in a fragmentary manner in papers by Faraday, Thomson, Maxwell, Joule, Siemens, Matthiessen, Clark, Varley, Culley, and others. The science of the schools is so dissimilar from that of the practical electrician that it has been quite impossible to give students any sufficient, or even approximately sufficient, text-book. It has been necessary to refer them to disjointed treatises in the Reports of the British Association, in the 'Cambridge Mathematical Journal,' the 'Phil. Trans.' and the 'Phil. Magazine.' A student might have mastered Delarive's large and valuable treatise and yet feel as if in an unknown country and listening to an unknown tongue in the company of practical men. It is also not a little curious that the science known to the practical men was, so to speak, far more scientific than the science of the text-books. These latter contain an apparently incoherent series of facts, and it is only by some considerable mental labour that, after reading the long roll of disjointed experiments, the student can even approximately understand any one experiment in its entirety; the explanation of part of the very first phenomenon described cannot be given until one of the very last experiments has been mastered.

The author has found it quite impossible, for this very reason, to write his treatise on the ordinary plan of beginning with simple experiments and gradually building up a science by the description of a series of more and more complex phenomena. Not a single electrical fact can be correctly understood or even explained until a general view of the science has been taken and the terms employed defined. The terms which are employed imply no hypothesis, and yet the very explanation of them builds up what may be called a theory. The terms cannot be explained by mere definitions, because they refer to phenomena with which the reader is unacquainted. The mere explanation of the terms, therefore, requires some rapid description of facts, the truth of which the reader must at first take for granted. Many of the assertions cannot be proved to be true except by complex apparatus, and the action of this complex apparatus cannot be explained until the general theory has been mastered.

The plan followed in the book is therefore as follows:—First, a general synthetical view of the science has been given, in which the main phenomena are described and the terms employed explained. This general view of the science cannot be made very easy reading, although it will probably be found easier by those who have no preconceived notions about tension, intensity, and so forth, than by students of old text-books. If this portion of the work can be mastered, the student will then be readily able to understand what follows, viz., the description of the apparatus used to measure electrical magnitudes and to produce electricity under various conditions. The difference between the Electricity of schools and of the testing office has been mainly brought about by the absolute necessity in practice for definite measurement. The lecturer is content to say, under such and such circumstances, a current flows or a resistance is increased. The practical electrician must know how much current and how much resistance, or he knows nothing; the difference is analogous to that between quantitative and qualitative analysis. This measurement of electrical magnitudes absolutely requires the use of the word and idea potential, and of various units each with an appropriate name, in terms of which each electrical magnitude can be expressed. On a proper choice of units depends the simplicity of the expression for the laws which connect electrical phenomena. After describing these laws and measurements, the author has given their chief practical application to telegraphy and a few examples of the construction of telegraphic apparatus. These fluctuate in form from year to year, and the special forms now in use will soon become antiquated; but the general theory of Electricity on which the construction and use of these depends is permanent, depending on no hypothesis, and it has been the author's aim to state this general theory in a connected manner and in such a simple form that it might be readily understood by practical men.

The above introduction is allowed to stand unaltered because it correctly describes what the author aimed at. He feels that the actual book falls very far short of the ideal he had conceived; he perceives only too well that the arrangement might be very greatly improved, and the statements made in much clearer language. The book has been unfortunately written in intervals snatched from professional engagements at irregular periods, but the author would rather claim indulgence on the score that the effort made has at least been in the right direction, although far from fully successful.

He has to acknowledge having received very kind assistance from his friends Sir W. Thomson, Professor J. C. Maxwell, Mr. Culley, and Mr. C. F. Varley; as well as from three of his assistants, Mr. W. Bottomley, Mr. W. E. Ayrton, and Mr. W. F. King, who kindly examined the proofs.

Mr. Latimer Clark and Mr. Culley have allowed free use to be made of extracts from their valuable handbooks.

CONTENTS.

CHAPTER I.

ELECTRIC QUANTITY.

CHAPTER II.

POTENTIAL.

CHAPTER III.

CURRENT.

CHAPTER IV.

RESISTANCE.

CHAPTER V.

ELECTRO-STATIC MEASUREMENT.

CHAPTER VI.

MAGNETISM.

CHAPTER VII.

MAGNETIC MEASUREMENTS.

CHAPTER VIII.

ELECTRO-MAGNETIC MEASUREMENT.

CHAPTER IX.

MEASUREMENT OF ELECTRO-MAGNETIC INDUCTION.

CHAPTER X.

UNITS ADOPTED IN PRACTICE.

CHAPTER XI.

CHEMICAL THEORY OF ELECTROMOTIVE FORCE.

CHAPTER XII.

THERMO-ELECTRICITY.

CHAPTER XIII.

GALVANOMETERS.

CHAPTER XIX.

ELECTRO-STATIC INDUCTIVE MACHINES.

CHAPTER XX.

MAGNETO-ELECTRICAL APPARATUS.

CHAPTER XXI.

ELECTRO-MAGNETIC ENGINES.

CHAPTER XXII.

TELEGRAPHIC APPARATUS.

CHAPTER XXIII.

SPEED OF SIGNALLING.

CHAPTER XXIV.

TELEGRAPHIC LINES.

CHAPTER XXV.

FAULTS IN TELEGRAPHIC LINES.

CHAPTER XXVI.

USEFUL APPLICATIONS OF ELECTRICITY OTHER THAN TELEGRAPHIC.

CHAPTER XXVII.

ATMOSPHERIC AND TERRESTRIAL ELECTRICITY.

CHAPTER XXVIII.

MARINER'S COMPASS.

LIST OF TABLES.

ELECTRICITY AND MAGNETISM.

CHAPTER I.

ELECTRIC QUANTITY.

§ 1. A PIECE of glass and a piece of gutta percha, or other resinous material, after being rubbed together, will be found to attract one another slightly. One piece of resin thus rubbed repels another similarly treated piece of resin, and one piece of rubbed glass repels another piece of rubbed glass; it is also found that either the rubbed resin or the rubbed glass attracts any light body in its neighbourhood. The properties acquired by the glass or resin are not permanent.

ELECTRICITY is the name given to the supposed agent producing the described condition of bodies. It seems to have been natural to regard this agent as a kind of very subtle fluid, and the nomenclature adopted in treating of electricity is based on this idea. There has been much wrangling as to the hypotheses of one and of two fluids. It is quite unnecessary to assume that the phenomena are due to one fluid, two fluids, or any fluid whatever; but in this treatise the names employed will be chiefly those which have been suggested to men of science by thinking of electrical phenomena as due to the presence or absence of a single fluid.

The stick of resin or glass, while retaining the properties

described above, is said to be electrified or charged with electricity; it carries electricity with it if moved from place to place. If these electrified bodies are wiped with a wet cloth, a damp hand, or with metal foil, they cease to be electrified. The electricity is then said to have been conducted away and the bodies which allow it to run off the glass or resin are called conductors of electricity. Metals, water, the human body, damp wood, and many other bodies are conductors.

The air must be a non-conductor, or it would have removed the electricity as well as the wet cloth.

Similarly, the resin and glass themselves are non-conductors, for when the electrified pieces are simply laid on a conductor they do not lose all their electricity, but remain electrified for some time in those portions which are not in the immediate neighbourhood of the conductor.

Non-conductors are also called insulators. Glass, gutta percha, india-rubber, air, are examples of insulators.

§ **2**. If a small piece of metal, supported by an insulating rod, be allowed to touch the electrified piece of glass or resin, it will be found to be in an electrical condition, similar to that of the glass or resin which it has touched.

The insulated conductor which has touched the resin repels the resin itself or any other insulated conductor which may have touched the electrified resin: it may be said to be electrified as the resin was, or charged with resinous electricity; it attracts the electrified glass, or any insulated conductor electrified by the glass or charged with what is sometimes called vitreous electricity.

It follows from these experiments that part of the electricity on the resin or glass is communicated to any conductor which touches either of the bodies. The electrical properties gained by the insulated conductor electrified by contact with the electrified resin have been gained at the expense of those possessed by the resin—the resin or glass loses what the metal gains; similarly, the electrified conductor can impart a portion of its properties to another conductor, losing that which it gives. We may, then, so far as can be yet seen, with propriety speak of a conductor as carrying a certain quantity of electricity, or as being charged with that quantity. .

The insulated conductor has acquired the special properties in virtue of which the resin or glass was said to be electrified, or charged with electricity; but the insulated and electrified conductor has some peculiarities which distinguish it from a similar piece of an electrified insulator. For instance, if the conductor be touched by the hand, or by the point of a wire held in the hand of a man not himself insulated, it will lose all its electricity in a time so short as to appear inappreciable; whereas the insulator can only lose its electricity gradually, when every part of its surface has been successively touched.

We may also expect that if from any cause the distribution of electricity in a body can be varied, even without its total amount being changed, this redistribution will take place almost instantaneously in the electrified conductor, and much more slowly in the electrified insulator.

§ **3**. The force exerted (other things being equal) by the electrified body on another similar body in its neighbourhood, is found to depend on the *quantity* of electricity. If I halve the quantity, distributing that electricity over two equal balls, which was previously contained on one, the force exerted by the electricity on each ball will, under any given circumstances, be halved. It is in virtue of this force only, that we have known the ball to be electrified, and we may therefore, with propriety, speak of the quantity of electricity on each ball after the redistribution, as half that on the first ball originally.

Resin and glass have been chosen as two typical materials, but any two different insulators rubbed together behave more or less as resin and glass do; thus relatively to a stick of shellac or resin, flannel behaves as a piece of glass would do.

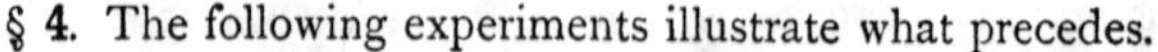

§ 4. The following experiments illustrate what precedes.

FIG. 1.

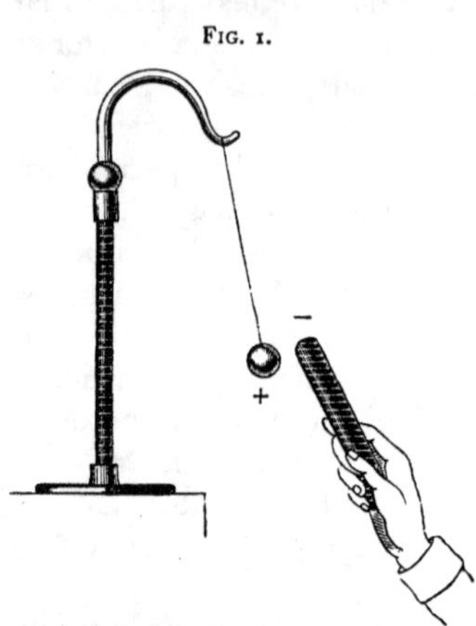

Suspend a pith ball by a silk thread (Fig. 1): pith, in order that the ball may be light; silk, in order that it may be insulated.[1]

1. A stick of shellac rubbed with flannel attracts the pith ball.

2. After contact with the shellac, the pith ball will by conduction become negatively electrified as the shellac is, and will be repelled by it.

3. Arrange the flannel, which is not a very good insulator, so that it may be insulated both while rubbing the shellac, and afterwards; this may be done by shaping it like a cup, and supporting it on a silk thread, or by gumming it on a metal disc fastened to a stick of vulcanite. Then the flannel, after rubbing the shellac, will be electrified with vitreous electricity, and will attract the pith ball electrified with resinous electricity.

Converse effects will be produced by electrifying the pith ball by means of the flannel. The silk threads, shellac, and flannel must all be very dry, or the moisture will form a conductor along which the electricity will rapidly escape. Sometimes the pith ball is gilt, to make it a better conductor.

Experiments, illustrating the proportion between the force observed and the charge of electricity, can be made by means of the pith ball.

4. Two pith balls electrified with different electricities attract one another (Fig. 2).

[1] The parts of the drawings shown dark, but crossed by thin white lines, are intended to represent insulators.

5. Two similarly electrified pith balls hung side by side repel one another (Fig. 3). The same effect may be observed by means of two pieces of gold leaf insulated, and hanging side by side. When these apparatus are arranged (as in Fig. 4)

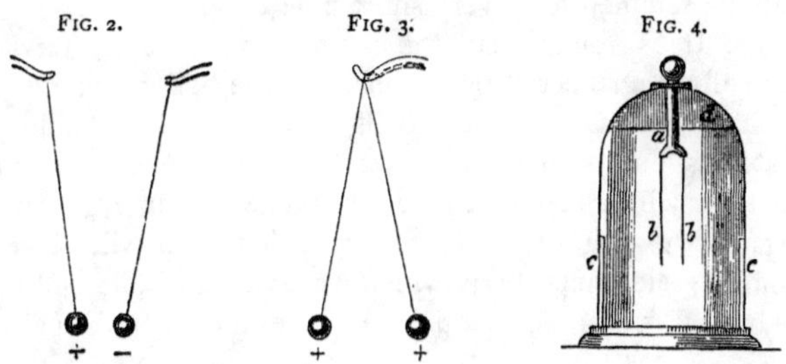

with glass cases and stands, and with means, such as the metal rod *a*, of readily communicating an electrical charge from any body the condition of which is to be examined, they are called electroscopes.[1] They indicate the presence of electricity by showing the existence of a force. They do not, strictly speaking, measure either the force or the quantity of electricity, but only indicate the presence of some force and some quantity. The little electroscope in Fig. 4 is furnished with a metal cap *d*, and two uninsulated strips of metal *c c*, the object of which is explained in § 14 and § 23.

In testing the laws of electrical quantity, it is convenient to use a more complex arrangement for producing electricity than is afforded by the mere stick of shellac or glass. The common electrical machine may be used to produce the electricity. This machine consists of a plate or cylinder of glass rubbed by flannel or some other semi-insulator while being turned, and having conductors conveniently arranged so as to gather either the vitreous electricity produced on the surface of the glass or the resinous electricity produced on the flannel. The best construction of these instruments will

[1] The name electro*meter* is often improperly applied to what is above described as an electro*scope*. Electrometers are described below, § 18.

be described when electrical laws have been more fully explained. The balls by which the foregoing laws are illustrated, may be held on glass or vulcanite stems, which must, however, be very dry and clean, or the electricity will only be retained for a very short time upon the balls.

§ 5. It is found that the distribution of electricity on the balls is unaffected by the mass of the ball, provided the surface remain constant. Balls made of wholly different materials but of the same size, if their surfaces be conductors, will behave in a precisely similar manner, so far as regards the quantity of electricity which each will abstract from any electrified body which it may touch: one ball may be wholly of brass, another a mere gilded pith ball, a

FIG. 5.

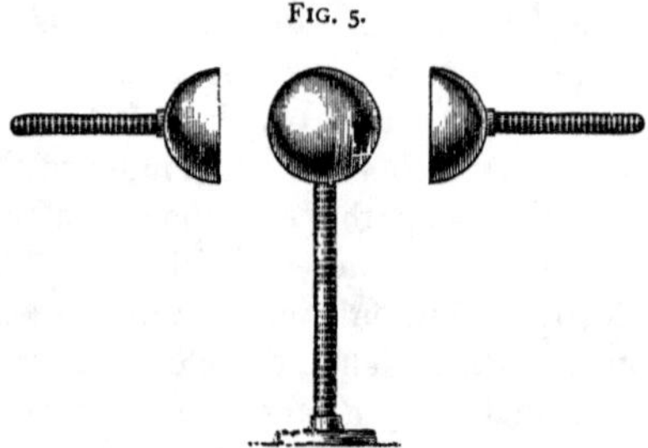

third a hollow iron ball; yet each will be found under similar circumstances to have what may be termed the same capacity for electricity. Moreover, let a ball (Fig. 5) be made of two hollow hemispheres, enclosing an independent conducting ball within them, and in contact with them, and let the system be electrified and the enclosing hemispheres removed by insulating handles. The internal ball will not be found electrified, and the two hemispheres, when placed in contact so as to form a complete ball, will, if the insulation has been perfect, be found to be as strongly electrified as at first. Electricity, while at rest, is therefore looked upon as residing in the surface only of the conductors. These statements may be verified with the assistance of the electroscopes before described.

Although electricity when at rest can only be detected on the surface of bodies, we shall presently see that, when in motion, it does not run over the surface only; it will pass more readily from one conductor to another along a solid rod than along a hollow rod of equal external dimensions and the same materials, vide § 3, Chapter IV.

§ **6**. Let one insulated conducting ball A be electrified by contact with rubbed resin, and another exactly similar ball B by contact with rubbed glass. If the two balls be now put in contact with one another, they will assume an electrical condition which is the same in both. If the ball A had most electricity at first, the whole system will be electrified as by rubbed resin; if B had most electricity at first, the whole system will be electrified as by rubbed glass; and in all cases the quantity of electricity on the two balls after contact will be equal to the difference of the charge on the two balls at first (it being remembered that the quantity of electricity is assumed to be measured by the force, which, if contained on a given conductor, it would be capable of exerting).

The distinction between the electricity due to rubbed glass and that due to rubbed resin is therefore analogous to that between positive and negative algebraic quantities, and justifies the use of the epithets positive and negative in place of vitreous and resinous. When positive and negative electricities are summed, the result is equal to the difference between the arithmetical values of the quantities. If the two quantities of electricity of different kinds were equal on the two balls, the result of the contact would be wholly to put an end to all electrical charge. The two bodies would be discharged and would be unelectrified, which we shall find to mean no more than that they will be in the same condition as all surrounding uninsulated bodies.

§ **7**. The electricity appearing on the rubbed glass is called positive, that appearing on the rubbed flannel or gutta percha is called negative; and the algebraic signs + and − are often used to denote the two different electrical conditions.

+ positive, vitreous } are three synonymous modes of
− negative, resinous } describing electrical conditions.

The symbols + and − have already been used on the foregoing figures showing attractions and repulsions, + repels +; − repels −; + attracts −.

§ 8. When electricity is produced, it is found invariably that equal quantities of positive and negative electricity are produced. True, the glass when rubbed becomes positive only, but the material with which it is rubbed becomes negative, and the quantity on the glass is precisely equal and opposite to that upon the rubber. If the rubber be not insulated, the electricity upon it will be at once conducted to the earth, and will for the time being make the rest of the earth more negative than before; but the earth, including the rubbed piece of glass, contains as a whole neither more nor less electricity than it did before; the distribution only has been altered.

When the whole surfaces of the two substances which have been rubbed together are thoroughly connected, either through the intervention of the mass of the earth or by any other conductor, the positive and negative electricities disappear, being neutralised as before. No substance is found to insulate so perfectly as to possess the power of keeping the two electricities asunder for more than a limited time. A perpetual leakage is always occurring from the one to the other through the mass of the insulator, until the combination or neutralisation is complete and all signs of electricity disappear. In elementary electrical experiments the one kind of electricity only is made manifest, because the one kind is concentrated in a small conductor and the other is probably diffused over the earth in the neighbourhood; the quantity at any one spot being too small to produce appreciable effects. Thus, when a stick of sealing-wax (being one kind of resin) is rubbed by a cloth, the sealing-wax alone appears electrified, simply because the positive electricity diffuses itself over the earth from the cloth, through the hand of the person holding it.

§ 9. When one insulator is rubbed against another, one of them becomes charged with positive and the other with negative electricity; and with any given pair of materials, one invariably becomes positively and the other negatively electrified; but whereas glass rubbed with silk or flannel becomes positively electrified, when rubbed with a cat's skin it becomes negatively electrified. It follows from this that the positive or negative electrification of the material does not depend absolutely on the substance of that material, but depends on some peculiar *relation* between the two substances in contact. It is proved by experiments that all insulators can be arranged as in the following list, which is such that those first on the list invariably become positive when rubbed by any of the substances taking rank after them, but negative when rubbed by a substance preceding them. This list is given on the authority of M. Ganot.

Cat's skin.
Glass.
Ivory.
Silk.
Rock crystal.
The hand.
Wood.
Sulphur.
Flannel.
Cotton.
Shellac.
Caoutchouc.
Resin.
Gutta percha.
Metals.
Gun cotton.

Those bodies which stand far apart on the list are distinctly and decidedly positive or negative relatively to one another, but those bodies which appear near together on the list may possibly be misplaced. A very trifling difference in the composition of the body, or even in the state of its surface or of the colouring matter employed, will raise or lower the place of the body in the list. A rise in temperature lowers the body in the list, i.e. a hot body rubbed by a cold one identical with it in chemical composition becomes negatively electrified. Generally it may be said that no difference between two insulators can be so trifling as not to necessitate the production of electricity when they are rubbed together. The relative position of two bodies on the scale can be readily tested by rubbing two insulated discs together and observing their action on a pith ball charged with electricity of a known character or sign.

§ **10**. The word *potential* will now be substituted for the general and vague term electrical condition. When a body charged with *positive* electricity is connected with the earth electricity is transferred *from* the charged body to the earth; and, similarly, when a body charged with *negative* electricity is connected with the earth electricity is transferred from the earth to the body. Generally, whenever two conductors in different electrical conditions are put in contact electricity will flow from one to the other. That which determines the direction of the transfer is the relative *potential* of the two conductors. Electricity always flows from a body at higher potential to one at lower potential when the two are in contact or connected by a conductor. When no transfer of electricity takes place under these conditions the bodies are said to be at the same potential, which may be either high or low. The potential of the earth is assumed as zero. *The potential* of a body is the difference of its potential from that of the earth. Potential admits of being measured and this measurement is fully described with the conditions tending to produce a given potential in Chapter II. Difference of potential for electricity is analogous to difference of level for water. From the above definition it follows, that all parts internal and external of any conductor in or on which electricity is at rest must be at one potential.

A body is said to be *uninsulated* when connected by a conductor with the earth. The potential of any uninsulated body is neither negative nor positive. There is in this view nothing to prevent our regarding the earth as an electrified body; indeed, we know that any one part of the earth is seldom or never in exactly the same electrical condition as any other part in the neighbourhood. We simply assume as our zero the condition of the earth in our neighbourhood for the time being; just as we may assume, in measuring heights, any arbitrary level, such as Trinity high-water mark: a point above this is a positive height, a depth below it may be written or regarded as a negative height.

§ **11.** It is frequently said that positive electricity attracts negative electricity, but that positive repels positive and negative repels negative. We have stated that electrified bodies do present attractions and repulsions of this kind, and by a slight extension of language the electricity itself may be spoken of as attracting or repelling; but there is a further phenomenon called *statical induction*, which does appear more distinctly to represent an attraction or repulsion of electricity, besides the attraction and repulsion of the bodies charged with electricity. A body A brought into the neighbourhood of a body B at a different potential immediately produces a distribution of electricity over the surface of B, such as would be produced by the system of attractions and repulsions enumerated in § 7. If A be charged positively it attracts negative electricity to that end of the body B which is near it, and repels positive electricity to the remoter portions of B. If the body B be insulated, it neither loses nor gains electricity, but its ends are competent to produce electrical phenomena of opposite kinds. Separating the two ends we may retain each charged with its positive and negative electricity. Or if we connect the further end of B with the earth even for a moment, the positive electricity will be driven off to the earth, and a permanent negative charge will then be retained on B. Otherwise when A is removed the + and − electricities on B recombine and exactly neutralise one another. By induction, as in the case of electricity obtained by friction, precisely equal quantities of positive and negative electricities are simultaneously produced. It will be convenient to represent the distribution of electricity on the surface of bodies by dotted lines, the distances of which from the surface are proportional to the quantity of electricity per square inch at that point; then, if the electricity be positive the dotted line will be shown outside the body; if negative, the dotted line will appear inside the body. Along one line separating the positively charged portion from the negatively charged portion there will be absolutely no charge. The annexed Figure (6) represents an original and an induced charge represented to the eye according to this plan. The dotted line on A shows the original charge

FIG. 6.

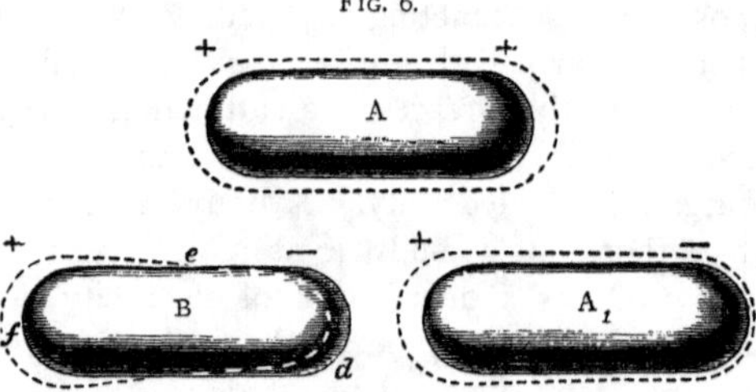

when A was at a great distance from B. When brought into the position A_1 near B the original distribution is disturbed, and at the same time positive and negative electricities are induced at the two ends of B; at the point *e* there is no charge.

§ 12. This induction of electricity must take place in the space surrounding every electrified body. In a room containing a ball electrified positively, the surface of the walls, the furniture, the experimenter himself must necessarily all be charged negatively in virtue of this induction. Where does this negative electricity come from? If the electrified body has been charged positively by rubbing, and the negative electricity has been allowed free access to the earth, it may be said that this negative electricity has been attracted to the surface of the walls, furniture, &c., distributing itself according to definite laws which must be separately studied. If both rubber and glass have been insulated, then each induces on all surrounding surfaces positive and negative electricities equal each to each, but these induced quantities are now not necessarily equal to the amount on the glass or on the rubber, unless these be removed very far apart from one another. If the two oppositely electrified bodies are kept close together, their inductive actions are spent almost entirely on each other and their action on the surrounding walls of the room is almost nothing, for where the one tends to induce a positive, the other tends to induce a negative charge; as the insulated electrified bodies are removed farther apart each produces its independent effect more completely. It will be found impossible rightly to understand electrical phenomena without always recognising the presence of this induced charge of electricity opposite in character to the first or original charge. The very existence of the original charge implies the induced charge.

§ **13**. Induction always takes place between two conductors at different potentials separated by an insulator. If the conductors are at the same potential, whether this be high or low, there is no induction.

If the wall of the room and an insulated body inside the room are at the same potential, the insulated body will be found to produce no electrical effects. The walls of the room and the insulated body might both be insulated from the earth and at a high potential, but none of the electrical effects hitherto described could be produced by an experimenter in the room. The insulated body would not attract light bodies; it would induce no charge or redistribution of electricity on a conductor held in its neighbourhood, and would not itself be charged with electricity or electrified. To produce all these phenomena we require not only that the insulated body in the room be at a high potential, but that the surrounding walls be at a different potential. If the insulated body at a high potential were connected with the earth electricity would run from it to the earth, and then a negative charge would appear on the surface of the body and a positive charge on the inside of the room. The body would then become electrified.

§ **14.** Viewed in the light given by these facts the attraction which an electrified body A exerts on uncharged bodies in the neighbourhood is simply due to the induced electrification which it produces in those bodies. The light uninsulated body B (Fig. 7) is attracted to the negatively

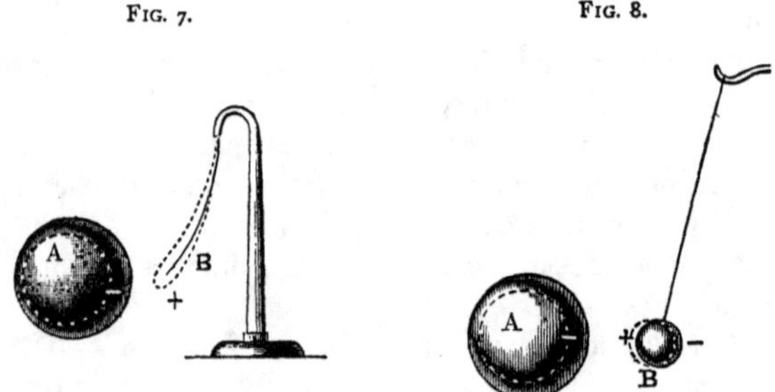

FIG. 7. FIG. 8.

electrified body A in virtue of the positive charge on B; this positive charge is also repelled by the walls of the room which will be positively electrified by induction from A. The light insulated body B (Fig. 8) is attracted because its charge at the near side is attracted. The charge on the far side of B is repelled, on the contrary, by the body A, but less repelled than the near side is attracted, because it is more distant. The charge on the near side of B is again repelled from the walls of the room towards the body A; the charge on the far side is attracted towards the walls and from A, but less than the near side is attracted, because the far side is nearer the walls. It is not until all these actions are taken into account that the forces set in action can be fully calculated; moreover, unless B be very small, it disturbs the distribution of electricity on A very sensibly.

In the electroscope shown in Fig. 4, § 3, the metal strips *c c* are inductively electrified by any charge on the gold leaves *b b*. They attract the gold leaves and increase their divergence. They also make the action of the instrument more regular than it could be if glass were opposite *b b*, for the glass would always be liable to have an electrical charge of its own, independently of any charge on *b b*.

A similar complicated series of actions occur when a positively electrified ball is brought into the neighbourhood of another positively electrified ball : each ball repels its neighbour and is attracted by the negative induced electricity on the surrounding walls. If the walls were positive also they would repel the balls back to one another, and if all were at the same potential the two positive balls would be in equilibrium and would not be electrified.

The phenomenon of induction allows us to examine the electrical condition of any body without abstracting electricity from it. If I hold a positively electrified body over the knob on the electroscope (Fig. 4), the knob will be negatively charged and the gold leaves positively charged by induction ; the gold leaves will therefore be deflected. On the removal of the inducing body, the electricities recombine and the deflection ceases. It is easy, however, by touching the under side of the knob or plate used for this purpose with an uninsulated conductor such as the hand, to allow the one electricity to run to earth, and then we have the electroscope permanently charged with electricity of the opposite kind to that contained on the inducing body.

§ **15.** The distribution of electricity can be examined in two ways, the first of which is the following. We may touch the surface of the body which we believe to be electrified with a small insulated disc called a *proof plane*, and then remove this conductor, and observe whether it is competent to produce any of the electrical attractions and repulsions or inductions. If the conductor be small, and if it be held on a long insulating stem of small size also, it will not much disturb the distribution of the electricity over the surface to be tested though some disturbance will always be produced by induction. While touching the body, it will sensibly form part of the surface of that body, and will be charged as the body is charged at that point, or nearly so. When removed, it will therefore retain a charge nearly proportional to what is termed the density of the electricity at that point, and this density may therefore be tested by observing the attracting or repelling force which the *proof plane* is in each case capable of exerting directly or by induction on some body assumed to be at a constant electrical potential—for instance, on the pith ball electroscope. By experiments of this nature, the distribution of electricity has been studied, and it is found that no electricity can be detected inside a hollow and empty conductor. A proof plane introduced (as in Fig. 9) into the interior of a highly electrified ball withdraws no sensible charge of electricity unless by accident it touches the edge of the aperture while being withdrawn. This distribution is a necessary consequence of the law that each elementary portion of a charge of electricity repels every other similar portion with a force inversely proportional to the square of the distance separating them. We shall study hereafter a few of the laws of distribution of electricity on the surface of conductors of regular form, on the assumption that they are so far from all neighbouring conductors, that the distribution depends only on the form of the electrified surface. These laws

FIG. 9.

will show that electricity tends to accumulate on all projections, and that the density at points is necessarily very large. Next we must study the distribution of electricity over two conducting surfaces opposite each other. The distribution in this case depends not only on the form of each surface, but on their proximity. For instance, the inside of a hollow conductor will be inductively charged by any electrified and insulated body placed there, and the charge on the internal surface will be greater the closer the two surfaces are placed. The charge is also affected by the insulator separating the conductor.

A second mode of testing the distribution of electricity is to remove the portion of the body the electricity of which is to be tested from the system of which it forms part, by insulating it from that system; its electricity may then be tested by the proof plane or by its direct effects.

§ **16**. It follows from what has already been stated (§ 11) that an electrified conductor may at certain portions of its surface have little or no charge. If those parts are touched by the proof plane no electricity will be removed by it. Thus, if a cylinder be electrified by induction, so that one end is positive, the other end negative, as shown on the body B, Fig. 6, some point near the middle at *e* will not be charged. It will not electrify the proof plane or any other small conductor, and even if a portion of the cylinder itself be removed it will give no signs of electricity. If it be touched by a large conductor, the whole distribution of electricity will be changed by induction before the contact takes place. Thus, if I connect the point *e*, Fig. 6, with the earth the whole distribution of electricity on B will be changed, for although *e* is no more charged with electricity than the earth itself the potential of the whole body B has been raised by induction from A on B; the approach of the connecting wire alters the distribution of electricity, positive electricity accumulates opposite the wire even before the contact is made, and the result of connecting *e* with the earth would be to leave the body B charged with negative electricity only and at the potential of the earth.

There are distributions of electricity such that the electrified conductor may actually be in contact with the largest conductor or with the earth without losing its electricity or the distribution being in any way changed, the conductor being at the potential of the earth; for instance, consider the positively electrified conductor A, Fig. 10, insulated and separated from the conductor B by a thin dielectric C. Let there be a negative charge on the conductor B equal to the positive charge on A, then no sensible charge will be found upon the external surface of either A or B, supposing them held far away from other conductors. I can produce this distribution by electrifying A while B is in contact with the earth. The positive charge on A will induce a negative charge on B, as shown by the dotted lines. The charge on A will be on the surface opposite B: the charge on B will be on the surface opposite A. I may then allow either A or B to be in connection with the earth without sensibly disturbing the charge on A or B. If I allow *both* to be in connection with the earth or with one another, the electricities will combine and neutralise one another. The dielectric need not be solid, as in Fig. 10, but may consist of air only, as in Fig. 11. The distribution of electricity described is that which occurs in a charged *Leyden jar* (Fig. 12). The outside coating A has a large charge of electricity almost equal to the charge of the internal coating B; nevertheless none of the electricity runs from the outer coating to the earth. The potential of the outer coating is zero. It is often said that electricity in this case is latent or fixed—in truth it is no more latent or fixed than any other charge of electricity. The distribution in this case is such that no sensible charge is on the outside of the outside coating, the whole quantity being on the inside of the outside coating.

If we were to form a Leyden jar with an opening admitting the introduction of a proof plane between the inner and outer coatings, we might take off from either coating a quantity proportional to the charge at each place. This, in fact, is what we do when by the proof plane we remove a portion of the charge from a conductor inside a room, or from the walls of a room inside which an electrified body is

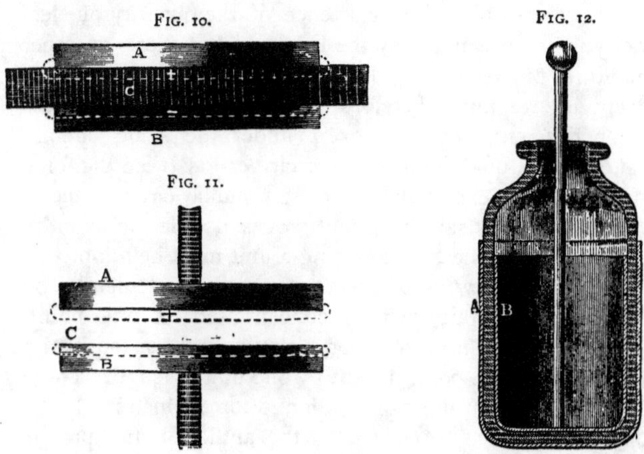

FIG. 10.

FIG. 11.

FIG. 12.

placed. There is no difference in theory between the inner and outer coatings of the Leyden jar; the outside of the inner coating, the inside of the outer coating are charged. From these electricity can be withdrawn by the proof plane; from the other faces of either coating none can be taken.

Whenever a conductor is charged a kind of Leyden jar is necessarily formed. The conductor is the inner coating, the air the dielectric, and the nearest surrounding conductors, such as the wall of the room or the person of the operator, form the outer coating; but the name of 'Leyden jar' is reserved for those cases in which the two opposed conductors are brought very close together purposely. The arrangement is also called a *condenser* or *accumulator*. The difference of potential between the two coatings of the Leyden jar remains constant whichever coating is in connection with the earth. If the original charge on the inside be positive, the outer insulated coating will be at a negative potential when the inner coating is put to earth.

§ 17. The quantity of electricity on a given conductor may be measured. The existence of the quantity of electricity is proved merely by the force which it exerts on other quantities of electricity. In order to measure quantities of electricity we must therefore measure the relative forces which different quantities exert under the same circumstances: if a quantity A of electricity exerts twice the force that quantity B exerts under precisely similar circumstances, we may properly say that quantity A is double the quantity B. In order to measure anything a unit must be adopted.

The *unit quantity* of electricity may conveniently be called that quantity which, concentrated at one point, would exert the unit force upon a similar and equal quantity concentrated at a point distant by one unit of length. There are many different units of length and force which might be adopted. The units chosen by the author in the present work are the centimètre for the measure of length; and the force capable of giving in one second a velocity of one centimètre per second to a gramme mass for the unit of force. The unit quantity of electricity upon this system, known as the electro-static system, is that which if concentrated at one point would repel an equal quantity at a point one centimètre distant with such a force as would, after acting for one second, cause a gramme to move with a velocity of one centimètre per second. Another unit of electricity might be defined as that which would repel a similar unit with the force of one grain at a distance of one foot. The idea at the root of both definitions would be identical, but the apparently more complex definition leads to greater simplicity in calculations.

§ **18**. The practical measurement of quantities of electricity can in many cases be made by directly measuring the electrical forces in action; the apparatus in which these forces are weighed is called an *absolute electrometer*. Any apparatus in which the forces produced by different quantities under the same circumstances are numerically compared but not actually measured in units of force is termed an *electrometer*. Indirect methods of measuring quantity are often more convenient for practical purposes, but these measurements can and ought to be all made in units of the kind described. In studying the distribution of electricity under various conditions, we must not be satisfied with merely knowing generally that at certain points there will be more, at others less, electricity; we must not even be satisfied with knowing the relative amounts on various points of a given conductor; we must aim at knowing exactly the quantity of electricity per square unit of surface, which is termed the density of the electrical charge. The electrometers employed in comparing quantities of electricity on different portions of any surface or surfaces must give us the relative amounts on various points, or they will not be measuring instruments. An absolute electrometer does more, it gives not only the relative but the absolute amounts.

§ **19**. Hitherto electricity has been spoken of as produced directly by friction and indirectly by statical induction only; there are several other modes by which electricity is produced:—1. The simple contact of two insulated pieces of dissimilar metals results in charging one metal with positive, the other with negative electricity in precisely equal amounts; or it may be more correct to say that after contact the metals are found to be thus dissimilarly charged. The charges so produced or observed are very small. 2. If a metal be dipped in a liquid a similar effect occurs, the liquid and the metal being electrified in opposite ways. A difference of potentials is produced by the contact. The amount of electrification differs with different metals and different liquids, but is always very small compared with that which might be produced by friction. 3. When two dissimilar metals are plunged side by side into a liquid, such as water or a weak solution of sulphuric acid, they do not exhibit any signs of electrification. The three materials remain at one potential or nearly so.[1] A further description of this curious fact is given Chapter II. § 22. 4. If while the two dissimilar metals are in the liquid they are joined by metallic contact to terminal pieces of one and the same metal, these terminal pieces will be brought to the same difference of potentials as that which would be produced by direct contact between the dissimilar metals. Thus, though zinc, water, and copper in an insulated

FIG. 13.

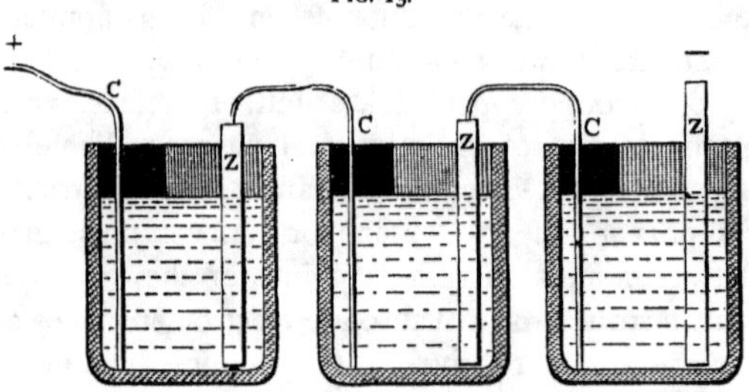

jar are all at one potential, if I join a copper terminal to the zinc, then this copper terminal will become positive relatively to the zinc, water, and second copper, which all remain at one potential.

The name of galvanic cell is given to an insulating jar containing two dissimilar metals plunged in a liquid composed of two or more chemical elements, one of which at least tends to combine with one or other of the two metals, or

[1] The Voltaic theory of the galvanic cell is adopted in this treatise. The above statement is in direct contradiction with many treatises on electricity, which generally state that the metals become one positive and the other negative. *Vide* Chapter II. § 23.

both in different degrees. But whereas in the single cell no charge of electricity is given to either metal, if we insulate successive jars of the liquid one from another, and plunge successive pairs of metals, C and Z, joined as in Fig. 13, into these jars, very considerable charges of electricity will be communicated to conductors in contact with the final plates of metal; thus, if coppers and zincs be used, the liquid being water or a weak solution of sulphuric acid, the last copper plate will charge a conductor positively, the last zinc plate an equal conductor negatively. Sulphate of zinc will be formed during the process, and this chemical action is found to be essential to the production of any considerable quantity of electricity in this manner, which is therefore often said to be due to chemical action as distinguished from friction. The charge of electricity obtained in this way may be looked upon as wholly due to the chemical action; but, on the other hand, it may be looked upon as due to the successive junctions between the zincs and coppers, and it is found that the amount of charge obtained in this manner on a given conductor is simply proportional to the number of these junctions, and that it depends on the metals in contact, not upon the liquids. In other words, the difference of potentials produced is proportional to the number of junctions. These two views are called respectively the chemical theory and the contact theory of the galvanic cell, and have been supposed to be incompatible. They are both true.

§ **20.** There is no difference whatever in kind between the electricity produced by friction and that produced by chemical reaction. It is worthy of remark that, in each case, the electricity requires for its production the contact of dissimilar materials. This contact requires to be supplemented by friction in the case of insulators, by chemical reaction in the case of conductors. The friction between two dissimilar insulators invariably produces electricity. The difference of the chemical action of any conducting liquid compound on two dissimilar metals produces electricity. The analogy between friction and chemical action is not known. Electricity in each case is produced so that equal quantities of positive and negative electricity are simultaneously produced. This is sometimes expressed by saying that all bodies are always electrified, and that the contact and friction, or contact and chemical action, produce merely a redistribution of electricity.

§ **21.** Electricity may also be produced by the simple pressure, or indeed contact, of two dissimilar insulators. The electricity will be retained by the insulators after their separation This is precisely analogous to the production of electricity by the contact of two conductors.

§ **22.** Certain minerals when warmed acquire an electric charge, differing in sign at different parts of the mineral; thus, one end of a heated crystal of tourmaline will be positively electrified, while the other is negatively electrified. This electricity is sometimes called pyro-electricity. The phenomenon has not been much studied; the electrical charge is probably due to a polarity in the structure of the tourmaline at different parts, which virtually makes in one crystal a system like that of a magnet having opposite properties at opposite ends. The electrical phenomena produced by the contact of dissimilar metals are produced even when the dissimilarity consists merely in the difference of temper in one and the same piece of metal. A soft and a hard piece of brass wire behave as dissimilar metals, although their chemical composition may be identical. If this view be correct, we may say that, wherever electricity is directly produced, it requires the contact of two dissimilar materials.

§ **23.** The attractions and repulsions produced by electricity have hitherto been spoken of as absolute, or as being produced under all circumstances; but if an uninsulated metal plate D (Fig. 14) be interposed between an electrified body A and the insulated suspended pith ball B all attraction or repulsion will cease, just as if the metal plate were opaque to the electric influence. If, however, the metal plate or screen be insulated (as in Fig. 15) it will increase the attraction or repulsion instead of destroying them. These two apparently different effects are due to the different distributions of electricity produced in the two cases.

Let A be electrified positively, and the plate D be uninsulated, then on the side next A a negative charge will be induced, diffused over a considerable surface; the effect of this diffused negative charge is very nearly to neutralise all attractions or repulsions due to A, on the farther side of the screen. The metal cap of the electroscope (Fig. 4) is intended to screen the gold leaves from inductive effects, and should not be insulated. The whole glass case should be coated with an open wire case for the same reason.

When, however, the metal plate D is insulated the farther

FIG. 14. FIG. 15.

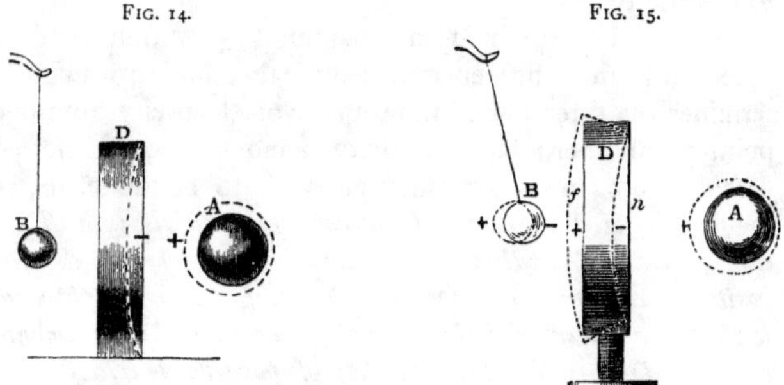

side of D becomes positively electrified as A was, the charge on the side *n* near to A and the charge on A nearly neutralise one another as before; but the positive charge on the far side *f* of D is thus left free to attract or repel, and the result is the same as if the body A had been advanced in the direction of the screen by an amount equal to the thickness of the screen.

We can now understand the reason why a Leyden jar containing a very large quantity of electricity neither attracts nor repels light bodies in its neighbourhood. The effect of the more concentrated inner charge and more diffused outer charge is such that one precisely neutralises the other. This statement is here made as of a fact ascertained by experiment. It can also be theoretically demonstrated.

CHAPTER II.

POTENTIAL.

§ **1.** The word *Potential*, introduced by Green, has only lately been generally adopted by electricians, and is still often misunderstood; it expresses a very simple idea, and one quite distinct from the meaning of any other term relating to electricity.

As already explained in Chapter I. § 7 difference of potential is that difference of electrical condition which determines the direction of the transfer of electricity from one point to another; but electricity cannot be so transferred without doing work or requiring work to be done, hence the following definition. *Difference of potentials is a difference of electrical condition in virtue of which work is done by positive electricity in moving from the point at a higher potential to that at a lower potential, and it is measured by the amount of work done by the unit quantity of positive electricity when thus transferred.* The idea of potential essentially involves a relative condition of two points, so that no one point or body can be said simply to have an absolute potential but for the sake of brevity.

The potential of a body or point is used to denote *the difference between the potential of the body or point and the potential of the earth.*

These definitions require considerable illustration before they can be fully understood.

Electrified bodies repel and attract one another, and by a slight extension of language we say that a quantity of positive electricity attracts a quantity of negative, but repels a quantity of positive electricity. If, therefore, we move a quantity of positive electricity towards another similar quantity we meet with a resistance capable of measurement, equal, for example, to the weight of so many grains. In overcoming this resistance work must be done, precisely as work must be done to lift a pound or a grain. The work done in moving a body from A to B is measured by the product of the distance multiplied into the force overcome; if the weight of a grain be the unit of force and the foot the unit of distance, the unit of work will be the foot grain. If, then, in moving a certain quantity of electricity from A to B we overcome a resistance of ten grains through a space of five feet we do work equal to fifty foot grains during the operation. On the other hand, the repulsion or attraction of electrified bodies tends to perform work; for the body just brought to B may be driven back to A by the force of electricity alone. In the one case, work is said to be done *upon* the electrified body in consequence of its electrification; in the other case it is done *by* the electrified body in virtue of its electrification; less accurately we might say the work was done by the electricity, or performed upon the electricity; the measure of the work is the same in the two cases, which are analogous to letting a body fall from the level A to the level B, and raising it up again from B to A.

§ 2. An electrified body moving from one point to another may at one time require work done upon it in order to overcome the resistance; at another part of the journey it may pull in the direction it is going and then work is done by it. I speak here only of the work done or required in consequence of the electrical condition of the body.

The whole work which has been required in consequence of electrical attractions or repulsions to move it from any point A to any point B will be the algebraic sum of the work done by and done upon the electrified body, the first being called positive and the second negative work.

Thus, if in moving the electrified body from A to B, we first have to overcome a resistance, and do work upon it equal to 10 foot grains, whereas afterwards it pulls towards B, doing work equal to 30 foot grains, then in the whole passage from the point A to the point B the work done by the body may be said to be 20 foot grains; it is true that during one part of the passage it did more than this, but only after having required aid previously.

The path followed in going from A to B will be a matter of indifference so far as this total work done by or upon the body is concerned. We have a precisely analogous case in gravitation: a body of a pound weight in falling from a height of 40 feet to a height of 20 feet above the sea, will do necessarily 20 foot-pounds of work in virtue of that fall, no matter what path it follows. We may lift it above A and do work upon it by lifting it before letting it fall, still the whole work done by the body in its passage from A to B and in virtue of that fall will be 20 foot-pounds; it may fall by the most roundabout or the most direct road, the work done will be the same; it may fall below the level of A, and bound up to A: the whole sum of the work will be unchanged, depending merely on the difference of level between the first and second spot. This work may indeed be represented in various ways: thus, if the body fall direct through a vacuum the work appears in the form of what is called actual or kinetic energy; that is to say, it is wholly represented by the motion of the mass. If, on the other hand, the body falls slowly, lifting another weight, the work will be represented partly by the weight lifted, partly by the heat due to the friction of the mechanism; but the work done by a body due to its fall from one level to another is constant in amount however various in form. The work done in overcoming electrical force or done by electrical force is subject to the same law

§ **3.** In moving a weight from a point A to a point B on the same level no work on the whole is either done upon or by the body in respect of its weight; and similarly in moving a small electrified body from a point A to some other point B, it may happen that the point B is so situated that on the whole no work is either done upon or by the body in respect of the electrical forces in action on the body. In that case the two points might be at the same electrical level or height, but the recognised term in respect of electrical forces is potential; the points A and B are at the same *potential.* If our small electrified body, for instance, be moved round another large electrified body, neither approaching nearer nor receding farther from it, and so far from all other conductors as not to be sensibly attracted or repelled by them, it will pass along a path every point of which is at the same electric potential.

In moving any actual body from spot to spot some work must always be performed to overcome friction, but as in moving a heavy body from one point to another, at the same gravitation potential or level, no work is required in respect of its gravitating properties, so in moving an electrified body from one point to another at the same electrical potential no work is required in respect of its electrical properties, although of course work will certainly be required to overcome friction and may be required in respect of gravitation if the body be raised or in respect of inertia if we accelerate the motion of the mass.

§ **4.** *The potential* of a body is the excess or defect of its potential above or below that of the earth in the neighbourhood—the potential of the earth at that point being arbitrarily assumed as *nil.*

The potential increases in proportion to the increase of work done by any given quantity of electricity in moving from the point to the earth; and since the potential is proportional to the work and to the quantity of electricity transferred, and to no other quantity, *the potential of a point is measured by the work which a positive unit of electricity does in passing from that point to the earth.* The unit quantity of electricity might, so far as this definition is concerned, be chosen arbitrarily, but there is a certain convenience for many calculations in choosing the unit as defined in Chapter I. § 17. Every point everywhere may be said to be at a certain electric potential, just as every point everywhere may be said to be at a certain level above or below a datum line arbitrarily chosen, such as the Trinity high-water mark. In speaking of the potential at a point it is as unnecessary to conceive of the presence of any electricity at that point as it is to think of the presence of a heavy body at a point when we speak of its height above the sea.

§ 5. The electric potential at the point depends on the electrical condition of all bodies in the neighbourhood; that is to say, sufficiently near to exercise any sensible force on a small electrified body at the point. Moreover, in testing the equality of the potential at two points by the work done upon or by an electrified body in its motion from one point to the other we must remember to choose a body containing only a very small charge of electricity, which we will call the test charge; otherwise, the mere presence of this test body or test charge of electricity would sensibly change the potential at the point at which it was at the time of the experiment; increasing or decreasing for the time being the work which must be done in order to bring any other small quantity of electricity to that point. At first it might appear as if the analogy of gravitation deserted us here, but that is not so; for if I say that two points A and B shall, relatively to the earth, be at the same level when no work is done upon or by a heavy body in passing from one to the other, I must remember that in placing a heavy body at the point A, I do change for the time the gravitation level of that point if the body be of sensible size compared with the earth; for its presence at A has increased the attraction of all other heavy bodies to A, so that for the time being a small weight passing from A to B would do work; the position of the centre of gravity of the earth having been changed.

§ 6. The *difference of potential* between two points A and B, being the difference of condition in virtue of which electricity does work in moving from one to the other, is measured by the work required to move a unit of electricity against electric repulsion from A to B, or, what is the same thing, it is measured by the work which a unit of electricity would do while being impelled from B to A.

The point A is said to have a higher potential than B if a unit of positive electricity in passing from A to B performs work. It is assumed that the unit of electricity does not disturb the distribution of electricity in the neighbourhood.

The conception of the work which must be done upon or by electricity in passing from one point to another must be grasped as the only idea which can explain difference of potential. When bodies are spoken of as being in the same electrical condition we mean that they are at the same potential. Difference of potential can therefore be expressed in foot grains or any other recognised unit of work.

In this paragraph the work is spoken of as being done by the unit of electricity simply to avoid the awkward periphrasis 'done by a small electrified body charged with one unit of electricity' and 'done in consequence of the electric charge only.' That is to say, it is the extra work which must be done in moving the body from the one place to the other in consequence of its being electrified.

§ 7. Let us apply our definition to special cases. First, take an electrified conductor on which electricity is at rest, having assumed that distribution which is determined by its own shape and the shape and position of neighbouring conductors. All points on the surface of such a conductor are at the same potential. If any one point A were at a higher potential than another B, the electricity at A would as surely run to B as a weight would fall from a higher level to a lower unless resisted by some force; whereas, on the conductor there is no impediment to the free motion of electricity. One end of the conductor may be positively electrified, the other end negatively electrified; the centre may have no sensible charge as in the body B, Fig. 6; nevertheless all points of the surface are at the same potential, for I might move any little electrified body all over the surface without its being retarded or impelled in any direction by electrical forces. All points in the interior of the conductor are also at the same potential as the surface, although no charge of electricity is ever found at any internal point.

The little test charge of electricity, when introduced into any cavity in the interior of a body, would be equally ready to move in all directions, and would be in perfect equilibrium. At first it might seem that inside or outside the body a unit of positive electricity *a* (Fig. 16) would be attracted by that end N of the conductor N P which was negatively charged, and would be repelled by the other end P; but in thinking thus we forget the influence of the external neighbouring conductor M, which has already produced the arrangement of the charge upon N P. The test charge, wherever applied, will not tend to move in one direction more than another, but to subdivide itself over the large conductor N P, in the same manner as the original charge is distributed.

FIG. 16.

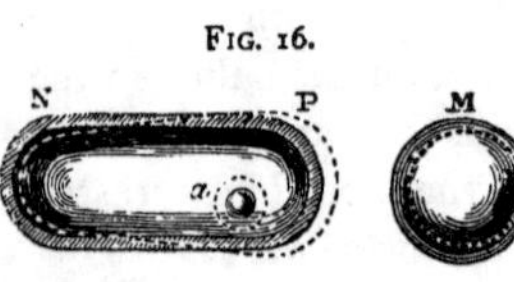

§ 8. Let us next consider the space round a charged conductor, this space being necessarily filled with air or some other insulator. First, conceive the conductor to be uniformly charged with one kind of electricity, as a sphere might be in the centre of a spherical room (Fig. 17). Then the space close to the sphere would be very nearly at the same potential as the sphere, for our test charge of electricity would do very little work in moving up to the sphere if attracted to it, and would require little work to be done upon it to move it up to the sphere against the repelling force. Let us conceive the potential of the sphere to be positive and the test charge positive also. Then the potential of the space round the sphere falls, or becomes less positive as we recede from the sphere. The work required to bring the test charge to the ball increases as it is removed farther and farther from the ball, although the force with which it is repelled diminishes. Again, the case is analogous to that of gravitation. The work which a body will do falling to the earth increases as the height increases from which it falls, although the attraction between the earth and the body diminishes as it recedes from the earth.

FIG. 17.

§ 9. As the test charge approaches the wall of the room surrounding our positively charged sphere, it approaches a negative charge of electricity, and is more and more attracted by it; this attraction further increases the work required to bring the test charge back to the electrified sphere, and the potential falls faster and faster. The fall continues until the test charge touches the wall of the room, which is thus shown to be necessarily at a lower potential than the charged sphere. Had we begun with a negative charge on the internal sphere, we should have found that the wall of the room would have been at a higher potential than the sphere. Thus we find that there is a necessary difference of potential between the inner and outer coating of a Leyden jar, or generally that any two conductors between which induction is taking place must be at different potentials.

The potential diminishes gradually from the internal sphere to the surrounding conductor, and all concentric spherical surfaces will be at one potential, i.e. we might, so far as electrical forces are concerned, without doing or receiving any work, move the test charge all over any concentric spherical surface, indicated by the lines *p*1, *p*2, *p*3, in Fig. 17. Whatever be the shape of the internal electrified body, I may conceive in the dielectric surrounding it equipotential surfaces of this kind, the form of which will depend on the form of the internal and external conductors.

We may further conceive successive equipotential surfaces separated by such distances that the same amount of work would be done by the test charge in moving from any one to the next. An equal amount of work would be required to move the test charge back from any one of these surfaces to that adjacent to it and at a higher potential.

§ 10. Consider the more complex case of a body charged partly with positive and partly with negative electricity but all at one potential. This involves a complicated distribution of electricity in neighbouring conductors, such, for instance, as is shown in the annexed diagram (Fig. 18).

FIG. 18.

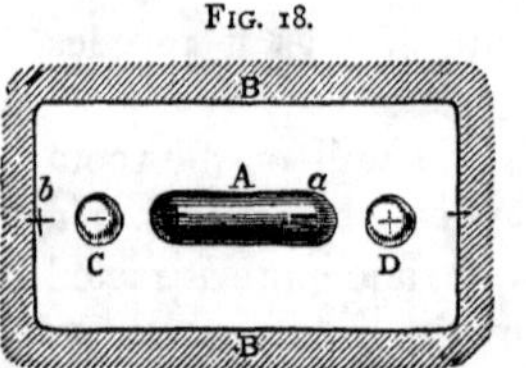

Very near the surface of the conductor A, the potential of the dielectric will be sensibly the same as that of A, and there is nothing here to indicate whether the potential of A is positive or negative relatively to the general enveloping conductor B; but receding from A towards C the potential of the space falls, whereas, as we pass from A towards D, it rises; again, receding from D towards the envelope B, the potential falls, but as we pass from C to the envelope the potential rises, so that close to B the potential is the same at all points, but whether higher or lower as a whole, there is nothing in the diagram to tell us. All these conclusions are deduced from the simple conception of the work required to move our imaginary test charge from place to place. Nor can any simpler conception be suggested. We see from the above diagram, that a body charged with negative electricity might have a positive potential relatively to a point charged with positive electricity, and *vice versâ.* For the body A may all be at a positive potential relatively to the body B, notwithstanding the fact that the part *a* of this conductor is negatively charged while some point of B, such as *b*, is positively charged.

§ **11.** The charge induced between two opposed conductors separated by a dielectric, implies a difference of potential between the conductors as shown above. Moreover, as the difference of potentials increases, so must the induced charges increase, for in order to make it more and more difficult to move the test charge from one surface to the other, the repulsion from one side and attraction to the other must increase, and this additional attraction and repulsion can only be increased by increasing the quantities of electricity. On the other hand, so long as the difference of potentials between the surfaces remains constant the charge on the opposing surfaces must remain constant; both potentials may rise and fall together, but the constant difference of potential implies a constant internal charge. An example will make the meaning of this statement clear. Suppose an ordinary Leyden jar to be charged with negative electricity and to have its outer coating in connection with earth. The potential of the inner coating will be negative, relatively to the earth; and calling the potential of the earth zero, as is usually done for brevity, we may, as stated in § 4, simply say that the potential of the inner coating of our jar is negative. There will be a positive charge of electricity on the inside of the outer coating of the jar equal to the negative charge within.

Insulate the outer coating, and electrify it with a positive charge. Its potential will be raised, but the potential of the inner coating will be raised by a like amount. The negative charge will remain inside the jar undisturbed in amount;

opposite to it will remain the positive charge on the *inner* side of the outer coating, the only change being that on the outer side of the outer coating we have now a positive charge. The effect of this additional positive charge will be to increase the work required to bring our test charge from a distance up to the jar, or to any point inside the jar, i.e. the potential both of inside and outside and of all adjacent points has been raised.

§ **12.** Next, suppose that two jars (Fig. 19), having their inner coatings in electrical connection, are charged with negative electricity, the outer coatings being uninsulated, i.e. at the potential of the earth. The potential of the inner coatings will be negative, and if the two jars are equal in all respects, the negative charge in each will be equal. Insulate jar A, and increase the potential of its outer coating by electrifying it positively. The negative charge will now redistribute itself between the two jars.

FIG. 19.

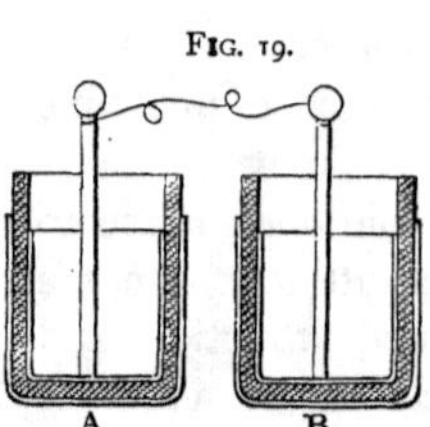

The potential of the outer coating of B remains constant. The potential of the inner coatings of A and B must be uniform throughout, since they are in metallic connection. Their potential as a whole will be somewhat raised, but not so much as that of the outer coating of A; hence the difference of potentials between the coatings of A, will have been increased, and its internal charge will have increased, and this will have occurred at the expense of B, where the difference of potentials between the inner and outer coatings will have diminished.

§ **13.** If one coating of any Leyden jar be kept at a constant potential, such as that of the earth at the spot is generally assumed to be, the quantity of electricity which the other coating contains is simply proportional to its potential, a fact determined by experiment. Thus, if I have the means of producing a constant potential, or rather a constant difference of potential, from that of the earth, I shall also have the means of collecting a constant quantity of electricity. The charge assumed by any insulated conductor inside a conducting envelope, however far remote, is simply proportional to the difference of potentials between the envelope and insulated conductor; and as a limit we may say that the charge on any insulated conductor, when there are no electrified bodies in the neighbourhood, is simply proportional to the potential of the conductor; that is to say, the difference of its potential from that of the earth. The force it exerts is proportional to the quantity, and the work required to overcome that force is proportional to the force.

§ **14.** In a Leyden jar it is immaterial which of the coatings is in connection with the earth, or whether either of them be so. Connection with the earth is merely a device for keeping the potential of that particular coating constant, or nearly so, by maintaining it in connection with a very large conductor. Thus, the inner coating of a jar when in connection with the earth, will take a negative charge if the outer coating be positively electrified, and, the difference of potentials being the same, this charge will be precisely the same in amount as if the outer coating had been in connection with the earth, and the inner coating had been directly electrified by negative electricity.

FIG. 20.

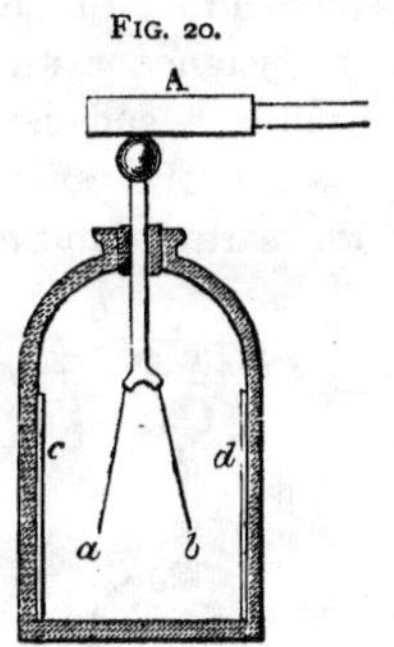

§ **15.** Let us consider the construction of electroscopes by the light of the knowledge we have now acquired; for instance, the gold leaf electroscope, Fig. 20.

The repulsion between the two gold leaves *a* and *b* depends on the quantity of electricity with which they are charged. But upon what does this quantity itself depend? Merely on the difference of potential between the gold leaf and the conductors *c* and *d* immediately surrounding it. When the gold leaves *a* and *b* are connected with the electrified body A to be examined, *a b* and A assume the same potential; then the quantity of electricity accumulated on the gold leaf depends on the difference of that potential from the neighbouring conductors *c* and *d*; let *c* and *d* be insulated, and at the same potential as A, then, no matter how much electricity there may be on A, none will come to *a* and *b*, and no divergence will occur in the leaves. In the ordinary construction of electroscopes, some parts of the surrounding conductors *c* and *d* are glass, and their potential depends on conditions over which we have no control; *c* and *d* should be in a metal case with openings, to allow *a* and *b* to be seen; for instance, a wire cage round glass, the meshes of which approach sufficiently near to keep the whole surface of the glass at one potential; then, if *c* and *d* be in connection with the earth, *a* and *b* will be charged with electricity whenever there is a difference of potential between A and the earth. Exactly similar reasoning applies to the Peltier electroscope, Fig. 21. In this instrument instead of the gold leaf we have a rod *a b*, free to move on a vertical axis *v*, and repelled at each end by a fixed conductor *c d* in electrical connection with it, but placed on an insulating support D; the rod is directed by a small magnet *m n*; the instrument is so placed that when *c d* has no charge of electricity, the magnet places the rod just clear of these fixed conductors *c* and *d*; then when B with *a b* are all charged with electricity, the rod *a b* is repelled until the force of electric repulsion is just balanced by the directing force of the magnet. The

FIG. 21.

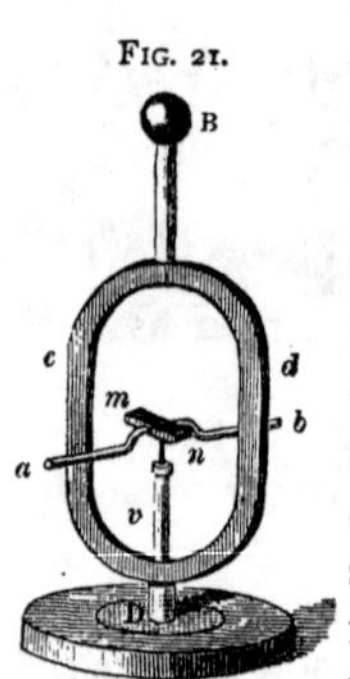

force depends on the quantity of electricity on the rod and balls, but this quantity is proportional to the difference of potential between the system B *c d*, &c., and the enveloping conductor A, which is not shown in the drawing but which encloses the whole insulated system B *c d*. This electroscope, therefore, like the preceding one, and like all others, indicates difference of potential by means of the quantity which that difference causes to accumulate on an insulated conductor.

In the instruments usually made there is a divided ring to show how far the rod *a b* is deflected. The instrument indicates more conveniently than the gold leaf electroscope whether a given potential be higher or lower than another; but inasmuch as the deflections are not proportional to the difference of potential between *a b* and the case A, and are not even connected by any simple law with this difference of potential, the Peltier electroscope cannot be used to *measure* difference of potentials, i.e. to compare two potentials or differences of potentials accurately, so as to allow us to say that one is distinctly two, three, or four times as great as another. For this purpose we require much more complex arrangements, electrometers or instruments in which the attractions and repulsions produced by given differences of potential between the parts can be calculated definitely.

All electrometers measure directly differences of potentials, and measure quantities only indirectly.

§ **16**. If two electrified conductors A and B, which are at the same potential, be joined by a wire, no disturbance in the electric distribution on the system will take place, unless indeed the wire be of sensible size relatively to the other conductors, and at a different potential; but, assuming the wire to be small, or at the same potential as A and B, the electricity on the bodies after being joined will be in equilibrium as before, the necessary condition of equality of potential throughout being satisfied. If, on the other hand, A be at a higher potential than B, positive electricity must, when the connection is made, flow from A to B, to re-establish electric equilibrium. The amount of the electricity thus transferred must be such as will restore the equilibrium; it will be great when the difference of potential is great and when the size of the bodies is large, and small under the opposite conditions. The existence or continuance of the flow of electricity from one point to another depends solely on the difference of potential between the points. The magnitude of the conductors has only one influence in the result, by requiring that a larger quantity of electricity shall flow to re-establish equilibrium. We may illustrate this by an experiment with water. If we join two reservoirs of water, big or little, by a pipe, no flow takes place from one to the other if the surfaces of the water in both are at the same level. If they be not, the flow will take place from the higher to the lower; the quantity of fluid transferred depends on the capacity of the reservoirs and original difference of level: it continues until the level is the same in both. Substitute potential for level, electricity for water, conductor for reservoir, and the above statements are all true for electricity.

§ **17.** If I put one end of a wire in connection with the earth and the other at a point X in the air (which may be at a very high potential) no electricity flows through my wire from the point to the earth, simply because at the point in question there was no electricity to flow, its capacity and charge being zero; but the potential of the point will have been changed by the mere presence of the wire to that of the earth. For this purpose, while the wire was approaching the point, a redistribution of electricity on its surface has been going on under the influence of the induction to which the potential of point X was due.

If the wire has a sharp point so that a very small quantity of electricity will produce a great density, electricity will actually flow from the air to the earth; successive particles of air negatively charged will fly from the point, and be replaced by particles of air positively charged, each of which will be discharged through the wire. If the potential of the point be sufficiently high the phenomenon is accompanied by noise and a brush of light. A lighted match on the end of the wire also allows the transfer of electricity to take place; the burnt particles fly off with the charge of one sign and the air about to be burnt brings electricity of the other sign to the wire.

§ **18.** By definition the difference of potential was declared to depend on the work done by or upon electricity in moving from one point to another. The nature of the work done by or to a quantity of electricity moved on a conductor by or against a force of attraction or repulsion is clear enough—a tangible force is used or overcome; a solid body is either put in motion, or its motion is resisted; but when electricity moves along a wire from a body at one potential to a body at another, no solid body is moved at all, and no equivalent work appears at first sight to have been done. The equivalent is found, however, in heat generated in the wire by the passage of the electricity. It is well known from the research of Joule that 772 foot-pounds of work are equivalent to the quantity of heat which raises 1 lb. of water 1° Fahr., and although no visible mechanical work is done, where a quantity Q of electricity passes along a wire from A to B, heat is generated precisely equivalent in amount to the work which the attractions and repulsions of the electrified bodies A and B would have done when acting upon the same amount of electricity Q, conveyed on a small moving conductor from the body A to the body B. We shall find that electricity in motion is capable of doing work in other ways, but in whatever way work or its equivalent is produced by electricity moving from A to B, the amount will always be equal to the quantity of electricity transferred multiplied into the excess of potential of A over B.

§ **19.** Difference of potential may be produced by mere induction. A small insulated conductor placed at any point in space where, owing to the neighbourhood of electrified bodies, the potential was x, will itself assume the potential x, without losing or gaining any electricity. Then if this body be connected with the earth, electricity will flow from the body to or from the earth sufficient in amount to bring the body to the potential of the earth; if x be positive, the current will be to the earth; if x be negative, the current will be from the earth to the body.

§ **20.** Difference of potential is produced by friction between insulators followed by separation. Two insulators rubbed against each other become oppositely charged, and there is a difference of potential between them. It is probable that for each pair of substances rubbed together there is a certain maximum difference of potential which cannot be exceeded. The list already given, Chapter I. § 9, showing the order in which some materials stand, so that each becomes positive when rubbed by any of the substances placed after it, necessarily shows also the order in which materials must be classed, so that when one is touched or rubbed by another following it in the list, the potential of the former may become positive relatively to that of the latter. Moreover, a greater difference of potential is produced by friction between substances far apart on the list than between substances close together on the list. It is possible that the law which will in the next paragraph be enunciated for conductors may also hold good for insulators.

§ **21.** When two dissimilar conductors touch one another, a difference of potential is produced between the conductors charging them, as mentioned Chapter I. § 19. The difference of potential is constant with constant materials, i.e. copper and zinc at a given temperature touching one another are invariably at potentials differing by a constant measurable amount. The same may be said of any two metals. Moreover, all metallic conductors may be ranged in a list, such that any one of them in contact with any of the conductors later in the list will have a potential positive relatively to that conductor. Moreover, calling these bodies A B C D, &c., the difference of potential between A and C is equal to the difference of potential between A and B added to the difference of potential between B and C, or generally if these bodies were all in contact one with another in the order A B C D . . . N, &c., and if we call $a\ b\ c\ d \ldots n$, &c., the potentials of these bodies, $a - n = (a - b) + (b - c) + (c - d) \ldots + (m - n)$. Thus if three bodies be in contact, as in Fig. 22, the difference of potential between the ends A and B may be calculated from the two end metals only; in the example given, it does not matter what the difference of potentials between gold and copper alone would be, for call that a, and call the difference between gold and zinc c, and that between copper and zinc b, then $(a - b) + (b - c) = a - c$, as if gold and zinc had been directly in contact. It may be stated quite generally that in any series of metallic conductors thus placed in contact, the difference of potentials between the ends depends on the extreme conductors of the series. The following is a list of conductors, ranged in such an order that each becomes positive when touched by those which follow. Zinc, lead, tin, iron, antimony, bismuth, copper, silver, gold. The earlier metals on the list are called *electropositive* to those which follow. The exact relative differences of potential have as yet been experimentally ascertained only in a few cases.

FIG. 22.

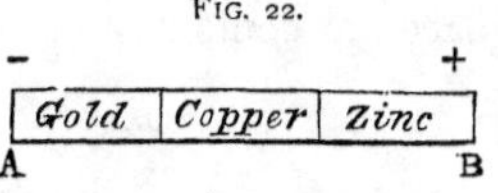

§ **22.** It is believed that all compound solid bodies which are conductors behave in the same way as simple metallic conductors so far as the production of a difference of potential due to mere contact is concerned, and this is certainly the case in many instances. Liquid conductors also appear relatively to one another to form a series of the same kind. But compound liquids and solids do not admit of being arranged relatively to one another in the simple order described as applicable to metals.

This difference between the compound liquid and the simple metallic conductor, appears to be intimately connected with the fact, that electricity in passing through these compounds decomposes them, a phenomenon to be more especially described hereafter. The compounds which are thus decomposed are called *electrolytes*. The following series of phenomena occur when metals and electrolytes are placed in contact :—

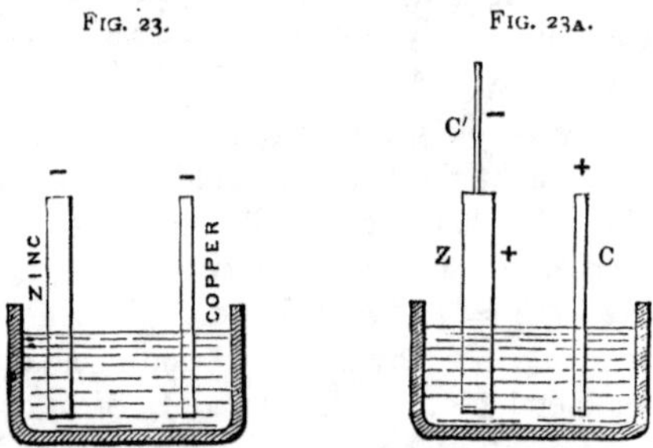

FIG. 23. FIG. 23A.

1. When a single metal is placed in contact with an electrolyte, a definite difference of potentials is produced between the liquid and the metal. If zinc be plunged in water, the zinc becomes negative, the water positive, as in Fig. 23. Copper plunged in water also becomes negative but much less so than zinc.

2. If two metals be plunged in water (as copper and zinc, in Fig. 23), the copper, the zinc, and the water forming a galvanic cell, all remain at one potential and no charge of electricity is observed on any part of the system. (It may be proper to remark that this statement is in direct contradiction to the popular impression.) If the copper and zinc had been joined by a metal they would have assumed the same difference of potential as if in direct contact. If a piece of copper be now joined to the zinc (as in Fig. 23_A) C will

become positive and C′ negative, the difference of potentials being that due to the direct contact between C′ and Z only, the water having the effect of simply conducting the charge from Z to C, and of maintaining C and Z at one potential.

3. If a series of galvanic cells be joined (as in Fig. 13), there will be a difference of potentials between the first copper and the last zinc equal to the sum of the differences produced by the two joints between zinc and copper; or, taking the difference of potential which a single junction can produce as one unit, the arrangement in Fig. 13 would give a difference of potential = 2; but if we join another piece of copper to the last zinc, this extra piece of copper will differ in potential from the copper at the other end by the amount 3. This is Volta's theory of the galvanic battery. The difference of potential produced by these arrangements is so small that its direct observation presents considerable difficulty until a large number of galvanic cells be joined in series, when they will be found to be capable of producing a difference of potentials, which can be indicated by electroscopes. It will be observed that the electrolyte is an essential element of the series, for we cannot accumulate differences of potential by simply joining metals in series, the difference which any series produces being simply the same as if the first and last metals were in contact. The electrolyte behaves as if it tended to produce no difference of potentials when in contact with metals, and Volta believed that it did not. We know that it does; but its apparent passivity when two metals are plunged into it instead of one is proved by the following experiments.

Place a metal disc B (Fig. 24) under a light suspended flat strip of metal or needle A, maintained at a high positive potential by connection with a highly charged Leyden jar D. When the disc is of uniform metal the needle A is not deflected to right or left by the presence of B. A charge accumulates on A and B when they are brought close, but the charges are symmetrically distributed relatively to A, so that A is simply attracted to B and does not tend to turn round on the axis or suspending wire E. But if the disc B be made of two metals, such as zinc and copper, with their

FIG. 24.

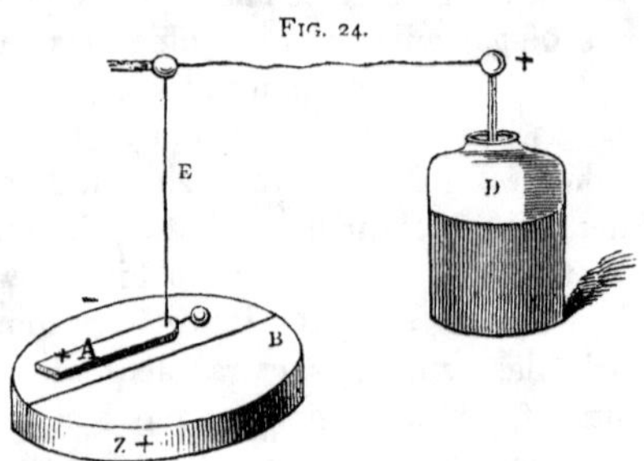

junction placed under the needle A, this needle no longer remains in equilibrium, but deflects towards the side on which the copper is placed, showing that now the charge on B is not symmetrically distributed but that there is a greater

FIG. 25.

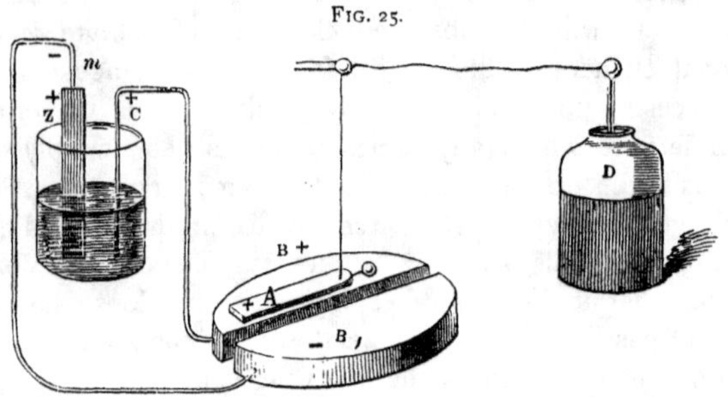

induced charge on the copper than on the zinc. This can only be due to the fact that there is a greater difference of potential between the needle and the copper than between the needle and the zinc; in other words, there is a difference of potential due to contact between the zinc and copper, the zinc being positive relatively to the copper. If the potential of A be negative instead of positive the deflection will be in the opposite direction. The two half discs may be separated from one another by a narrow opening as in Fig. 25. The needle will not deflect if the two halves are of the same metal. It will deflect to a definite amount if the discs are of different metals but in metallic connection by a wire, and the deflection *d* will, when A is positive, be as before from the zinc to the copper, if these are the metals employed for B and B_1. In making this experiment care must be taken to ensure that the half discs are symmetrically placed on the two sides of A, otherwise deflections occur due to charges induced on the two sides of A even when B and B_1 are at one potential. If when the potential of A is reversed being made alternately + and − to equal amounts, we obtain equal deflections in opposite directions, we may be certain that this symmetry is attained. Let two such half discs of copper be carefully adjusted under A; when these are joined by metallic contact there should be no deflection however high the potential of A may be. Then connect the side B with the copper pole of a galvanic cell, and the side B_1 with the zinc pole (Fig. 25); the needle A will deflect towards the side B_1 which is in connection with the zinc pole, and the amount of the deflection will correspond to the same difference of potential as that already observed as due to the simple contact of zinc and copper. Remark that, whereas in Fig. 24 A was attracted to the *copper* half disc it is in Fig. 25 attracted to the half disc in connection with the *zinc*. We know from the first experiment that the junction *m* has made the zinc in the water positive and the copper above *m* with the half disc B_1 negative. We find that the copper C and the half disc B are positive to just the same extent as Z must be, and therefore conclude that the water has simply brought the copper strip and disc B to the potential of the zinc. The experiment is a delicate one, and it can hardly be said to be proved that the difference of potentials between B and B_1 is *exactly* equal to that produced by the simple metallic contact of zinc and copper; there may be a slight difference due to the liquid, and different liquids may possibly augment or decrease this small difference. Another experiment, hitherto unpublished, still more strikingly proves the Voltaic theory. When the two half discs of copper and zinc (Fig. 24) are connected by a metallic wire, it is impossible to find any position of A such that a reversal of its potential does not cause a deflection, and if A is in a symmetrical position relatively to those discs a reversal of the potential of A will always give equal deflections to right or left. When this symmetrical position has been found connect the zinc and copper by a drop of water instead of by the metallic wire. The needle A will remain undeflected in its central position whether its potential be high or low, positive or negative. The two half discs of different metals behave as if they were of one and the same metal in metallic connection. This experiment, which has been carefully made by Sir William Thomson, appears to be absolutely conclusive. The surface of the metals should be polished and clean, for the experiment will not succeed if they are tarnished. Oxides on the surface of the metals introduce complex actions.

The reason why the erroneous statement has so long remained unchallenged undoubtedly is because whenever the two poles of a galvanic cell are connected with electrodes or wires of one metal, as in at least nine hundred and ninety-nine cases out of a thousand they are, the difference between the two electrodes or wires really *is* what it has always been supposed to be. Thus the difference of potential between the two copper wires attached to the zinc and copper of a Daniell's cell is that which has hitherto been generally attributed to the zinc and copper plates of the cell.

§ **23.** The property of producing a difference of potential may be said to be due to a peculiar force, to which force the name of *electromotive force* is given. When we say that zinc and water produce a definite electromotive force, we mean that by their contact a certain definite difference of potentials is produced. A series of the galvanic batteries or cells (Chapter I. § 16) produces a definite electromotive force between the terminal metals plunged in the solution which, if the law above stated held good, would depend on the metals only and not on the solution employed. This, as stated in the last section, is approximately at least found to be true. The electromotive force of a cell or the difference of potentials between the metal poles or *electrodes*, as they are often called, is constant so long as constant metals and a constant solution are used. The words *electromotive force* and *difference of potential* are used frequently one for the other, but they are not strictly speaking identical. It must be remembered that electromotive force is not a mechanical force tending to set a mass in motion, but a name given to the supposed force which causes or tends to cause a transfer of electricity. Wherever difference of potential is found there must therefore be an electromotive force; but we shall find (Chapter III. § 22) that there are cases in which electricity is set in motion, from one point to another, between which that difference of condition does not exist which we have defined as difference of potential. Electromotive force is therefore the more general term of the two, and includes difference of potential as one of its forms.

§ **24.** The electromotive force exerted between two dissimilar metals is altered by every change in their temperatures, but the connection between the change of temperatures and the change of electromotive force has not been thoroughly investigated. Two parts of one and the same body at different temperatures are probably always at different potentials. This has been verified only in certain cases, as in the crystals of tourmaline.

§ **25.** Electromotive force may also be produced by electricity in motion, and by magnetism in ways which we cannot even describe, until the simpler phenomena of electricity in motion and of magnetism have been described;

but it may be said generally that all causes which have the power of altering the distribution of electricity can produce electromotive force or difference of potential. Every source of electricity must as such be able to produce a difference of potential; since no charge of electricity whatever can be made sensible without some difference of potentials between the charged body and the earth or neighbouring conductors. Friction between insulators is found to produce a great electromotive force, producing a large charge on even a small conductor, whereas the galvanic cell or the contact of conductors produces a very small electromotive force, giving a small charge only if the conductor be small. On the other hand, when the conductor is large the galvanic cell will almost instantaneously charge the whole to the maximum potential it can produce, developing by chemical reaction an immense quantity of electricity; whereas the quantity developed by friction from the contact of insulators is so small that if it be allowed to diffuse itself over a large conductor the potential of the conductor will be very little raised. For instance, if we connect a brass ball of a few inches diameter with the conductor of a frictional machine, a few turns of the machine raise its potential so much that its mere approach to the knob of an electroscope will cause the gold leaves to diverge. If we touch the same ball with one electrode of a galvanic cell, the other being connected with earth, the brass ball will indeed receive a charge, but its quantity will be so small and its potential so low that instruments to detect it must be perhaps a thousand times more sensitive than any I have yet described. But if we connect the conductor of a very large condenser or Leyden jar with the galvanic cell, we shall communicate to it such a charge that although its potential would be insensible on the electroscopes hitherto described, its quantity is such that it would sensibly heat a wire in its escape to earth, and would produce many other effects which could not be obtained without the greatest difficulty from the same Leyden jar charged by a frictional machine. A frictional machine charges a *small* Leyden jar with a much greater charge than could be obtained in the same jar even from 1000 galvanic cells ranged in series as in § 16.

§ **26**. Difference of potential or electromotive force must be measured in terms of some unit adapted to measure work. Every unit of work must be represented by the operation of a force overcoming a resistance so as to move it through a distance; or, what is the same, it may be represented by the resistance overcome and moved through a distance. In other words, the unit of force exerted through the unit of space is the unit of work. The most common unit of work is the foot-pound, being the weight of a pound overcome so as to be lifted through the distance of a foot, but the so-called absolute unit of work is that which leads to greatest simplicity in electrical calculations. This unit is the absolute unit of force (Chapter I. § 17) overcoming a resistance through the unit distance, say one centimètre. The absolute unit of work (centimètre, gramme, second) is equal to the foot-pound divided by 13,825 *g*, where *g* is the velocity acquired at the end of one second by a body falling in vacuo: taking this as 981 centimètres per second the absolute unit of work is equal to the foot-pound divided by 13,562,325. The unit difference of potential or electromotive force in electrostatic measure exists between two points when the unit quantity of electricity in passing from one to the other will do the unit amount of work.

The practical measurement of the difference of potentials between two points can in certain cases be made by observing the work done by definite quantities of electricity in passing from one point to the other; thus we may observe the total amount of heat generated in a wire by a given quantity of electricity passing between two points kept at a constant difference of potentials. From the heat we may calculate the work, and from the heat and quantity we may calculate the difference of potentials. [Similarly, if we wished to ascertain the difference of level between two points we might let a weight (a standard quantity of matter) fall from one to the other, measure the total heat generated by the concussion which brought the weight to rest, from the heat deduce the amount of work done, and from this work and the known quantity of matter, deduce the difference of level or of gravitation potential. Fortunately there are more direct methods available or engineers would have some difficulty in levelling.]

Difference of electric potentials is more generally ascertained indirectly by a knowledge of the laws connecting potential with other electrical magnitudes. Thus we know that the quantity of electricity with which two opposing surfaces of conductors are charged is simply proportional to the difference of potential between them, assuming the distance and dielectric to remain constant. Electrometers afford us the means of comparing such quantities as these, and therefore electrometers (as shown in § 16) afford us the means of comparing differences of potential. The measurement of currents and of resistances to be described in the following chapters give other means of comparing differences of potential.

CHAPTER III.

CURRENT.

§ 1. Electricity has already been frequently spoken of as redistributing itself over a given conductor, or moving from one conductor to another along a wire, and we may with propriety speak of the *current* of electricity by which the redistribution is effected. Bodies along which electricity moves acquire, so long as the motion lasts, very singular properties, and in order to avoid cumbrous phraseology the properties which are actually observed as belonging to the bodies through which a current of electricity flows, are spoken of as the attributes of the current of electricity itself. Some of the properties of electric currents are most conveniently observed in long uniform conductors, such as wires, along which the flow takes place in one simple direction. Currents in wires will chiefly be spoken of in the first instance, although identical properties are possessed by currents moving in any manner through bodies of any form. The direction of a current is assumed as the direction from the place of high potential to the place of low potential; in other words, it is the direction in which positive electricity flows. Thus, to recur to our earliest definition of positive and negative electricity, if one conductor A be electrified by contact with a stick of glass which has been rubbed with a resinous material, and another conductor B be electrified by contact with the resin used to rub the glass, then upon joining A and B, a current of positive or vitreous electricity will flow from A to B until they are brought to the same potential. By using two large conductors A and B, or two Leyden jars of large capacity, and electrifying them with a frictional electrical machine of considerable size to a high potential, a considerable quantity of electricity may be accumulated on A and B, and a considerable current will flow from A to B, when they are joined.

§ 2. A current of electricity thus produced will be transient, and even while it lasts it will not remain constant, for during its continuance the difference of potentials producing it will continually diminish; indeed, if the above were the only manner of producing an electric current, we might still be ignorant of its peculiar properties. When plates of zinc and copper not touching one another are plunged in water and the copper is then joined to the zinc by a wire outside the water, a current flows from the copper to the zinc along the wire, and from the zinc to the copper through the water. According to the theory of the cell explained in the last chapter, the zinc when it touched the copper became positive and the copper negative, the

FIG. 26.

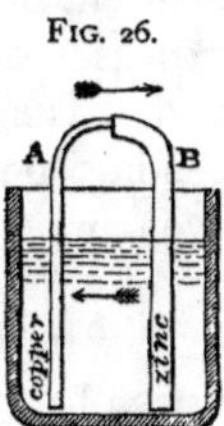

electricities being separated at the metallic junction, but there being no opposition to their recombining through the water, the current flows in the direction shown. The existence of the current is shown by the fact that if A and B be joined by a long copper wire, this wire acquires the same properties as if it joined two large conductors charged with opposite kinds of electricity. These properties are described in § 6 and the rest of this chapter.

§ **3.** The transfer of electricity from A to B involves the performance of work or its equivalent, and to perform work implies a source of power, or in other language an expenditure of energy. The mere contact of two dissimilar substances cannot be a source of power. It is found that while the current flows the water is decomposed, and oxide of zinc formed. This chemical reaction is a true source of power; the oxygen leaves the hydrogen of the water to join with the zinc, for which it has a greater affinity. The zinc is consumed in the process, as coal is consumed when it burns while combining with the oxygen of the air. The source of power is accurately described by saying: the intrinsic energy of a given weight of zinc and water is greater than that of the hydrogen gas and oxide of zinc produced by the combination, the difference is equal to the work done by the current of electricity produced. The work done by the current is therefore proportional to the amount of zinc consumed. The electromotive force of the cell is constant, depending on the metals in contact; the performance of a given amount of work by the transfer of electricity from one point to another, between which there is a constant difference of potentials or electromotive force, requires the transfer of a definite amount of electricity, hence the quantity of electricity produced by the galvanic cell is proportional to the zinc consumed. The effect described as occurring in the simple form of the galvanic cell is produced whenever we join two solid conductors A and B plunged in a compound liquid, one element of which tends to combine more strongly with A than with B, or with B than with A.

If we consider the liquid alone we find that positive electricity is produced apparently at the surface of contact between the liquid and one conductor, and is taken away as fast as it is produced to neutralise the negative electricity produced apparently at the surface, where the other conductor touches the liquid.

§ **4**. A bitter war raged for a long time between the electricians who maintained that in this case the electricity was due to contact, and those who maintained that it was due to chemical action ; like many other disputes, it turns much upon the use of words.[1] Both contact between dissimilar substances and chemical action are necessary to produce the effect; the laws regulating the potential and those regulating the current are intimately connected with the nature of the substances in contact, and with the amount of chemical action. Perhaps it is strictly accurate to say that difference of potential is produced by contact, and that the current which is maintained by it is produced by chemical action. As we shall see hereafter, the difference of potentials can be accurately determined from a consideration of the chemical action, but then this chemical action depends probably on the very properties which cause a difference of potential to be produced by contact. In cases where no known chemical action occurs, as where copper and zinc touch one another, the difference of potential is produced, and since this involves a redistribution of electricity, a small but definite consumption of energy must then occur; the source of this power cannot yet be said to be known.

§ **5**. The law described in Chapter II. § 21, by giving a contact potential series, or electromotive series, for metals, shows

[1] The opponents of contact electricity denied and falsely explained things now known to be true, and the original supporters of the contact theory were ignorant of dynamics.

why we have no hope ever to obtain a permanent current by any arrangement of metals, each at one temperature. The electromotive force at the joint C (Fig. 27) is necessarily equal to that at joint D, and opposed to it, i.e. the E. M. F. (as electromotive force may for brevity be written) at C tends to send the electricity round in the direction opposed to the hands of a watch, while the E. M. F. at D tends to send electricity round in the opposite direction, and the two forces being equal, electricity moves neither way.

FIG. 27.

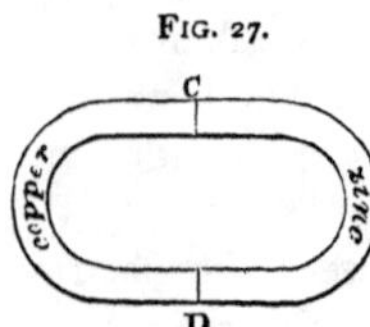

When instead of bringing the zinc and copper into contact at D they are plunged into water, the E. M. F. at the junction remains as before; but owing apparently to the electrolysis or decomposition of the water, little or no electromotive force is at first manifested at the surface where the water touches the metals, and the current can therefore flow as described in § 2. The arrangement of potentials in the cell, in the plates, and in the wire joining the plates, cannot be explained until after Chapter IV., and indeed has never been very fully or clearly established. What chiefly concerns us is that galvanic cells can be arranged so as to produce a permanent current conveying considerable quantities of electricity ; the *strength* of the current is simply proportional to the quantity conveyed in a given time.

§ **6.** The properties of electric currents are very important. Two parallel wires in which electric currents flow in the same direction, attract one another It is simpler to state this fact by saying that parallel currents in the same direction attract one another. Parallel currents in opposite directions repel one another.

When the wires conveying the currents are straight but not parallel, they attract one another if both currents flow to or from the apex of the acute angle which the wires make with one another.

The wires or currents repel one another if one current approaches and the other recedes from the apex of the angle.

§ 7. Consider a rectangle of wire E F G H (Fig. 28) held over a straight wire A B, each having currents circulating in them, as shown by the arrows, and let the rectangle be capable of turning on a vertical axis X X_1; it is found by experiment that E G is attracted towards A; F H, on the contrary, towards B; both therefore tend to turn the rectangle in the same direction round its axis; that portion of H G which is behind A B is attracted towards B, and repelled from A by § 6. On the contrary, the portion of H G which is in front of A B is repelled from B and attracted towards A; all these forces act therefore in one direction, and tend to place E F G H in

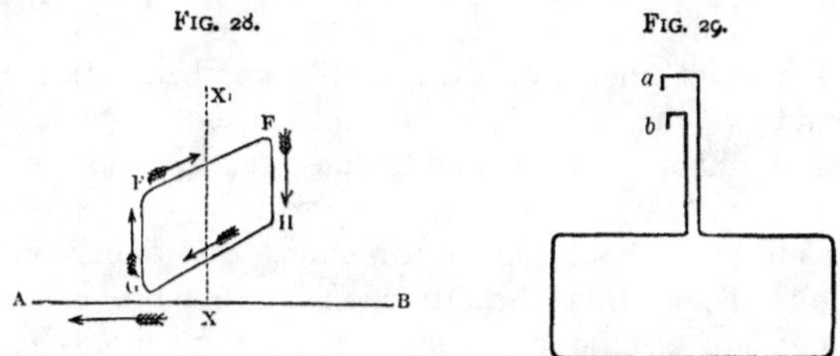

FIG. 28. FIG. 29.

a plane parallel to A B. The forces in E F are acting in the opposite direction, but E F being farther from A B than the other portions of the current, the forces due to it are weaker and are overpowered. These attractions and repulsions are easily verified with a rectangle of copper wire made as in Fig. 29, and supported by two pivots *a* and *b* resting in two mercury cups, which are connected by thick wires with a Grove's cell.

§ 8. If we conceive one rectangle A B C D (Fig. 30) inside another E F G H, all the actions described in the last will be strengthened, and the two rectangles will tend to place themselves in parallel planes, and moreover in such a position that the current is going in the same direction in both rectangles. The truth of this proposition is evidently not limited to rectangular systems, and generally any two closed wire circuits in which currents are flowing tend to arrange themselves in this manner. When the two are in the same plane they may be so arranged, as for instance where they are concentric circles, that the one does not attract the other at all but merely directs it, as described above. If the one circuit were not in the same plane as the other they would attract one another even after they had placed themselves in parallel planes, and if forced to remain in such a position that the currents were flowing in opposite directions in the two circuits, they would repel one another. If the two circuits were in one plane, but not concentric, there might be a resultant force tending to cause relative movement in that plane, due to the greater proximity of the wires at certain parts. All these attractions and repulsions are wholly distinct from the attractions and repulsions between charges of electricity at rest. They were discovered by Ampère.

FIG. 30.

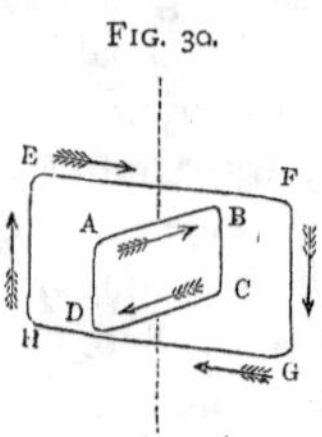

§ 9. All the actions of currents one upon another may obviously be multiplied by using, instead of a single wire, a coil of wires, through each winding or turn of which the same current is flowing. Thus, a circuit composed of twenty turns of wire on a reel would be acted upon with twenty times the force that a single turn would experience with the same current flowing through it; and again, if the second circuit be also composed of twenty wires, each with a current equal to the original one, the forces in action will be again multiplied twentyfold. So that a circular coil A (Fig. 31) of twenty turns of wire hung up by a fibre inside a fixed coil B of twenty turns of wire, will experience a directing force 400 times greater for any given current circulating in both than would be experienced by a coil with a single turn hung inside a coil with only one turn. This fact allows the construction of instruments called *electro-dynamometers*, adapted to show the presence of electric currents. A coil A of perhaps several thousand turns may be hung up inside a coil B, also consisting of a large number of turns, each turn being insulated from its neighbours by silk. A and B are, when no current is passing, maintained in planes at right angles to one another by a small directing force, such as the torsion of a wire. When a current is passed through both, the inner coil is turned in such a direction as to place it more parallel to B than before, and with the currents running in the same direction. The instrument may be modified, so that a known current is passing through A, and the one to be examined passed through B only. The direction of the unknown current is indicated by the direction in which A turns, and its magnitude or strength by the angle through which it is turned.

FIG. 31.

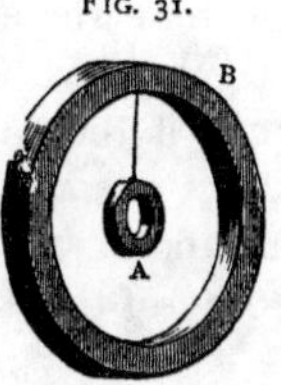

§ **10**. Other arrangements of a similar kind will suggest themselves to the reader. If the centre coil A, instead of resembling a ring, were a coil of small diameter as in Fig. 32, forming a cylinder of considerable length, so arranged that the current flowed in the same direction round all parts of the cylinder, the deflection of the internal cylinder would be more immediately visible, and the ends *a* and *b* might be considered as two poles, having a tendency to place themselves at right angles to the plane of the directing coil. When such a cylinder as this is placed wholly inside another, having similar coils parallel to it, it will be in stable or unstable equilibrium, as the currents flow in the same or in opposite directions.

FIG. 32.

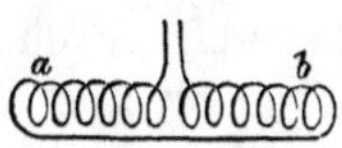

If the pole *a* were introduced inside the coil B A, as shown in Fig. 33, the coil *a* would be sucked in by the action of one current on the other. If, on the other hand, the currents flowed in such a direction that the pole *b* were placed inside or near the similar pole B, as in Fig. 34, the inner coil would be expelled or repelled from B. These actions are apparent whatever be the diameter of the coils. Conceive next that two flat spiral coils (Fig. 35) are placed face to face: if the currents flow in the same direction, they will attract one another; if in opposite directions, they will repel one another.

Any of these arrangements may be made use of to show

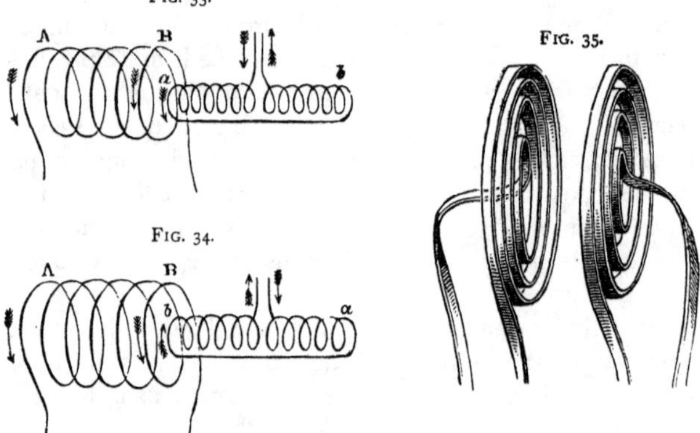

Fig. 33.

Fig. 34.

Fig. 35.

the presence, direction, and magnitude of a current in a wire. By using a large number of turns of fine copper wire insulated with silk, and suspended so as to turn with very small frictional or torsional resistance, it is easy to construct apparatus showing all the phenomena described in § 9 and § 10. The long cylindrical coil described in this section is sometimes called a solenoid.

§ **11.** Magnets are found to be influenced by electric currents almost exactly as solenoids are. In the presence of a current, they are directed so that if free to move, they stand across the current. This fact was first observed by Oersted. The end of the magnet which points to the south, when freely suspended, is similar to that pole of the solenoid in which the current is moving in the direction of the hands of a watch, holding the watch with its back to the coil; or, in other words, if the solenoid be like a right-handed corkscrew and the current enters at the point, the point will behave like the end of a magnet which points south. The solenoid and magnet have many properties in common. The solenoid may be directed by a single rectilinear current, and so may the magnet; but just as the directive action on the solenoid is increased by wrapping the directing coil all round it, by bringing the coils into close proximity, and by increasing the magnitude of the current flowing through the directing coil, so the directing force or couple acting on a magnet is greatly increased by sending the current in the directing coil round it many times, by bringing that coil very close to the magnet, and by using a powerful current. This property of the magnet allows us to construct instruments called galvanoscopes and galvanometers for the detection and measurement of currents without using a double coil of insulated wire. In galvanoscopes a magnet hangs inside a directing coil, each turn of which is placed north and south. The magnet hangs with its poles north and south so long as no current passes through the coil, but when a current passes, it is deflected more or less towards one side or the other, until the couple due to the directing action of the current is balanced by the couple due to the directing action of the earth. When the current in the directing coil (Fig. 36) flows from south to north in the top of the coil, the end of the magnet which pointed south, and which will

FIG. 36.

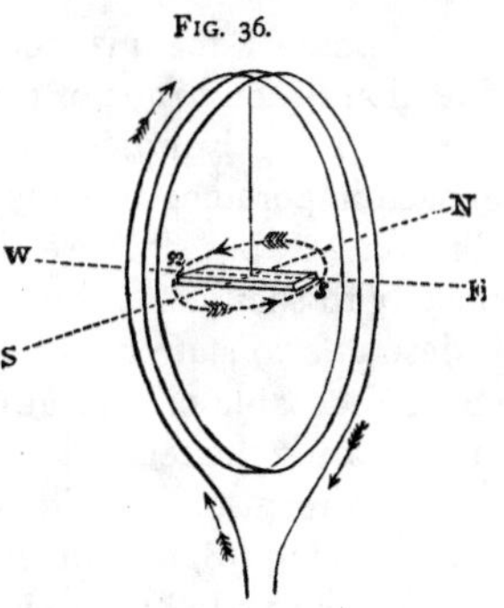

hereafter be called the south pole of the magnet, turns towards the east.

The direction in which a magnet tends to turn across a current may also be described as follows. Imagine a man lying on the wire which conveys the current, in such a direction that the current was from his feet towards his head, his face being turned towards the magnet; then, under the influence of the current, the pole of the magnet which, when free, turns to the south, will turn towards the right hand of the man. Or let a current be flowing through a copper corkscrew, and let the magnet take up its natural position inside the coils of wire; then if the corkscrew be turned the way of the current it will screw from south to north, through the compass needle considered as a cork.

The following is a third description of the direction in which a current deflects a magnet. Imagine a watch strung on the wire conveying a current so that this current goes in at the back of the watch and comes out at the face through the central pivots; then the south pole of the magnet is impelled by the current in the direction of the hands of the watch (Fig. 37).

FIG. 37.

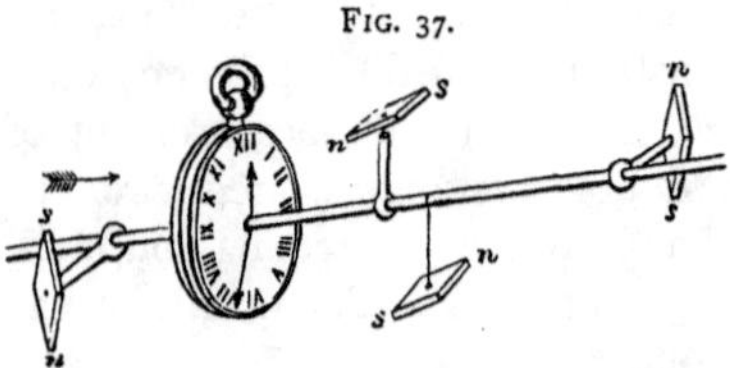

§ **12**. The galvanoscope and galvanometer are instruments of such importance that they will be described at length in Chapter X.; but since we shall have occasion in future continually to speak of electric currents and their properties, it is desirable to state how a galvanometer may be easily constructed capable of indicating the presence of a current and of comparing the relative strengths of various currents. Wind copper wire insulated with silk on a hollow brass cylindrical bobbin A (Fig. 38) with deep flanges B B_1, which may have feet at C by which the bobbin is supported on wood or vulcanite. Inside A fit a small brass plug D, having at one end a hollow chamber, closed by the lens E, with a focal distance of about 120 centimètres. In the little chamber suspend a mirror and magnet by a single silk fibre, such as may be drawn out of a cheap silk ribbon. This fibre must be so thin as to be nearly invisible. The mirror should be formed of microscope glass as truly plane and as thin as possible. The magnet may be attached to the back by a little shellac dissolved in spirits of wine. Care must be taken that the mirror is not drawn out of shape by the magnet. The silk fibre must also be attached with shellac varnish. It may then be threaded through a hole in the chamber by means of a needle of sealing wax or shellac, and secured with a little mastic or other varnish. The plug D can then be introduced or withdrawn from A at pleasure.

FIG. 38.

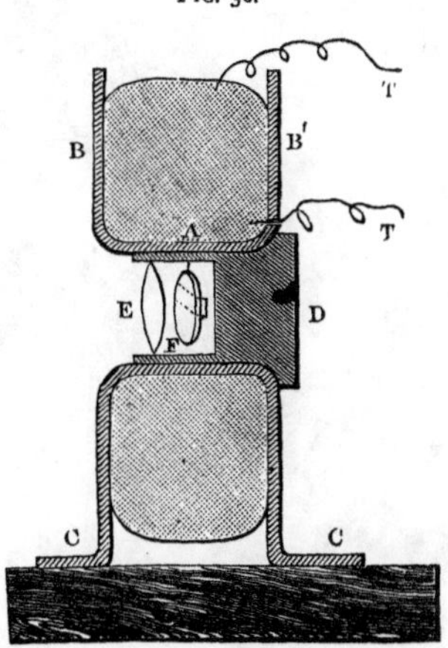

If currents are to be observed which are passing through circuits of great length or containing bad conductors the wire should be thin, say No. 40, and many thousand turns may be employed: the diameter of the chamber inside the plug D may be 1·5 centimètre, the length from B to B_1 3·5 centimètres, and the outside diameter of the flanges B B_1 6 or 7 centimètres. This size will contain many thousand turns of fine wire.

If currents are to be observed which are passing in short lengths of wire or other good conductors the space inside the flanges B B_1 may be filled with two or three dozen turns of stout copper wire, say No. 16 or No. 20. The two ends of this coil T T_1 may conveniently be connected to two

Fig. 39.

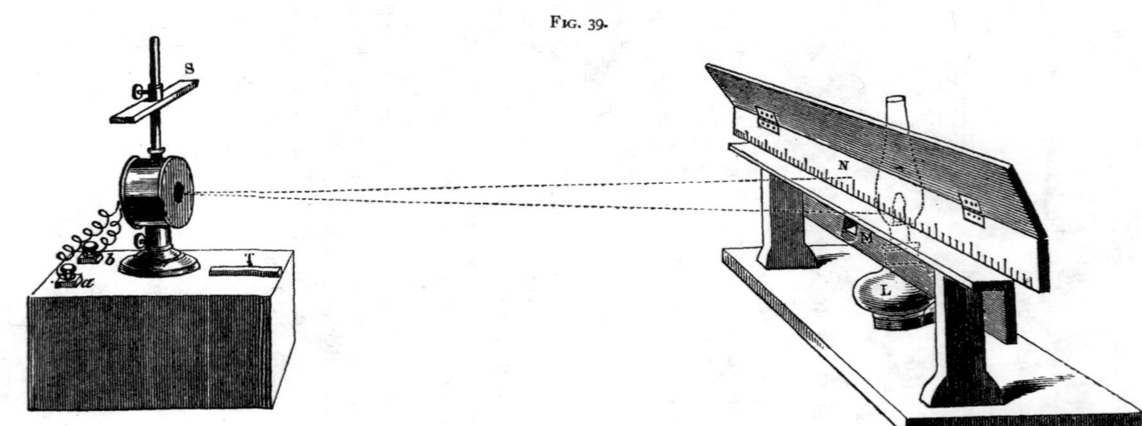

brass pieces (Fig. 39) well insulated by vulcanite and having screws by which other wires can be joined to the same terminals as they are called. The instrument is completed by a paraffin lamp L, placed behind a screen having a slit M in it about 60 centimètres in front of the coil and horizontal white scale N about 45 centimètres long.

When placed as in Fig. 39 the light from the lamp passes through the slit in the screen, through the lens E on to the mirror F, by which it is reflected back on to the scale. An image of the flame is seen on the scale. When the light falls perpendicularly on the mirror this image appears on the scale immediately above the slit in the screen. If by the passage of a current through the coil the magnet is deflected to the right or left, the image moves to the right or left along the scale, the angle formed by the reflected rays being twice the angle through which the magnet and mirror are deflected. A very small angle produces a great displacement of the image. With the dimensions named the horizontal displacement of the image is nearly proportional to the strength of the current. If the scale be bent so as to form part of a cylindrical surface having the axis of suspension of the mirror as its central axis, the reflected spot of light is more clearly seen through the whole range. This instrument is Sir William Thomson's mirror galvanometer. With its assistance the presence, increase, or decrease of a current can be observed. It is convenient to place a bar magnet S in the magnetic meridian immediately above the coil; by raising or lowering this magnet, the directive force of the earth may be increased or weakened. If the south pole of S is placed to the south the magnet may by trial be put at such a distance from the suspended mirror and magnet as almost exactly to counterbalance the effect of the earth's magnetism. The instrument will then be very sensitive, but the spot of light will never remain quite stationary. A second magnet T, placed perpendicular to the magnetic meridian, may be used to adjust the zero of the instrument, i.e. to bring back the spot of light to a fiducial mark at the centre of the scale when no current is passing. The direction of the magnetic meridian is that in which a free magnet naturally points.

§ **13**. A current not only acts on a piece of steel or iron which is already a magnet, but it converts any piece of non-magnetised steel or iron in its neighbourhood into a magnet having its poles so situated that they lie in the line along which a free magnet would place itself under the action of the current. This magnetising action is more powerful as the iron is placed nearer the current, as the current is more powerful, and as a greater length of the current acts in the same sense on the iron. Thus, a piece of iron placed inside a helix or bobbin (Fig. 40) of many coils is strongly magnetised by the current and has its north and south poles placed as shown in Fig. 40.

FIG. 40.

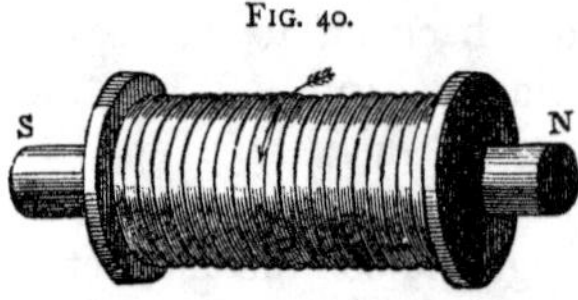

The magnetisation produced by the current is only temporary if the iron be soft or annealed, but a portion of the magnetisation produced in hard iron is retained long after the current has ceased to flow, and in a hard steel bar some portion of it is permanently retained. Work is done, and energy expended, in producing this magnetisation.

§ **14**. The current in the wire implies a transfer of electricity under the action of electromotive force; and by the very definition of electromotive force work in some form must be done during the transfer.

When a current flows through a simple wire and does not magnetise iron or set any mass in motion, the energy expended in producing the current is wholly employed in heating the conducting wire, the heat developed in any part of the wire being precisely equivalent to the work which would be done in bringing the same quantity of electricity from the one end of the wire to the other on a little conductor against the statical repulsion described in § 1, Chapter II. If any portion of the energy is employed in other ways, as described above, so much less heat is developed in the wire. The rise of temperature in the wire depends on the specific heat of the metal of which it is composed.

§ 15. When the current traverses a compound liquid conductor instead of a solid simple metal wire, the liquid is in many cases decomposed, one element or group of elements moves to the spot at which the current enters the fluid, and the other to the spot at which the current leaves the fluid. Faraday called the metal surface at which the positive current entered the fluid the *anode*, and the other surface the *kathode*. The compound decomposed by the electricity is called an *electrolyte*, the process of decomposition *electrolysis* and the products of electrolysis *ions*. Thus when two glass tubes (Fig. 41) C and D, filled with water, are inverted over a vessel of water, and the two platinum wires A B introduced into the vessel, then upon connecting A and B with a sufficiently powerful galvanic battery so that a current may pass from A to B, the water is electrolysed; oxygen is found in C and hydrogen in D, in the proportions forming water.

FIG. 41.

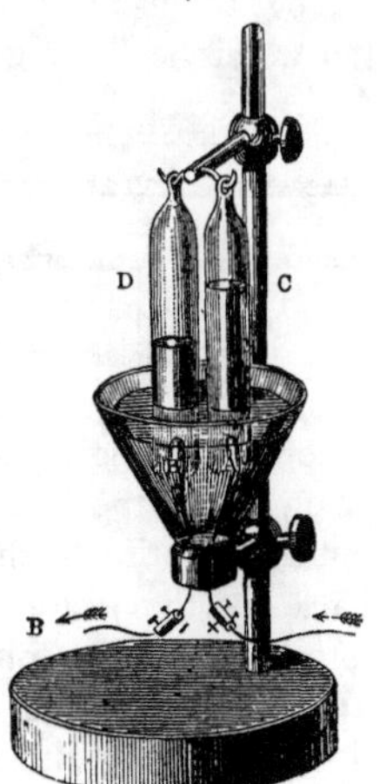

Energy is expended in decomposing any compound, just as energy is evolved in the combination of elements which have a chemical affinity one for another. The energy expended in the decomposition of an electrolyte is not available to produce motion or heat in the circuit.

§ 16. Currents traverse even very bad conductors, but the current is small, i.e. comparatively little electricity passes in a given time with a given E. M. F. Bad conductors are generally compound bodies. The resins, may be taken as examples. Feeble currents also traverse electrolytes without producing any sensible amount of electrolysis. It is certain that work of some kind is done by the current as it passes through these bodies; but it is not yet known by what action the work is represented, that is to say, it is not known whether the bad conductor is heated, or decomposed, or whether some other form of work represents the energy expended.

§ **17.** If a current be allowed to set a magnet in motion, for instance, to expel one pole of a magnet previously introduced into a helix, the current experiences a real resistance, and its flow is checked by the effort. The mere presence of the magnet if it is at rest does not check the current; a certain statical force exists between the current and the magnet, but so long as no motion occurs in consequence of this force or against this force no work is done, and the current flows as if the magnet were not there. A rough analogy to this might be found in the following arrangement. Let water be flowing through a pipe at one side of which there is a piston A (Fig. 42) held in position by a spring at B. The water as it flows through the pipe will press on the piston A, and by means of a piston-rod may exert a force at B. When this force just balances the force of the spring, the water in flowing past the piston does no work by means of it or on it, and the current proceeds as if no piston were there; but if the spring be then weakened or let go so as to be forced back by the piston, the lateral pressure of the water in forcing back the piston overcomes a resistance through a certain space and does work as the current of electricity does in moving the magnet. Moreover, the flow of water will be checked or diminished while the work of pushing back the spring is being done. When the spring has been

FIG. 42.

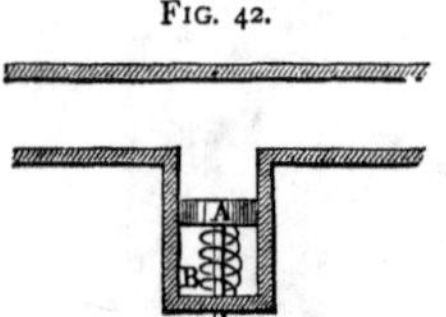

pushed back so far that its elastic force balances the pressure in the pipe, the current in the main pipe will flow on as before, unaffected by the presence of the spring B. In like manner the electric current which was checked in its flow while deflecting the magnet flows on as before after the magnet has come to rest. The analogy is imperfect, inasmuch as the diminution of the water current is accompanied by a change of capacity for the water, whereas the diminution of the electric current is unaccompanied by any increase of capacity. The water is only diverted, whereas the electricity is really retarded. This diminution of the current while it is doing work occurs not only when the work consists in moving a magnet, but also when the work consists in moving a wire or wires conveying currents, as in the electro-dynamometer, or in magnetising soft iron.

§ **18.** If the piston A in Fig. 42 be forced back towards the pipe containing water, it will produce a current, the effect being reciprocal to that which was produced when the current was diminished by forcing forward the piston; work is done by the piston as it is forced forward, and this work is expended in producing an extra current of water.

Similarly, if the magnet which has been deflected be forcibly moved back, energy is required to force it back against the resistance due to the electrical repulsion of the current, and this energy performs work represented by an increase in the current exactly corresponding to the diminution experienced when the current was expending energy in forcing back the magnet. The current is said to be induced in the wire by the motion of the magnet relatively to the wire. The case is one of energy stored and restored. When the current forced back the magnet the energy of the current was expended in such a manner as to be stored up in the system. When the magnet returns to its original position the energy is restored to the current. The example already given of water in a pipe forcing back water against a spring affords one instance of energy stored and restored; another is afforded by the common pendulum. The energy of the pendulum exists alternately in a latent or potential form due to the attraction of gravitation, and as actual energy due to motion. As the bob rises the actual energy is gradually transformed into potential energy, being thus stored up. As the bob falls the potential energy is reconverted into actual energy, being thus restored. Just so, if a current deflects a magnet and causes it to swing backwards and forwards, the energy alternately exists in the form of electric repulsion and actual energy of motion; but there is this difference between electric and gravitation examples: the force of gravitation is neither increased nor diminished by the motion of the pendulum, whereas when the magnet swings in obedience to the impulse given by the current, the current diminishes, and when the magnet swings back against the impulse of the current, the current is increased.

§ **19.** Motion of the piston in Fig. 42 would produce a current in the pipe, whether one existed before or not; if the piston were drawn back from the pipe it would suck water in at the mouth, if moved forward it would drive water out; quite similarly, the motion of a magnet in the neighbourhood of a conductor, the motion of a wire containing an electric current, or the increase or decrease of magnetism in a magnet near a conductor, will each of them cause currents to flow in that conductor; *the direction of the current in the conductor or wire will be such that it resists the motion of the magnet or of the current, or the change in the current, or the change of magnetisation.*

The following are examples of the application of this general principle, first enunciated by Lenz. Let there be a metallic ring A B (Fig. 43), a second ring C D, in which a current flows in the direction of the arrows, and a magnet N S; then, while the relative position of C D, A B, and N S do not vary, and while the current in C D and the magnetism in N S remain constant, neither increasing nor diminishing, no current whatever will flow in the ring A B.

but any change in any one of these conditions will produce a current in A B; thus:

1. If the ring C D moves nearer A B a current will be induced in A B in the direction of the inside arrows, and during this action the current in C D will be diminished.

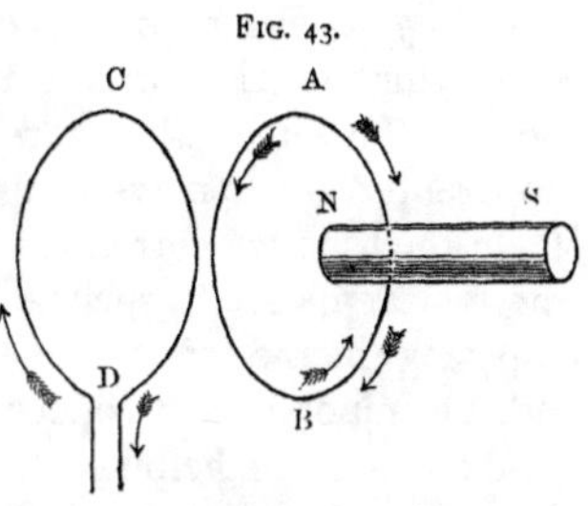

FIG. 43.

2. If the ring C D be removed farther from A B a current will be induced in A B in the direction of the outside arrows, and during the induction the current in C D will be diminished.

3. If the pole N of the magnet N S be pushed into the ring or nearer to it, a current will be induced in A B in the direction of the inside arrow, and the motion is resisted.

4. If the pole N of the magnet be withdrawn to the right hand, out of or away from the ring, a current will be induced in A B in the direction of the outside arrows, and the motion is resisted.

5. If the magnetism of the magnet be increased, a current will be induced in A B in the direction of the inside arrows, and the increase of magnetism is thereby resisted.

6. If the magnetism of the magnet be diminished, a current will be induced in A B in the direction of the outside arrows, and the diminution of magnetism is thereby resisted-

If instead of simple rings we have long thick coils of many turns, the effects will be much more sensible. The effects of induction between straight wires and magnets can with ease be deduced from the general principle enunciated above. *Induction* is the name given to this phenomenon, which, however, has nothing in common with the induction described in Chapter I. To distinguish between these phenomena, that described in Chapter I. must be designated electrostatic induction, and the induction of currents, electro-magnetic induction. Electrostatic induction is called 'influence' in French and German.

Owing to electro-magnetic induction magnets and wires conveying electric currents are not as free to move as other bodies. They may when at rest be in perfect equilibrium, and apparently free to move in all directions, but when we move them they induce currents in neighbouring conductors, and these currents are in such a direction as to produce a force opposing the motion of the first magnet or current. It is, indeed, impossible to conceive that by moving they should produce a force helping their own proper motion as in that case perpetual motion, or rather a perpetually increasing source of energy, would be the result.

§ **20.** A current which *commences* in a given circuit may be likened, so far as its effects on a neighbouring conductor are concerned, to a permanent current brought suddenly from an infinite distance to the spot where it stands. We know that by bringing a current C D (Fig. 43) from a distance to a position alongside a wire forming part of a distinct circuit A B, we should cause the induction of a current in A B opposite in direction to that flowing in the parallel wire C D. The beginning of a current in C D has exactly the same effect and induces a current in the opposite direction in A B; again, an increase of current in C D acts in the same manner as bringing C D nearer to A B. It induces a current in the opposite direction to that in C D. These induced currents cease as soon as the inducing current C D ceases to increase, just as the induced current in A B would cease as soon as C D, while conveying a permanent current, ceased to approach A B.

The diminution of a current in C D produces the same effect as removing C D from the neighbourhood of A B, i.e. it induces a current in A B in the same direction as that in C D. The total cessation of the current C D acts like the infinitely distant removal of C D with its current, and of course induces a current in A B in the same direction as that which flowed through C D. We may therefore add to the examples given in § 19 two more.

7. If the current in C D ceases or is diminished, a current will be induced in A B in the direction of the outside arrows, and the diminution of the current in C D is thereby delayed.

8. If the current in C D commences or is increased, a current will be induced in A B, in the direction of the inside arrows, and the increase of the current in C D is thereby delayed.

§ **21.** Induction is the unfailing accompaniment of the beginning or increase and termination or decrease of a current, for there are always conductors somewhere near in which the induced currents flow. The induced currents diminish for the time being the strength of the inducing current, and thus we see that neighbouring bodies change the rate at which a beginning or ceasing current comes to its permanent condition. If the whole or a large part of a circuit of small resistance is very near the inducing current, and so disposed that the induction tends to occur throughout in one direction, the induced current will be considerable, and its reaction on the inducing current will also be great, shortening the time it requires to reach the permanent condition. If the circuit in which the induced current flows is, on the contrary, far removed from the inducing current, or only exposed to induction for a small part of its length, or so placed that the current tends to flow in opposite directions at different parts of the circuit, or has a great resistance, then the induced current will be small and its reaction on the inducing current will also be small. The inducing current produces an electromotive force in the circuit conveying the induced current, and we may say that the induced current is due to the induced electromotive force. If the inducing current A be near a number of conductors B C D, the induced current in B tends to weaken that in C and D, inasmuch as a current beginning in B would induce currents in C and D in the direction of the original current A. Thus the induced current in B is less than it would have been if C and D had not been there, and the inducing current in A is less checked than it would have been if C and D had not been there.

An increasing or diminishing current not only induces an

FIG. 44.

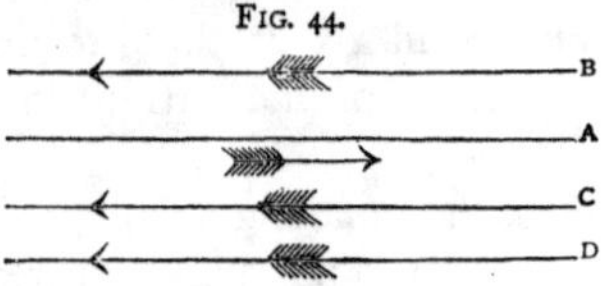

E. M. F in neighbouring conductors but also exercises an inductive action on the current in which it flows. Thus let us consider a circuit coiled back as in the annexed figure. An increasing current between A and B, flowing as shown by the arrow, tends to induce a current between C and D in the opposite direction. The E. M. F thus induced between C and D opposes the original current, and delays its increase. If the current between A and B is diminishing, it tends to induce a current between C and D in the same direction as it is flowing, and the result is to delay the decrease. Thus the action in both cases is to delay change. Even when the wire is straight a similar but much weaker effect occurs. A current flowing

FIG. 45.

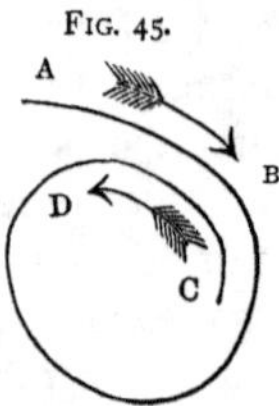

FIG. 46.

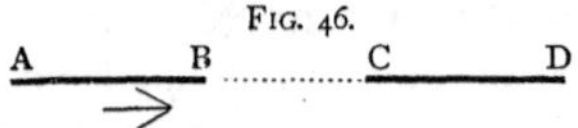

(Fig. 46) from A to B repels one flowing from C to D; if then a current increases in A B, it induces a current in front of itself in the direction in which it is flowing, and is checked in so doing. The effect is to diminish the abruptness of the increase.

§ **22**. The conductor in which the current is induced

need not form what is called a closed circuit, i.e. such a conductor as is formed by a ring of wire round which the current can continue to flow permanently if a permanent E. M. F. be kept up round it, as distinguished from a broken circuit, such as would be formed by a ring of wire incomplete at one or more points, where the presence of air or other non-conductors would stop any permanent current; but although the induced current will be very different in the two cases of a closed and open circuit it will be produced in both. In the closed circuit we may have a current induced without difference of potentials between the parts. We cannot have difference of potential between two parts of a conductor without a current ensuing, but we may have a current due to E. M. F. without any difference of potential. The analogy of water in a pipe will make this clear. If there be difference of level between two reservoirs in connection with one another, as in Fig. 47, the water will flow from the higher level to the lower. But even if the two reservoirs be at the same level, when a rope is rapidly drawn through the pipe from A to B, water will by friction be dragged along the pipe, and water will flow from A to B, causing B to rise in level or gravitation potential. Here the current cannot be said to be due to a difference of potential, and the difference of potential which finally results from the action is opposed to that which would have produced the current.

FIG. 47.

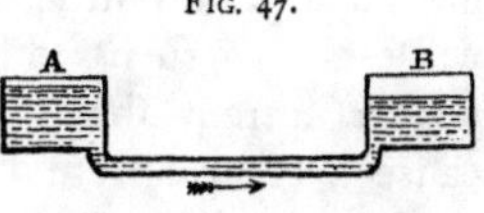

Again, if the water be enclosed in a circular pipe (Fig. 48), and an internal wire *a a a* be caused to rotate inside this pipe about the axis of the ring, it will set all the water in the pipe in motion, without causing any difference of pressure between two parts of the pipe; in this case there is no difference of gravitation or pressure potential causing the motion, nor is any difference of potential necessarily caused by the motion. The two cases of a closed and broken circuit are analogous to this. In the closed circuit the current may continue indefinitely so long as the motion of the inducing magnet continues, but no difference of potential need be produced between any parts of the circuit. In the broken circuit, on the contrary, the current is not produced by a difference of potential between different parts, but the E. M. F. drives positive electricity to one end of the wire, and negative electricity to the other, producing a difference of potentials which will send back a reverse current so soon as the inducing action of the magnet is over; the first current may be exceedingly small, even in cases where if the circuit were closed the current would be great, for a small quantity will in bodies of small capacity be quite enough to produce a difference of potential balancing the inductive action of the magnet. Just as in Fig. 47, if the reservoirs A and B are small, a very little water dragged from A to B by friction will establish such a difference of potentials as will stop all further current though the friction might be sufficient to cause a great current in the closed circuit (Fig. 48). As soon as the difference of potentials between A and B in the broken circuit is sufficient to cause a reverse current equal to that which the magnet moving as it does can induce, no further current will be induced in the broken circuit, precisely as under similar circumstances the friction of the rod would cease to produce a current of water; but no motion of the magnet or other inducing system can be so small as to fail to produce a continued current in the closed circuit, for no difference of potentials is necessarily created tending to reverse the action.

FIG. 48.

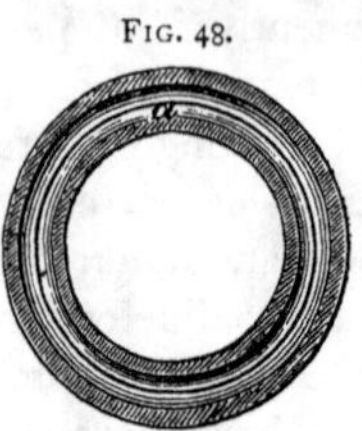

§ **23.** A complex case arises when the closed circuit is long and of sensible capacity while the inducing action takes place on one part only. This case is analogous to a long elastic pipe (as in Fig. 49), inside which a short rod is moving, producing a current by friction; here there may be accumulation of water in front of the rod and a deficiency behind. There may be, therefore, an increase of pressure in front of the rod and a defect behind, tending to reverse the current produced by the friction of the rod. Just so with the electric current, there may be at parts of the long circuit differences of potential produced tending to reverse the direction of the induced current; the potential being raised at the parts *into* which the positive current is flowing, and depressed at those parts *from* which it is flowing. This implies unequal currents in different parts of the circuit. Examples of this kind of action occur in submarine cables.

FIG. 49.

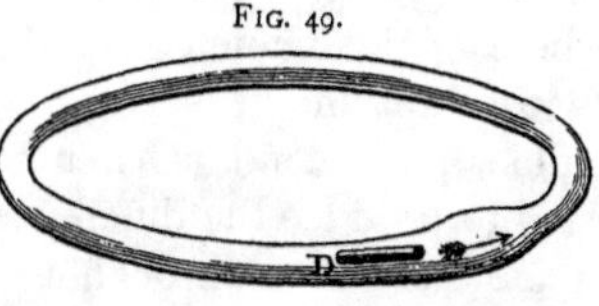

§ **24.** The strength of a constant current in any circuit is equal in all parts of the circuit. In this case, although one part of the circuit may be a thick wire and another part a thin one, a third part an electrolyte, &c., the quantity of electricity conveyed past each section is the same in the same time, i.e. the strength of the current is the same at each part. Equal lengths of current, whether conveyed in a thick or thin wire, will produce precisely the same effect in directing magnets and in producing magnetism, &c. This equal current in all parts of the circuit is independent of the capacity of each part, as it is independent of the difference of materials. There are not two kinds or qualities of current; a current has but the one quality of magnitude, meaning that it conveys a certain definite quantity of electricity past a given point in a given time. When the epithets great, strong, intense, are applied to currents they all mean the same thing, and mean that a large quantity of electricity is conveyed by them. The uniform current of electricity is analogous to the uniform current of water. If water be flowing from one reservoir to another through a succession of pipes of different diameters *all full*, the water will flow in a uniform current as defined above through all of them; that is to say, the same quantity of water per second passes through every pipe; the velocity of the water is different wherever the diameters of the pipes differ; but the current is constant in the sense that it is a current of so many gallons per second. When a good form of voltaic battery is used to produce the difference of potentials, and the current is allowed to flow through a metallic conductor, kept at rest at the same temperature and away from the neighbourhood of moving magnets or other moving currents, we obtain this simple uniform current in all parts of the circuit.

§ **25.** It will be obvious that this simplicity must be widely departed from, when even this uniform current is first started and when it ends, and that simplicity is still farther removed from the case in which currents are induced by moving magnets, &c.; these currents must vary at every moment in any one place, and differ at all parts of the circuit. To take the simplest case first: when the poles of the galvanic cell Z C are first joined at *n* and *m* to the wires A B C D electricity will rush from the cell into the wires; this electricity has to charge each portion of the wires statically: the current begins close to the cell some time before it reaches the remoter portions of the wire; it flows at different rates through different sections of the wire, according to their size, capacity, and material; it induces currents in all conductors in the neighbourhood, and is checked while doing so, and not until all this is over shall we have that permanent condition in which a constant current flows through all parts of the circuit. The series of phenomena just described occurs whenever an electric signal is sent along a wire. The earth generally forms one part of the circuit used for this purpose, and the circuit is completed or closed by making contact at one place only, as at *m*, the wire at *n* being already joined to Z;

FIG. 50.

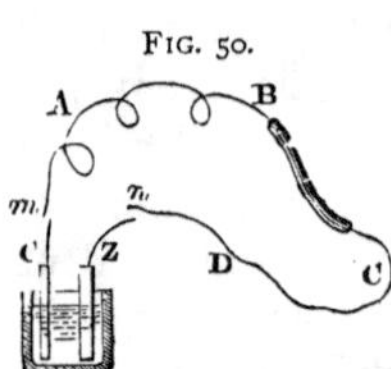

the phenomena are not made at all simpler by these changes. The speed of electricity is often spoken of, but what has now been said shows that these words without qualification can have no meaning; electricity starting from *m* does not reach A, B or C like a bullet, but in a gradually increasing wave, and the manner and rate of its arrival depend evidently on many circumstances, such as the size and material of the wire, its distance from surrounding conductors, &c. If the cell be connected with two long wires insulated at the further ends (as in Fig. 51), or if one pole be connected with the earth and the other with a large insulated conductor or long wire, we shall have a series of precisely similar phenomena, except that the final condition of equilibrium will be that in which all parts of the conductors being duly charged to the potentials which the cell produces, no further current will flow at all.

FIG. 51.

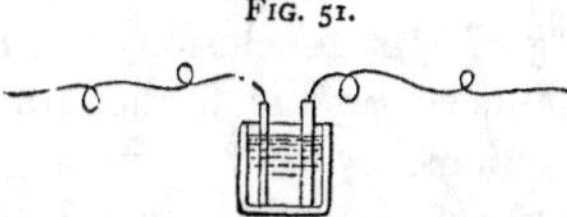

The laws according to which the varying induced currents flow in different parts of the circuit are subject to the still further complication, that the inducing system does not produce any constant difference of potential such as is produced by the cell, and that even the current which it induces in any one part of the circuit varies as the magnet or inducing system varies in its position relatively to the circuit.

§ **26.** When two dissimilar metals (Fig. 27) are joined so as to form a conducting circuit, and the junction C is at a different temperature from the junction D, an electric current is found to flow through the circuit, a difference of potential or E. M. F. occurring at both junctions. In both cases, taking iron and copper below 300° C. as an example, we should have the tendency to send the current from the iron to the copper across the junction, but that tendency is greatest at the cold junction, and therefore the current flows from the iron to the copper across the cold junction. The source of energy here is heat, which is absorbed at the hot junction, and given out at the cold junction; but less heat is given out at the cold than is absorbed at the hot junction by an amount equivalent to the work done by the electric current. This current is often called a thermo-electric current, but it differs in no quality from other currents. The E. M. F. produced is small.

§ **27**. In conclusion, we have found that currents are produced by the friction of non-conductors, by chemical reactions, by heat; by the approach, commencement, or increase of a current in any neighbouring conductor; by the removal, cessation, or diminution of any neighbouring current; by the motion of a neighbouring magnet relatively to a conductor and by the increase or decrease in the magnetism of this magnet.

Lastly, any change in the distribution of the statical charge of electricity on the surface of bodies produces currents until the redistribution is completed and equilibrium is restored. We find no difference of kind between all these currents; they all have the same properties, but combined in very varying degrees. In studying the laws which connect currents with other electrical magnitudes, we find that we must distinguish the case of the constant current which is uniform in all parts of the circuit, and at rest relatively to all other conductors and magnets, from that of the more complex varying currents, and of those which move relatively to other currents, conductors, or magnets.

CHAPTER IV.

RESISTANCE.

§ 1. BODIES have already been described as being bad or good conductors, and an imperfect conductor may be said to *oppose* the passage of an electric current. All known conductors oppose a sensible *resistance* to the passage of a current, by which we mean that if two bodies of any sensible capacity and at different potentials be joined, the current produced occupies a sensible time in passing between them, whatever material be employed to join the bodies, and however it may be shaped.[1] The strength of the current, or, in other words, the quantity of electricity passing per second from one point to another, when a constant difference of potentials is maintained between them, depends on the resistance of the wire or conductor joining those two points. A bad conductor does not let the electricity pass so rapidly as a good conductor, or, in other words, a bad conductor offers more resistance than a good one. When no electro-magnetic phenomena are produced, the current flowing from a point at potential A to a point at potential B depends simply on what is here called the *resistance* of the conductor separating them.

§ 2. With a given conductor joining two points, it is found by experiment that upon doubling the difference of potential between the points, twice as strong a current flows as before; in other words, with a constant resistance, the current is simply proportional to the E. M. F. or difference of potentials between the points. Again, it is found that keeping the difference of potential constant, and keeping the section and material of the conducting wire constant but doubling its length, we halve the current which flows, and

[1] The self-induction of a current would cause a delay in its passage between two points even if the conductor had no resistance, but the delay due to resistance is easily separated from that due to self-induction.

generally that if the E. M. F. and section and material of the wire be kept constant, the current will be inversely proportional to the *length* of the conductor. Again, keeping the E. M. F., length, and material all constant the current is halved by halving the area of the cross section of the wire. Consequently, if we define *resistance* as proportional to the length of the wire of constant section, and as inversely proportional to the cross section where that varies, we shall be justified in saying that with a given difference of potentials or E. M. F. between two points, the current which flows will be inversely proportional to the resistance separating these points; and, again, that with a constant resistance separating two points, the current flowing will be simply proportional to the E. M. F. or difference of potential between the points. If, then, we call C the current, I the electromotive force, and R the resistance of the conductor, we find that C is proportional to the quotient $\frac{I}{R}$, and is affected by no other circumstance, hence we have

$$C = \frac{I}{R}, \text{ or } R = \frac{I}{C}, \text{ or } I = C\,R.$$

This equation expresses Ohm's law, which may be stated thus:—

When a current is produced in a conductor by an E. M. F. *the ratio of the* E. M. F. *to the current is independent of the strength of the current, and is called the resistance of the conductor.*

This definition of resistance would not be justified if we did not always obtain one and the same value for R in any one conductor, whatever electromotive force may be employed to force a current through it. The electrical resistance of a conductor is not analogous to mechanical resistance, such as the friction which water experiences in passing through a pipe, for this frictional resistance is not constant when different quantities of water are being forced through the pipe, whereas the magnitude called electrical resistance is quite constant whatever quantity of electricity be forced through the conductor. This fact leads to much greater simplicity in the calculations of the distribution of electrical currents than in calculations of the flow of water. The accuracy of Ohm's law is most easily illustrated with a galvanometer having a short coil of thick wire. Take a Grove's cell and make a circuit through the galvanometer, and such a length of fine wire as gives a convenient deflection, it will be found that the deflection is nearly inversely proportional to the length of the fine wire; when this length is doubled, the deflection is halved. This would be strictly true if the deflections of the galvanometer were proportional to the current, and if the resistance of the galvanometer and of the cell were *nil.* Taking these resistances into account, then, with any cell or battery of constant E. M. F. and with any galvanometer, we shall find the deflections inversely proportional to the total resistances of the circuit.

§ **3.** *Resistance in a wire of constant section and material is directly proportional to the length and inversely proportional to the area of the cross section.* The form of the cross section is a matter of indifference, showing that the resistance is in no way affected by the extent of surface of the conducting wire or rod, and that although electricity at rest is found only on the surface, electricity when flowing as a current is propagated along all parts of the conductor alike.

The most easily explained manner of comparing two resistances is by means of the differential galvanometer. Let the coil of a galvanometer be formed of two insulated wires wound on side by side, so that each makes the same number of turns. Then if equal currents be sent round the two coils in opposite directions there will be no deflection; if the two currents be not equal, the stronger will produce a deflection. Let G_1 G represent the two coils in the annexed diagram, and let R_1 R be two resistances which are to be compared; join the two galvanometer coils at B and the two resistances at A connecting R_1 with G_1 and R with G, as shown; complete the circuit by connecting B with A, through a battery C Z. One portion of the current will pass through G R, the other portion through G_1 R_1. The magnitude of the current through both these conductors depends on their resistance and on the difference of potential between A and B which is the same in both cases. Hence the current through G and R will be equal to the current through G_1 and R_1 if the resistances of the two branches are equal. It is easy to make the resistance of G_1 equal to the resistance of G, by adding a little piece of wire to the coil which has the smallest resistance if there be any difference between them. If therefore we find no deflection caused by completing the circuit as above we may conclude that $R = R_1$. If R_1 be the greater, less current will pass through G_1 than through G and a deflection in one direction will follow; a deflection in the opposite direction would be produced if R_1 were the smaller. It is easy by successive trials to find the relative lengths of two wires R and R_1 which balance one another when different materials or different forms are used. By this instrument the law stated at the beginning of the paragraph is easily proved.

FIG. 52.

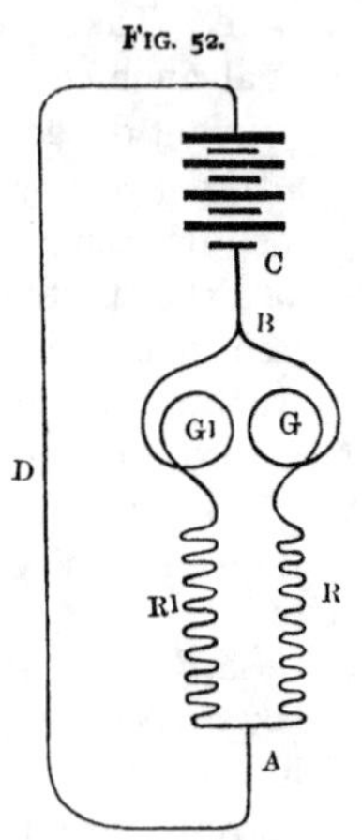

§ 4. Since the resistance of a wire of any given material is inversely proportional to the cross section of the wire, it will also be inversely proportional to the weight per unit of length; or, in other words, the resistance of a uniform wire of any material is inversely proportional to the weight per foot of the wire, i.e. a wire weighing twenty grains per foot has half the resistance of a wire weighing ten grains per foot. Inasmuch as all bodies have not the same specific gravity, the relative resistance of different materials will be different, according as we refer them to similar cross sections and lengths, or to similar weights and lengths. When treating of the measurement of resistance, a Table will be given in which the relative resistances of various materials are given, referred to both units; meanwhile, it may be sufficient to state that pure copper or pure silver have smaller resistances than any other known material; that alloys have a larger resistance than metals; electrolytes a considerably greater resistance than most alloys; that some liquids, such as oil, have so great a resistance as to become insulators, but that all known insulators, except gases, do permit the passage of electricity in a way differing rather in degree than in kind from the way in which metals permit the passage of electricity. Thus bad conductors or insulators will hereafter be frequently spoken of as bodies of great resistance. The difference in this respect between an insulator and a good conductor is enormous. Taking the resistance of silver at 0° C. as the unit, a wire of equal length and diameter of German silver would have a resistance of 12·82, and a rod of gutta percha of equal bulk and length about 850,000,000,000,000,000,000, or $8{\cdot}5 \times 10^{20}$; nevertheless, Ohm's law applies to the resistance of each material.

§ **5**. The resistance of all materials alters with a change of temperature. With the metals and good conductors, the resistance becomes greater with a rise of temperature; with electrolytes and bad conductors it diminishes. There is thus less difference between the resistances of these dissimilar bodies at high temperatures than at low. Inasmuch as the passage of a current through a wire heats it, the passage of a current tends continually to increase the resistance which it meets with. This can easily be seen with a differential galvanometer. After carefully balancing R and R_1, Fig. 52, alter the circuit so as to pass the current for some minutes through R_1 and G_1 only. On reconnecting R and G a deflection will be observed, and R will have to be increased to balance R_1, until the wires have been left to resume their former temperature. Wires of graduated length and section, insulated by silk and wound on bobbins are employed to represent certain definite resistances, and these bobbins of insulated wire are called resistance coils. It is essential that they should be made of a material, such as German silver, the resistance of which varies little with a change of temperature, and that in careful experiments the temperature of the resistance coil should be noted and allowed for.

§ **6.** A knowledge of the resistance of a conductor is essential to determine how much electricity will flow between two points in a given time when joined by that conductor; in other words, to determine the strength of a current which will under any given circumstances be produced; how much the current will be modified by a change in any given conductor; how a current will be subdivided and affected by having two or more paths open to it between the same points; to determine the effect of galvanic cells of different sizes and materials, since each kind of galvanic cell has an internal resistance depending on the size of the plates, on the distance between them, and on the solutions employed; to allow a comparison between the qualities of insulators; and to enable us to augment, diminish, and in all ways regulate any current at will.

§ **7.** The resistance of the materials of which any galvanic cell is made limits the current which it can produce. When the two metals are joined by the shortest and thickest wire practicable, the resistance of the circuit is practically the internal resistance of the battery, and in most forms this is very considerable. In a sawdust Daniell it is often more than the resistance of a mile of No. 8 iron wire, the size usually employed for land lines of telegraph : a quarter of a mile of such wire is a small resistance for a Daniell's cell. The resistance of the Grove cell is much smaller. The resistance of a battery decreases as the size of the plates is increased, because this is equivalent to increasing the area of the cross section of the liquids, the resistance of which is from 1 to 20 million times as great as that of metals of the same size.

Take two cells of any battery, join them as in Fig. 53, the copper being connected to the copper and the zinc to the zinc. Cells thus joined are said to be joined in multiple arc. The two cells are exactly equivalent to a single cell of double the size. The E. M. F. produced is that of one cell; the resistance is half that of one cell. Complete a circuit by inserting a galvanometer with a short thick coil between C and Z; the deflection obtained will be nearly double that which the one cell gives through the same galvanometer, because halving the resistance of the cell very nearly halves the resistance of the whole circuit. Next, make a circuit

FIG. 53. FIG. 54.

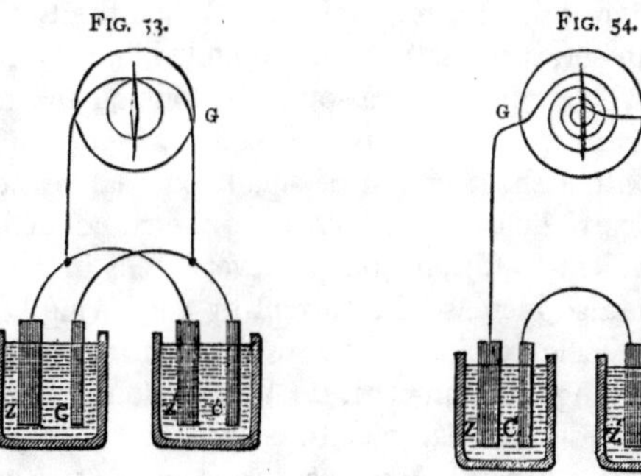

with one of the two cells and a galvanometer with a comparatively long coil of fine wire, reducing the current so as to have a convenient deflection by adding a resistance R if necessary. Add the second cell in multiple arc; no visible change will be produced in the deflection, because the resistance of the circuit is now chiefly made up of that of the galvanometer and resistance R. Diminishing the resistance of the battery hardly alters the whole resistance and does not sensibly alter the current. Thirdly, join the two cells in the manner described in Chapter I. § 19, the zinc being joined to the copper as in Fig. 13 or Fig. 54. This manner of joining is described by the words 'in series.' Now complete the circuit with the fine wire galvanometer and R, as in the second experiment. The deflection will be nearly doubled. The resistance has been slightly increased by adding the second cell in series, but the resistance of the batteries is only an insignificant portion of the whole ; while therefore the resistance of the circuit has hardly been changed, the E. M. F. has been doubled by doubling the number of metallic junctions, and twice the E. M. F. with a constant resistance gives twice the current and twice the deflection. Fourthly, return to the thick wire galvanometer, complete the circuit through it with the two cells in series ; the deflection will be almost exactly the same as when one cell only is used, and only half that obtained when the two cells are joined in multiple arc. When the two cells were joined in series the E. M. F. was doubled, but the resistance of the whole circuit was also nearly doubled and therefore the current remained nearly the same as before. Thus we see that with a short circuit of small external resistance we can increase the current by increasing the size of cells, or, what is equivalent to this, by joining several cells in multiple arc. We can also increase the current by employing liquids of smaller specific resistance, but we cannot increase the current by adding cells in series. With a long circuit of great external resistance large cells, or many of them joined in multiple arc, will fail to give us strong currents, but we may increase the current by joining the same cells in series.

When the resistance of the battery is neither excessively large nor excessively small in comparison with that of the rest of the circuit the current will be increased both by adding cells in series and by increasing their size or adding them in multiple arc. By the former process we increase the E. M. F. more than we increase the resistance. By the latter process we sensibly diminish the resistance of the circuit, leaving the E. M. F. unaltered.

Cells joined in series are sometimes described as joined for intensity, and cells joined in multiple arc as joined for quantity. These terms are remnants of an erroneous theory.

§ **8.** The resistance of the galvanometer employed to indicate a current in a circuit is a very material element in the circuit. A powerful current may be flowing from a large cell through a circuit of small resistance. If we introduce a galvanometer having a long coil of thin wire, we may by that very act diminish the current a thousand-fold. For circuits of small resistance galvanometers of small resistance must be used. For circuits of large resistance galvanometers of large resistance must also be used; not that their resistance is any advantage, but because we cannot have a galvanometer adapted to indicate very small currents without having a very large number of turns in the coil, and this involves necessarily a large resistance.

§ **9.** There are several forms of apparent resistance which are not resistances.

When a current passes to or from a metal to a liquid electrolyte, a great apparent resistance occurs, i.e. the current is diminished by the change of medium much more than by a considerable length of either material. This resistance is sometimes said to be due to the polarisation of the metals dipped into the solution. This word polarisation is sometimes very vaguely employed, but apparently here it means that the plates become coated with the products of the decomposition of the electrolyte, and that this coating produces a diminution of current. This diminution, which of course affects the current throughout its entire length, does not, however, appear to be due to anything analogous to resistance. The effect in question is due to something in the nature of a reciprocating force by which energy is stored up, i.e. when the original current ceases, a current in the opposite direction is set up at these surfaces of passage from liquid to solid by a kind of rebound. It appears, therefore, that the current has been diminished by the creation of an opposing electromotive force due to the arrangement of the elements into which the electrolyte itself has been decomposed. The term resistance is, however, continually applied to this cause of the diminution of a current even by those who are convinced that the diminution is not due to a true resistance. This false resistance or polarisation is easily observed. Make a circuit of a galvanometer, a copper wire, two Daniell's cells, and a couple of plates of one metal separated by water or any electrolyte. The deflection of the galvanometer during the first few minutes will be found to decrease rapidly; then if the cell be removed and the circuit closed, the two metal plates will send a current deflecting the galvanometer in the opposite direction; this current is strongest at first, and gradually ceases altogether.

§ **10.** When a current begins to flow across a solid insulator, such as gutta percha, a very similar phenomenon occurs; the current gradually and rapidly diminishes, as if the resistance of the gutta percha increased under the influence of the current. This apparent extra resistance is, however, no true resistance; when the original current ceases, the gutta percha sends back a gradually decreasing current in the opposite direction, and this current is of such magnitude and lasts for such a time as precisely to send back all the electricity which had, at first, apparently flowed through the gutta percha in excess of the quantity which would have passed in the same time through a constant resistance equal to the final resistance. The final resistance of the gutta percha is looked upon by some electricians as its true resistance, inasmuch as it is the only part of the apparent resistance which follows Ohm's law; the greater flow of current in the first instance is, according to this view, due not to a diminished resistance, but to an apparent absorption of electricity, as if by a number of condensers. Other electricians look upon this property of the solid insulator or electrolyte as quite analogous to the polarising property of the liquid electrolyte, and consider that the resistance of the material, as shown by the first current, is the true resistance and the subsequent diminution of current is due to an opposing electromotive force. The former view appears to the writer to be the more tenable.

This phenomenon is most easily observed with the aid of a considerable length of wire insulated with india-rubber or gutta percha. Take, say, a mile of such insulated copper wire as is used for submarine telegraph cables; place it in a tub of water; insulate one end *n* of the wire and connect the other *m* through a galvanometer G with one pole of a galvanic battery C Z of say 50 cells. Connect the other pole of the battery with the water by a copper plate, as in Fig. 55. The galvanometer must have a coil with some thousands of turns of fine wire. All the connections must be carefully insulated. When all the other arrangements have been completed the circuit may be completed by joining the wires at *m*; this will be followed by a violent throw of the galvanometer needle, due to the rapid rush of the electricity to charge the wire. When the needle comes to rest a steady deflection in the same direction will be observed, due to a current flowing from C through G and across the gutta percha sheath to the water and thus to Z. This deflection will gradually diminish, until after an hour it may be two-thirds or half the original deflection. Call this final deflection X and the deflections at each minute after the wires at *m* were joined

FIG. 55.

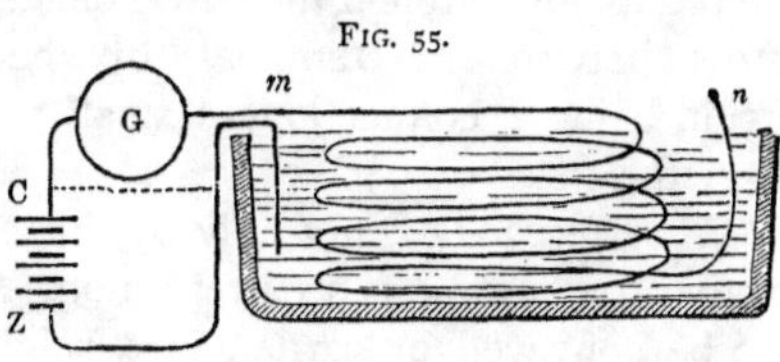

$$X + a_1, \quad X + a_2, \quad X + a_3 \quad . \quad . \quad . \quad X + a_{60}.$$

Now remove the cell C Z and substitute for it a metallic connection, as shown by the dotted line. This may be done by means of prearranged stops or keys so as not to disturb the insulation of any part. Then the charge in the wire will rush out through G, causing a violent throw in the opposite direction to that produced by the charge and equal in amount. After this discharge has taken place a steady deflection will be observed in the same direction as that due to the discharge, and this deflection at the end of each successive minute will be equal to $a_1\ a_2\ a_3\ .\ .\ .\ a_{60}$. It is assumed that a reflecting galvanometer is used, in which the deflections are proportional to the currents. The violence of the charge and discharge is such that in delicate experiments they are not allowed to flow through the galvanometer, but are conducted across between the terminals by what is termed a short circuit, being a connection of small resistance temporarily inserted.

§ **11.** Electricity is not only conducted from one body to another, by flowing as a current along a conductor; it may also be conveyed or carried in a solid conductor, through such an insulator as air, from one place to another. When two conductors charged to very different potentials are brought close together, the attraction of the electricity is such that it tears off the metal or material in fine powder, and this powder springs across the intervening space, carrying with it a charge of electricity. The air or gas itself is also electrified by contact with the conductor, and helps to convey the electricity. Light and heat are evolved in the process apparently much as light and heat are evolved when sparks are struck from steel. Electric sparks thus produced are said to overcome the resistance of the air, but this resistance has nothing in common with the resistance which is the subject of Ohm's law. The laws according to which sparks pass, and brushes, as they are called, form on points electrically charged, must be separately studied. The brush discharges, whether luminous or otherwise, are due to the accumulation of electricity in large quantities at points. The electricity has such a repulsion for itself, that if it accumulates sufficiently, the force becomes great enough to break down the pressure of the air, and highly electrified particles of the conductor and of air fly off the point. Every electrical spark seen is an illustration of this convection. Lightning is one example; another is the luminous brush which in the dark may be observed discharging the conductors of an electrical frictional machine. The air or gas heated by the spark probably conducts some electricity, so that only part of the electricity passing in the spark or brush is transferred by convection.

§ **12**. Rarefied gases are found to be tolerably good conductors. The laws of their resistance to the passage of electricity have only lately been investigated, and are but partially understood. It is uncertain how far their resistance can properly be said to follow Ohm's law. According to recent experiments by Mr. Varley, conduction in rarefied gases does follow Ohm's law, but there is a very large resistance at the surface of contact between the attenuated gas and the metal conductor. This resistance is constant and prevents any current from passing until the E. M. F. employed exceeds a certain definite magnitude, which is constant for each material and degree of rarefaction. This is very analogous to what takes place in electrolytes, except that through these some current apparently always passes whatever E. M. F. be employed, although no complete decomposition occurs until a certain definite E. M. F., constant for each electrolyte, has been reached. Experiments showing the action of a partial vacuum can be made with Geissler's tubes, which can be bought at any respectable optician's. These glass tubes contain highly rarefied gases, and electrodes leading through the glass are employed as part of the circuit. If a galvanometer and an electric battery form part of the circuit no current will be observed until perhaps two hundred cells are employed. Then the current passes with brilliant optical effects in the tube and the galvanometer is deflected. Induction apparatus producing high electromotive force, such as the well-known Ruhmkorff's coil, may be employed instead of the galvanic battery.

CHAPTER V.

ELECTRO-STATIC MEASUREMENT.

§ **1.** OUR knowledge of electricity and magnetism is derived from observation of certain forces, and the comparison of currents, quantities, potentials, and resistances are all effected by a comparison of forces acting under various circumstances. The measurement of forces requires fixed standards of length, mass, and time, which will also serve as fundamental standards for all electrical measurements. The

centimètre . . . gramme . . . second

are the three units of length, mass, and time which will be adopted in the present treatise.

As stated in Chapter I. § 17, the unit of *Force* adopted by us is the force which will produce a velocity of one centimètre per second in a free mass of one gramme by acting on it for one second.

This unit of force = ·00101915 × weight of a gramme at Paris. The weight of the gramme itself wherever we happen to be is the more common unit of force, but we shall find the so-called absolute unit more convenient in calculations, and any result can be readily reconverted into the more familiar measure by multiplying it into the above coefficient, or dividing it by the number 980·868.

The unit of work is the work performed by the unit force moving over a distance of one centimètre; it is equal to ·00101915 centimètre grammes; in other words, to lift the weight of one gramme through one centimètre at Paris requires an expenditure of work equal to 980·868 of the units of work.

§ **2.** In what is termed *electro-static measure* the unit quantity of electricity is that which exerts the unit force on a quantity equal to itself at a distance of one centimètre across air.

The unit difference of potential or unit electromotive force exists between two points when the unit of work is spent by a unit of electricity in moving from one to the other against the electric repulsion, described in Chapter I.

The resistance of a conductor between two points is a unit if it allows only one unit of electricity per second to pass from one to the other when the unit of electromotive force is maintained between them.

The system of electrical units as defined in this paragraph is called the electro-static absolute system, based on the centimètre, gramme, and second. No special names have yet been given to these units. They are the most convenient for use when dealing with the phenomena described in Chapter I. The equations expressing these definitions are given below in § 14.

§ **3.** It is found by experiment that the force f with which, at a given distance d, two small electrified bodies repel or attract one another, is proportional to the product of the charges, q and q_1, upon them; and further, that when the distance varies this force f is inversely proportional to the square of the distance d between them; it follows, from the definition adopted of force and quantity, that

$$f = \frac{q q_1}{d^2} \qquad (1)$$

from which equation, if we observe the force, and make q either equal to q_1, or to bear any known relation to it, we can determine the quantity q in absolute measure; or *vice versâ*, knowing q and q_1, we can determine what force they will exert at a given distance, as, for instance, in moving the index of an electrometer. The application of this equation is limited to *small* electrified bodies. In any body of sensible size the mutual induction between the two electrified bodies would disturb the distribution of electricity over the surface, and change that distribution at every distance.

§ **4.** The quantity of electricity with which a given conductor in a given place can be charged depends simply on the difference of potential between it and neighbouring conductors, and if these neighbouring conductors are uninsulated we may say that the charge will be simply proportional to the potential of the body charged; we may therefore speak of the capacity s of a given conductor for electricity, meaning thereby the constant quotient of the quantity on the conductor divided by its potential; or calling the quantity q, as before, and the potential i, we have

$$q = s\,i \qquad (2)$$

The capacity of a sphere at a distance from all conductors is equal to its radius; that is to say, a sphere one mètre in diameter will, when charged to the potential 6, contain 6 × 50, or 300 units of electricity.

The capacity s_1 of a sphere of radius x, suspended in the centre of a hollow uninsulated sphere, radius y, is

$$s_1 = \frac{x\,y}{y - x} \qquad (3)$$

The dielectric separating the two spheres is supposed to be air. The capacity of the internal conductor would change if any other dielectric were used. The capacity of a metal conductor is independent of the metal employed. The phenomenon is more complex when either solid or liquid electrolytes or insulators are used as the bodies to be charged.

Equation (3) shows that when the distance between the two opposed conductors is diminished, so that $y - x$ becomes small, the capacity of the system is very much increased. This is equally obvious from the formula for the capacity of a large flat plate one side of which is near a similar uninsulated flat plate, and separated from it by air, while the other side is far removed from all conductors; in such a case, let a denote the distance between the metallic surfaces and let s be the capacity *per unit of area*, then

$$s = \frac{1}{4\,\pi\,a} \qquad (4)$$

(π here and elsewhere always means the ratio of the circumference to the diameter of a circle = 3·1416. The surface of a sphere of radius unity is equal to 4π.)

and in order to find the total capacity of the plate we may multiply s into the area of the plate with sufficient accuracy for practical purposes, whenever a is small in comparison with this area; a must be measured in centimètres, and the surface in square centimètres. This method is not absolutely accurate, because at the edges of the plates the electricity will not be uniformly distributed, as it will in the middle of the plate. By increasing the surface and diminishing a, we may increase indefinitely the quantity which the plate or conductor will contain when raised to a given potential. The quantity with which the plates will be charged with a given potential is $q = s\,i$ as before.

§ **5.** The arrangement of opposed conductors intended to give a large capacity with comparatively small surface is termed a *condenser*. The capacity of a condenser depends on the dielectric separating the conductors. If for air we substitute gutta percha, the capacity will be increased about four and a quarter times. The coefficient by which the capacity of an air condenser must be multiplied in order to give the capacity of the same condenser when another dielectric is substituted for air is constant for each substance, and is called the *specific inductive capacity* of the dielectric. It is a quantity of much importance in telegraphy, and will in this treatise be designated by the letter K. It has been approximately determined for a few substances. The following table gives the numbers for these:

Values of K.

Substance	K	Substance	K
Air	= 1	India rubber	= 2·8
Resin	= 1·77	Hooper's vulcanised india rubber	= 3·1
Pitch	= 1·80	W. Smith's gutta percha	= 3·59
Beeswax	= 1·86	Gutta percha	= 4·2
Glass	= 1·90	Mica	= 5
Sulphur	= 1·93	Paraffin	= 1·98[1]
Shellac	= 1·95		

§ **6.** The numbers are approximate values only, and, in-

[1] Gibson and Barclay.

deed, extreme accuracy is unattainable on account of the following peculiarity observed in all solid dielectrics. When one plate A of the condenser is first raised to the desired potential by contact, say with one electrode C of a galvanic battery, the other electrode Z being in connection with the earth or second plate of the condenser as in Fig. 56, a charge rushes in with great rapidity, but the entrance of the electricity does not instantly cease, as is the case with an air condenser; on the contrary, although it decreases very rapidly, the flow of electricity into the condenser does not cease for many hours. This phenomenon has already been described in Chapter IV. § 10 in its bearing on currents. Similarly, when the two plates are joined by a wire so as to be brought to one potential, the electricity is discharged very rapidly at first; but this discharge is so far from being completed immediately that electricity continues to flow out for precisely as long a time as it ran in, and with precisely the same rapidity after each interval of time; i.e. if, upon maintaining a difference of potential x between the plates, *coatings, or armatures* (as they are often called) of the condenser, a quantity q per second is found flowing into the condenser at the expiration of thirty minutes, then thirty minutes after the two armatures have been joined, or, in ordinary language, after the condenser has been discharged, the same quantity q per second will be found flowing from one armature to the other. The effect produced is as though the dielectric were a kind of sponge absorbing electricity at a certain rate when subjected to a certain difference of potential, and yielding it all up again when the two plates were brought to one potential. A condenser with glass or paraffin between the armatures has not, therefore, the same definite

FIG. 56.

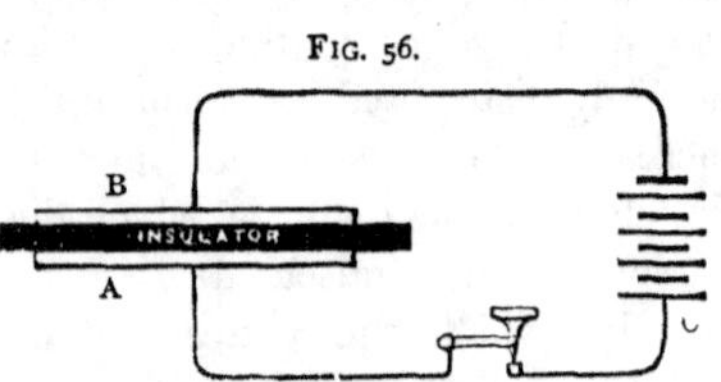

capacity as an air condenser; the capacity is generally understood to mean the capacity for receiving electricity from the first contact. When a condenser is discharged, if contact be not maintained between the armatures, the gradual restoration of this *quasi* absorbed charge raises the potential of the armature which had previously been highly charged, and accumulates upon it, so that on again making contact between the armatures a second considerable discharge is given, and a succession of discharges of this kind can be obtained from a large condenser for several hours. These are called residual discharges. The same law holds as to charges; after charging the armature to a given potential, and leaving it insulated, the potential gradually falls, owing to the absorption by the glass or gutta percha; then, on raising the potential of the armature afresh, by connecting it with the electrode cf a battery, a fresh charge can be poured into the condenser. This apparent absorption of the electricity by the dielectric is said by some writers to be due to polarisation caused by the continued electrification of the dielectric; the word polarisation, like induction, is applied to a great variety of phenomena having little in common.

These phenomena are readily observed in a condenser consisting of a mile of telegraph wire insulated by gutta percha; the copper wire is the one armature; if the gutta percha be covered with lead or tinfoil, as is sometimes done, this forms the other armature; or, if the gutta percha covered wire be placed in a tub of water, that water will be the second armature. With a sensitive galvanometer and a battery of 50 cells, or even less, all the phenomena described are easily observed. Condensers of smaller bulk and equal capacity can be obtained from the makers of telegraphic apparatus. When the condenser is like a common glass Leyden jar of small capacity, and insulated with a hard material, the residual discharges may be observed in the form of a succession of sparks after the jar has been charged to a high potential by a frictional machine.

§ 7. Let a small flat movable plate f, supported by the torsion of a wire $m\ n$ in Fig. 57, be placed flush with a much larger flat fixed plate $h\ h$ surrounding it on all sides, and let both plates be placed opposite and parallel to a third uninsulated plate g, then if a permanent difference of potentials be established in any way between g and the plates f and h, the quantity of electricity

FIG. 57.

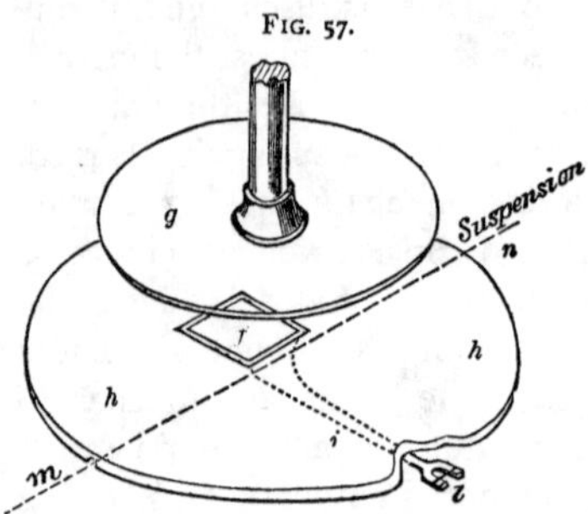

per unit of area on the plate f will be $\frac{i}{4\pi a}$, and the electricity will be uniformly distributed over the plate f, and the electricity of the opposite sign will also be uniformly distributed over the opposing surface of the plate g. The total force with which the plate f is attracted by g will be

$$f = \frac{i^2\, M}{8\,\pi\, a^2} \qquad (5)$$

Where M is the surface of the plate in square centimètres.[1] Apparatus can be constructed by which this force is actually measured, by weighing or otherwise, and this apparatus forms an absolute electrometer (Sir William Thomson's guard ring electrometer) by which we can determine the difference of potential i between the plates: $i = a\sqrt{\frac{8\pi f}{M}}$; f must of course be expressed in absolute measure, Chapter V. § 1.

§ 8. Measured by apparatus of this kind, the ordinary Daniell's cell (one form of galvanic battery) is found to produce a difference of potentials between its electrodes equal to ·00374. Experiment showed the attraction to be

[1] Vide paper 'On the Mathematical Theory of Electricity in Equilibrium, by Sir W. Thomson. *Phil. Mag.* 1854, second half-year, and republished in 1872 in a volume entitled *Electrostatics and Magnetism.*

·057 grammes per square centimètre between discs separated by ·1 centimètre, with a difference of potentials produced by 1000 Daniell's cells.

Hence, in equation (5), if the weighings had taken place in Paris, we should have had $f = 980·868 \times ·057$; but in Glasgow the force with which a gramme mass weighs is less than in Paris, so that $f = 981·4 \times ·057 = 55·94$; $a = ·1$, and $M = 1$; substituting these values in our equation, we obtain $i = 3·74$ for 1000 Daniell's cells.

Using this value in equation 4, we find that an air condenser, with a square mètre surface and the plates one millimètre apart, electrified by a thousand cells, would take a charge of $10000 \dfrac{3·74}{4\pi \times 0·1} = 2976$ units. If the plates had been separated by gutta percha instead of by air, the charge on the plates would be $4·2 \times 2976 = 12499$, the coefficient 4·2 being the specific inductive capacity of the material taken from § 5.

A ball, one centimètre in diameter, electrified by 1000 Daniell's cells, would take a charge of $·5 \times 3·74$, or 1·87 units of electricity.

From a knowledge of this quantity we may calculate the force on a similar ball similarly electrified, but so far off that the electricity on each ball would remain almost uniformly distributed. Two such balls similarly electrified at a distance of one mètre would repel one another with a force $= \dfrac{1·87 \times 1·87}{10000} = ·00035$ absolute units of force (equation 1) or ·000000357 grammes weight. When the balls are brought closer, the calculation of attractions or repulsions between them become exceedingly complicated, owing to the redistribution of the electricity on their surface.

§ **9.** The capacity of a long cylindrical conductor of the diameter d and length L enveloped by a concentric cylindrical conductor of the diameter D, and separated from it by an insulator with the specific inductive capacity K is

$$s = \frac{K L}{2 \log_e \frac{D}{d}} = \frac{K L}{4{\cdot}6052 \log \frac{D}{d}} \qquad (6)$$

($\log_e$ signifies that natural logarithms are to be employed instead of Napierian logarithms.)

The length of the cylinder is assumed to be so great that the capacity of the ends may be neglected; this formula is applicable to the insulated wire used for submarine cables. The capacity of one knot of the English Atlantic cable is

$$s = \frac{4{\cdot}2 \times 6087 \times 30{\cdot}48}{4{\cdot}6052 \times \log 3{\cdot}28} = 328000 \text{ (centimètres).}$$

6087 is the number of feet in a knot, and 30·48 the number of centimètres in a foot; 3·28 is the ratio between the diameter of the gutta-percha and that of the wire conductor. It follows from the above, that the charge per knot of this cable when electrified by 100 Daniell's cells is ·374 × 328000 = 122670 and every time the cable is charged or discharged this quantity per knot flows in and out; thus if ·01 second be occupied in charging 200 knots the mean strength of the current flowing for ·01 second will be $\frac{122670 \times 200}{100} = 245340$ units of current.

§ **10.** The term *electric density* signifies the quantity of electricity per square centimètre on a charged conductor. The equations (2), (3), and (4) allow us to calculate this for spheres and condensers with flat plates; equation (4) is applicable to any form of condenser in which the curvature is not considerable relatively to the thickness a of the dielectric. It is applicable, therefore, to the ordinary Leyden jar, with the simple modification that the value obtained from it must be multiplied by the number expressing the specific inductive capacity of the dielectric. The electrical attraction or repulsion, exerted on a *small* quantity q of electricity close to an electrified surface, is easily calculated when the electric density on the surface ρ is known. It is perpendicular to the surface, and in air is equal to

$$4 \pi \rho q = R q \qquad (7)$$

where R is the *electrostatic force* close to the surface, i.e. the force which the charge would exert per unit of quantity on the small charge q.

Between two parallel opposed conducting surfaces, differing in potential by the amount i, and separated by a small distance compared with their size, the resultant electrostatic force R tends to impel any small quantity of electricity straight across from one surface to the other, in a direction perpendicular to the surface, with a force f which is constant in amount. Retaining the previous notation we have

$$f = \mathrm{R}q = \frac{i}{a}\,q \qquad (8)$$

The work done on q in crossing is $fa = i\,q$.

The electric density on a small sphere at a given potential is much greater than on a large one, for the capacity increases only as the radius, while the area increases as the square of the radius; hence an infinitely small sphere charged to any sensible potential would have an infinitely great electric density on its surface, and the force it would exert on electricity in its immediate neighbourhood would be infinitely great; it would, in fact, repel its own parts infinitely, and we may therefore infer that it would be impossible to charge a very small sphere to a very high potential. This inference is justified by experiment. The distribution of electricity over bodies which have points or angles is such that the electric density becomes very great on these points, as it would on a very small sphere, even when the potential is not high. The result is a great repulsion of the electricity for itself, or rather a great repulsion between neighbouring parts of the matter charged with it; we then frequently see the electrified matter passing off in the condition known as an electric spark, or as what is termed an electric brush. Anything tending to produce a great density at any part of the surface of a charged conductor tends to produce the spark. Thus by approaching a finger to a charged conductor, the density is increased by induction opposite the finger, and may be increased sufficiently to produce the spark. Increased electric density by no means necessarily implies increase of potential unless the form and position of the conductors are constant.

§ **11.** There is a real diminution of air-pressure against the surface of a charged conductor, due to the repulsion of the electricity for itself. This mechanical force can be made evident by electrifying a soap-bubble, which expands when electrified, and collapses when discharged. If the air-pressure per square centimètre be called p, we have

$$p = 2 \pi \rho^2 \qquad (9)$$

The diminution of air-pressure required before a spark takes place between two slightly convex parallel plates has been tested by Sir William Thomson, with the results shown in the following table :

Length of sparks in centimètres = A	Electrostatic force R close to surface in absolute units. Vide § 10.	Electromotive force = R × A, or difference of potential, which produced a spark of length A.	Pressures of electricity from either surface immediately before disruption in grammes weight per square centimètre = $\frac{R^2}{8\pi \times 981 \cdot 4}$.
·00254	527·7	1·34	11·290
·00508	367·8	1·87	5·49
·0086	267·1	2·30	2·89
·0190	224·2	4·26	2·04
·0408	151·5	6·19	·931
·0688	140·8	9·69	·806
·1325	131	17·35	·696

It is curious to observe that the electrostatic force is not constant, as might have been expected; and that the electromotive force required to produce a spark does not increase in simple proportion to the length of the spark, being less per unit of distance between the opposed surfaces for long sparks than for short ones. It follows from the measurement in § 8, that 2,600 Daniell's cells would produce a spark of ·0688 centimètres between two very slightly convex surfaces: by observing the length of spark, which, under similar circumstances, can be obtained, say, from a Leyden jar, we may roughly estimate the potential to which it has been charged.

§ **12.** The brushes or sparks which fly off from points charged to high potentials, show that in all apparatus intended to remain charged at a high potential, every angle and point must be avoided on the external surfaces. It is easy to draw off, by a silent and invisible discharge from a point, by far the greatest part of the charge of a conductor without any direct contact with the discharging conductor: points are also spoken of as collecting electricity from any electrified body held in the neighbourhood; their action is as follows: If attached to an insulated conductor, and held near an electrified body A, they become charged by induction with the opposite kind of electricity. This flies off in sparks, or by a silent discharge, and leaves the insulated conductor charged with the same electricity as that contained in A. This property of points explains the action of lightning conductors. Lightning is an enormous electric spark passing between two clouds, or from a cloud to the earth; in the latter case the electrified cloud is attracted towards any prominence or good conductor, which becomes electrified by induction, and the spark of lightning passes when the difference of potentials is sufficient to overcome the mechanical resistance of the air. If the electrified prominence on the earth be armed with a point connected, by good conductors, such as large copper rods, with the earth, then, as soon as the potential of the point is raised even slightly, the electricity passes off from the point into the air; the prominence can no more be electrified highly by induction than a leaky bucket can be filled with water; the electrified clouds are not attracted to the neighbourhood, and even should they be driven there in such quantity that the electricity flying off from the point is insufficient to prevent a spark from passing, the spark will pass from the cloud to the point, inasmuch as the electric density and attraction will be greater there than anywhere in the neighbourhood. Electricity conveyed by a good metal conductor to the earth does no harm, and leaves no trace of its passage; whereas, a spark driven through an insulator or bad conductor, tears it to pieces on its passage; this fact may be verified by sending a spark through glass, which will be cracked and shivered, or through paper, which will have a hole torn in it. The electromotive force required to produce mechanical results of this character is much greater than that required to open a passage through a corresponding thickness of air. We may, therefore frequently, prevent the passage of sparks between two conductors, by covering one of them with ebonite, glass, or other hard insulator.

Sir W. Thomson has found that if a conductor with sharp edges or points is surrounded by another, presenting everywhere a smooth surface, a much greater difference of potential can be established between them without producing disruptive discharge, when the points and edges are positive, than when they are negative.

§ **13.** The distribution of electricity over opposing surfaces, when these are not of the simplest description, offers problems of extreme complication. Generally, we know that the density will be greatest where the opposing conductors are close together, and where they have angles or points, that it will increase directly as the difference of potential and directly as the specific inductive capacity of the dielectric. We must especially remember that the charge or electric density on opposed surfaces depends on difference of potential, and not on absolute potential, so that on electrifying the outside of a charged insulated Leyden jar, we shall raise not only the potential of the outer, but also that of the inner coating; from the same cause the charge of any condenser due to contact with two electrodes of a battery will be the same, whether one electrode of the battery be uninsulated or not, i.e. the quantity which will flow from one armature to the other when joined will be unaltered, but the quantity flowing from either armature to the earth will depend on the potential of that armature. Any change in the total quantity of a charge on a conductor will change its potential as a change of its potential will change the charge. Putting a charged conductor in contact with another conductor at the same potential, will not alter the distribution of the charge in either; but if two conductors at different potentials are brought into contact, there must necessarily be a redistribution of the charge due to the intermediate potential assumed by the two bodies. Mr. F. C. Webb in his treatise on electrical accumulation and conduction, has given many instructive examples of the distribution of electricity under different circumstances. The theory already stated explains his results.

§ **14**. The simple laws connecting potential, quantity, capacity, density, and electrical attraction or repulsion have now been stated, and the nature of the measurements has been indicated, by which potential, quantity, and capacity can be defined in terms of length, mass, and time; a current is necessarily measured by measuring the quantity which passes per second, and resistances are expressed in terms of that resistance which would allow the unit current to pass in the unit of time, with the unit electromotive force acting between the two ends of the wire. To give some idea of the material representation of units of this kind, it may be stated that this resistance would be represented by about one hundred million kilomètres of mercury at 0° Centigrade, in a tube, the sectional area of which was one thousandth of a square millimètre; electricians when they measure the resistance of the gutta percha envelope of a mile of cable, observe resistances of about $\frac{1}{5000}$th of this magnitude; approximately the insulation resistance of one foot of gutta-percha covered wire is of about this magnitude. The unit of current is such as would be given by a battery of about 268 Daniell's cells through the above resistance. The unit quantity of electricity is that on a sphere two centimètres diameter, electrified by one pole of 268 Daniell's cells in series, the other pole being in connection with the earth. This quantity discharged per second would give a current equal in strength to that flowing through the long mercury conductor or gutta percha envelope from the 268 cells. This series of units is called the electrostatic absolute centimètre-gramme. second series. This electrostatic series is the most convenient for calculations concerning electricity at rest, but when treating of currents and magnets, a distinct series of units is used; this double series of units involves no greater inconvenience than the use of the chain, acre, and rod for surveying, while the inch, foot, and square inch are used to describe machinery.

§ **15.** The four principal electrostatic units are directly determined by four fundamental equations; from equation (1), § 3, we have $f = \frac{q q_1}{d^2}$, from which, if $q_1 = q$, we directly find the unit of quantity in terms of the unit of force; we know by the definition of potential that the work so done in conveying the quantity q of electricity between two points at potentials differing by the amount i is equal to $q\,i$ or

$$i = \frac{w}{q} \qquad (10)$$

This gives the unit difference of potentials in terms of q and the unit of work; by definition § 14 the current $c = \frac{q}{t}$ where q is the quantity passing in the time t, and from this equation we obtain the unit of current in terms of q, and the unit of time; from Ohm's law $r = \frac{i}{c}$, by which we obtain the unit of resistance in terms of i and c.

Finally, the unit of capacity is directly derived from that of potential and quantity; the unit of density from that of surface and quantity.

If the capacity of a conductor be called s, we have $s = \frac{q}{i}$, where q is the quantity with which it is charged by the electromotive force i.

CHAPTER VI.

MAGNETISM.

§ **1**. A MAGNET in the popular acceptation of the word is a piece of steel, which has the peculiar property, among others, of attracting iron to its ends. Certain kinds of iron ore called loadstone have the same properties.

If a magnet A be free to turn in any direction, the presence of another magnet B will cause A to set itself in a certain definite position relatively to B. The position which one magnet tends to assume relatively to another, is conveniently defined in terms of an imaginary line, occupying a fixed position in each magnet, and which we will call the magnetic axis. The greatest manifestation of force exerted by a long thin magnet, is found to occur near its ends, and the two ends of any one such magnet possess opposite qualities; this peculiarity has caused the name of poles to be given to the ends of long thin magnets. These poles are commonly looked upon as centres of force, but except in the case of long, infinitely thin, and uniformly magnetised rods they cannot be considered as simple points exerting forces; nevertheless, the conception of a magnet as a pair of poles, capable of exerting opposite forces, joined by a bar exerting no force, is so familiar, and in many cases so nearly represents the facts that it will be employed in this treatise. The magnetic axis, as above defined, is the line joining the two imaginary poles.

§ **2**. Every magnet, if free to turn, takes up a definite position relatively to the earth, which is itself a magnet. The pole, which in each magnet turns to the north, will by us be called the north pole of the magnet. The other pole will be called the south pole. The two north poles of any two magnets repel one another; so do the two south poles; but any north pole attracts any south pole. Hence, the north pole of a magnet is similar in character to the south end of the earth. The pole which is similar to the south end of the earth is sometimes called the positive pole; the other, which we call the south pole of the magnet, is the negative pole. When a magnet is broken each piece forms a complete magnet with a north and south pole.

§ **3.** The *strength* of a pole is necessarily defined as proportional to the force which it is capable of exerting on another given pole; hence the force f exerted between two poles of the strengths m and m_1 must be proportional to the product $m\,m_1$. The force f is also found to be inversely proportional to the square of the distance D, separating the poles, and to depend on no other quantity; hence, choosing our units correctly, we have

$$f = \frac{m\,m_1}{D^2} \qquad (1)$$

The strength of a pole is a magnitude which must be measured in terms of some unit. When in the above equation we make f and D both equal to unity, the product $m\,m_1$ must also be equal to unity hence from equation (1) it follows that *the unit pole is that which at the unit distance repels another similar and equal pole with unit force.*

f will be an attraction or a repulsion according as the poles are of opposite or similar kinds. The number m is positive if it measures the strength of a north pole and negative if it measures the strength of a south pole; hence an attracting force will be affected with the negative sign, and a repelling force with the positive sign.

§ **4.** We observe that the presence of the magnet in some way modifies the surrounding region, since any other magnet brought into that region experiences a peculiar force. The neighbourhood of a magnet is often for convenience called a *magnetic field*; and for the same reason the effect produced by a magnet is often spoken of as due to the magnetic field instead of to the magnet itself. This mode of expression is the more proper, inasmuch as the same or a similar condition of space is produced by the passage of electric currents in the neighbourhood, without the presence of a magnet. Since the peculiarity of the magnetic field consists in the presence of a certain force, we may numerically express the properties of the field by measuring the strength and direction of the force, or, as it may be worded, the *intensity of the field*, and the *direction of the lines of force.*

This direction at any point is the direction in which the force tends to move a free pole; and the intensity H of the field is defined as proportional to the force f, with which it acts on a free pole; but this force f is also proportional to the strength m of the pole introduced into the field, and it depends on no other quantities; hence,

$$f = m\,\mathrm{H} \qquad (2)$$

and therefore the field of unit intensity will be that which acts with unit force on the unit pole.

§ 5. The *lines of force* produced by a long thin bar magnet near its poles radiate from the poles; the intensity of the field is equal to the quotient of the strength of the pole divided by the square of the distance from the pole; thus the unit field will be produced at the unit distance from the unit pole.

In a *uniform* magnetic field, the lines of force will be parallel; such a field can only be produced by special combinations of magnets, but a small field at a great distance from the pole producing it will be sensibly uniform. Thus in any room unaffected by the neighbourhood of iron or magnets, the magnetic field due to the earth will be sensibly uniform: its direction being that assumed by the dipping needle. The dipping needle is a long magnet supported in such a way as to be free to take up its position as directed by the earth, both in a horizontal and vertical plane; it requires to be very perfectly balanced before being magnetised, otherwise gravitation will prevent it from freely obeying the directing force of the earth's magnetism.

§ **6.** We can never really have a single pole of a magnet entirely free or disconnected from its opposite pole, but from the effect which would be produced on a single pole it is easy to deduce the effect produced by a magnetic field on a real bar magnet. In a uniform field, two equal opposite and parallel forces act on the two poles of the bar magnet, and tend to set it with its axis in the direction of the force of the field. This pair of forces tending to turn the bar, but not to give it any motion of translation, constitutes what is termed in mechanics a *couple.* When the magnet is so placed that its axis is at right angles to the lines of force in the field, this couple G is proportional to the intensity of the field H, the strength of the poles m, and the distance between them l; or

$$G = m\,l\,H \qquad (3)$$

The product $m\ l$ is called the *magnetic moment* of the magnet; and from equation (3), it follows that the moment of any given bar magnet is measured by the couple which it would experience in a field of unit intensity, when it is placed normal to the lines of force. A couple is measured by the product of one of its forces multiplied into the distance between them. The *intensity of magnetisation* of a magnet is the ratio of its magnetic moment to its volume.

§ **7.** When certain bodies (notably soft iron) are placed in a magnetic field they become magnetised, the axis joining their poles being in the same direction as would be assumed by the axis of a free steel magnet in the same part of the field. Thus if the small pieces of soft iron $n\ s$ are magnetised by the action of the magnet N S producing the lines of force shown in Fig. 58, the north pole will be near *n*, the south pole near *s* in each case. Magnetisation when produced in this way is said to be induced, and the action is called magnetic induction. The intensity of the magnetisation (except when great) is nearly proportional to the intensity of the field. We have seen in Chapter III. § 13, that soft iron, round which a current of electricity circulates, becomes magnetised. When, therefore, we can calculate the intensity of the magnetic field which we now see is produced by the electric current, we shall be able to calculate the intensity of magnetisation of the soft iron core. When the magnetisation approaches the limiting intensity which the soft iron is capable of receiving, it always falls short of that calculated on this principle.

Bodies in which the direction of magnetisation is the same as that of the field are termed *paramagnetic*. Iron, cobalt, and nickel, chromium and manganese, are paramagnetic; some compounds of iron are also paramagnetic. Some of these bodies retain their magnetism, so that we can

FIG. 58.

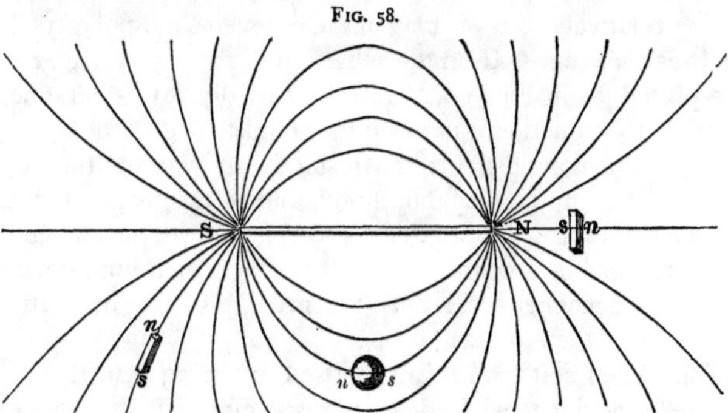

have an independent nickel magnet. Iron is capable of much more intense magnetisation than nickel, but nickel approaches iron in this respect more nearly than any other material. Certain other materials, such as bismuth, antimony, and zinc, are magnetised by a magnetic field, so that the direction of magnetisation is opposite to that of the field: they are called *diamagnetic.* None of these bodies can be so intensely magnetised as iron, nor do they retain their diamagnetism when removed from the field.

§ 8. One consequence of magnetic induction is that when a number of similar magnets are laid side by side we obtain a compound magnet stronger indeed than any of the component magnets, but much less strong than the sum of the strengths of the separate magnets used. For when a magnet N S is brought near another N′ S′, as in Fig. 59, the north pole N tends to induce a south pole at N′ and similarly N′ tends to induce a south pole at N. The result is that N and N′ by

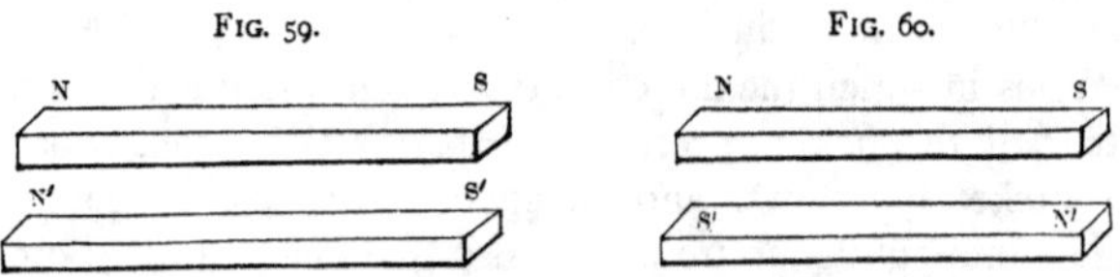

FIG. 59. FIG. 60.

their mutual action weaken one another, if N be sufficiently strong relatively to N′, it may actually reverse the polarity of the weak magnet. If on the other hand two equal magnets are placed, as in Fig. 60, N and S′ mutually strengthen one another by induction, but since they tend to induce opposite and equal magnetic fields the result is to weaken the resultant field in the neighbourhood, and if the magnets are allowed to touch, the strength of the field will be reduced to an insensible amount. When the magnets are not equal the weaker magnet will reduce the strength of the magnetic field due to the stronger.

§ 9. When soft iron is magnetised by being placed in a magnetic field a sensible time elapses before it assumes the maximum intensity of magnetisation which the field will produce. Similarly, when the bar of soft iron is withdrawn from the field it does not lose its magnetism instantly; the magnetism decreases as gradually as it increased, and in almost all cases some traces of magnetism will remain for hours or perhaps for ever after the iron has been withdrawn from the magnetic field. This remnant of magnetisation is often called *residual magnetism*; most ordinary pieces of iron show residual magnetism very distinctly, especially in large masses; but very perfectly annealed iron of certain qualities shows very little, and is valuable on that account in the construction of telegraph instruments. The cause of this phenomenon is called *coercive force.* The slowness with which iron in any mass gains or loses its magnetism is a serious impediment to the construction of quick-working telegraphic apparatus. The term 'soft iron' is applied to denote iron which loses its magnetism rapidly, or in other words iron which has little coercive force.

§ **10.** The conception of electric potential has been explained at length in Chapter II. *Magnetic potential* is an analogous conception. If we move a single magnetic pole from one point to another of the magnetic field, we shall find that the forces in the field perform work on the pole, or that they act as a resistance to its motion according as the motion is with, or contrary to, the forces acting on the pole; if the pole moves at right angles to the force, no work is done. The *difference of magnetic potential* between any two points of the field is measured by the work done by the magnetic forces on a unit pole moved against them from the one point to the other, supposing the unit pole to exercise no influence on the field in question. A point infinitely distant from the pole of any magnet must be at zero magnetic potential, and hence the *magnetic potential* of any point in the field is measured by the work done by the magnetic forces on a unit pole during its motion from a point infinitely far off from all magnets to the point in question, with the same limitation as before.

An equipotential surface in a magnetic field is a surface so drawn that the magnetic potential at all its points shall be the same. By drawing a series of equipotential surfaces, corresponding to the potentials 1, 2, 3 . . . n, we may map out any magnetic field so as to indicate its properties. The unit pole in passing from one such surface to the next against the magnetic forces will always perform one unit of work.

The direction of the magnetic force at any point is perpendicular to the equipotential surface at that point; its intensity is the reciprocal of the distance between one surface and the next at that point; i.e. if the distance from surface to surface be $\frac{1}{4}$, measured in units of length, the intensity of the field will be 4.

§ **11.** The magnetic field may be mapped out in another manner: this second method is due to Faraday.

Let a line whose direction at each point coincides with that of the force acting on the pole of a magnet at that point be called a line of magnetic force. By drawing a sufficient number of such lines we may indicate the *direction* of the force in every part of the magnetic field; but by drawing them according to a certain rule we may also indicate the *intensity* of the force at any point as well as the direction. It has been shown[1] that if in any point of their course the number of lines passing through a unit area is proportional to the intensity there, the same proportion between the number of lines in a unit of area and the intensity will hold good in every part of the course of the lines.

If, therefore, we space out the lines so that in any part of their course the number of lines which start from unit of area is numerically *equal* to the number measuring the intensity of the field there, then the intensity at any other part of the field will also be numerically equal to the number of lines which pass through unit of area there; so that each line indicates a constant and equal force.

The lines of force are everywhere perpendicular to the equipotential surfaces; and the number of lines passing through unit of area of an equipotential surface is the reciprocal of the distance between that equipotential surface

[1] Vide Maxwell on 'Faraday's Lines of Force,' *Cambridge Phil. Trans.* 1857.

and the next in order—a statement made above in slightly different language.

§ **12.** In a uniform field the lines of force are straight, parallel, and equidistant, and the equipotential surfaces are planes perpendicular to the lines of force, and equidistant from each other.

If one magnetic pole of strength m be alone in the field its lines of force are straight lines, radiating from the pole equally in all directions, and their number is $4 \pi m$. The equipotential surfaces are a series of spheres whose centres are at the pole and whose radii are m, $\frac{1}{2}m$, $\frac{1}{3}m$, $\frac{1}{4}m$, &c. In other magnetic arrangements the lines and surfaces are more complicated.

Since a current exerts a force on the pole of a magnet in its neighbourhood it may be said to produce a magnetic field, and we may draw magnetic lines of force and equipotential surfaces depending on the form of the circuit conveying the current, and the strength of that current. When the current is a straight line of indefinite length like a telegraph wire, a magnetic pole in its neighbourhood is urged by a force tending to turn it round the wire, so that at any given point the force is perpendicular to the plane passing through this point and the axis of the current. The equipotential surfaces are therefore a series of planes passing through the axis of the current and inclined at equal angles to each other. If the unit current be defined as *that current, the unit length of which acts with unit force on the unit magnetic pole at the unit distance,* then the number of the equipotential planes surrounding the wire is $4 \pi c$ where c is the strength of the current. Thus if the strength c were 1·909 we should have 24 such planes; if πc is not a whole number, c must be expressed in units so small that the error involved in taking the nearest whole number may be neglected. The lines of magnetic force are circles having their centres in the axis of the current and their planes perpendicular to it. The intensity R of the magnetic force at a distance k from the current is the reciprocal of the distance between two equipotential surfaces; we have therefore $R = \frac{2c}{k}$. . . . . 4°.

§ **13.** In most telegraphic instruments magnets or soft iron armatures are moved by forces due to the passage of electric currents in certain wires. The apparatus should be sensitive so that it may be worked even by feeble currents; in designing the apparatus it should therefore be our endeavour so to arrange the wire conveying the current as to produce the most intense magnetic field which that current is capable of producing, and to place the magnet or soft iron acted upon in the most intense part of that field. By so doing, and by reducing the forces opposing the motion of the soft iron or magnets as much as possible, we render the apparatus as sensitive as it can be made.

When the magnet to be moved or acted upon is large it will occupy a large portion of the magnetic field, and will therefore experience a larger force than if it were small; but the force which it experiences per unit of volume can seldom if ever be made so great as when the magnet itself is small, for a small and intense magnetic field can be produced with a much less current than a large and equally intense magnetic field. Hence, we find all very sensitive apparatus characterised by small moving parts. The inertia of large masses is also injurious in all rapidly moving parts, for not only are the large masses acted upon with less force, but owing to the increased distance of the greater portion of their bulk from the axis on which they must oscillate their moment of inertia is increased even more than their bulk.

Similarly, when a wire conveying a current, or a magnet, or a soft iron armature is to move under the influence of a magnet, it must be our aim so to arrange that magnet as to produce the most intense magnetic field possible at the spot where the moving piece is placed.

The mapping out of magnetic fields due to different

forms of magnet and different arrangements of wires conveying currents has therefore a great practical interest for the electrician.

§ **14.** The poles of a magnet are not at its extremities, but generally a little way from the end. It is not necessary that a magnet should be magnetised in the direction of its length; a bar may be magnetised transversely or indeed in any direction. Some magnets have more than one pair of poles.

If a long thin magnet be broken each part becomes a distinct magnet having its axis in the direction of the old axis; from this it appears that all parts of the magnet are in some peculiar polarised condition, and the actual poles of any given magnet are simply the result of the combination of all these polarised parts.

A piece of soft iron which is a magnet by induction can again induce magnetism in another piece of soft iron: thus, a magnet may sustain a long string of nails each hanging to its neighbour. This chain of nails has its pair of poles near the ends of the first and last nails in the series, and affords an example of what is meant by saying that all parts of a magnet are in a polarised condition; each nail when detached from the series will remain a magnet for some little time in virtue of its coercive force § 9. If a magnet be plunged in iron filings and withdrawn, these adhere most abundantly near the poles. They stand out from the magnet in tufts, largest where the field of force is strongest, that is, near the poles, and the direction of the chains or strings which they form corresponds to the direction of the lines of force; each separate filing becomes a small magnet for the time being.

§ **15.** Magnets are made from one another by taking advantage of this coercive force, which is found to be greatest in hard steel. A piece of steel may be magnetised by being stroked once or twice in the same direction by a powerful magnet, or even touched at one end by that magnet. Better results are obtained by placing the two opposite poles of equally strong magnets in the centre of the bar to be magnetised, and drawing them simultaneously away from the centre to the two ends. This operation is repeated two or three times, and the bar then turned over and treated in a similar way on the other face. The bar magnets may, with advantage, incline from one another while being dragged apart. A still more complete magnetisation is given by placing the bar A B between two powerful magnets N S and N′ S′ as shown, and then drawing the oppo-

FIG. 61.

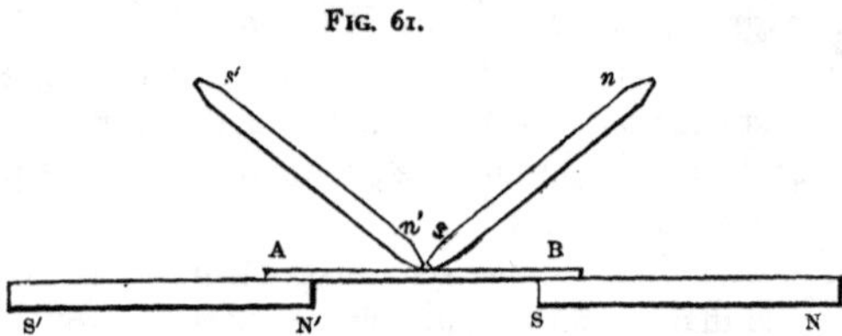

site poles of two other magnets from the centre of A B towards the ends. There are other methods of preparing magnets but they all consist in placing every part of a bar of steel in the strongest possible magnetic field and trusting to the coercive force of the steel to retain the induced magnetism.

§ **16.** The name electro-magnet is given to a magnet formed of a rod or bundle of rods of wrought iron, round which an electric current circulates in a coil of wire, as in Fig. 40. The electric current so arranged produces an intense magnetic field, and the most powerful magnets are produced in this manner. It is found that there is a limit to the amount of magnetism which in this way or any other can be induced in soft iron; when this limit is approached, the soft iron is said to be saturated with magnetism. Steel is sooner saturated than wrought iron; and as it resists the acquisition of magnetism more than soft iron does, so it retains more of the magnetism it acquires. This resistance to magnetisation is also attributed to coercive force. Electro-magnets can be made of any form. The two most common are the straight bar, in which the poles are as far apart as possible, and the horse-shoe, in which they are brought close together.

A piece of soft iron joining the poles of a magnet is called an armature; it adheres to the poles and diminishes very much, while in its place, the intensity of the magnetic field in the neighbourhood. An electro-magnet formed as a complete ring produces no sensible magnetic field in its

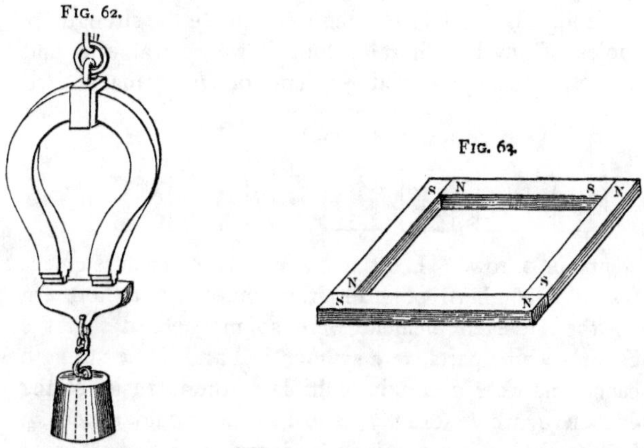

FIG. 62. FIG. 63.

neighbourhood; nevertheless, although without poles, it is certainly a magnet, and produces many of the magnetic phenomena. A series of equal magnets arranged (as in Fig. 63) so that the north pole of each is in contact with the south pole of its neighbour will also produce no magnetic field. An armature is found to diminish sensibly the loss of magnetism which is continually taking place in ordinary steel magnets. The armature is used to suspend weights from horse-shoe magnets, as in Fig. 62.

§ **17.** The strength m of the poles of a long soft iron bar of one square centimetre section held horizontally in the magnetic field due to the earth alone in England will be equal to about ·175 × 32·8 or 5·74 units, each pole would attract a pole of opposite name with a force $f = \frac{m\,m}{d^2}$, so that if the distance between the poles were one mètre, the force exerted would be $\frac{5{\cdot}74 \times 5{\cdot}74}{100^2}$ $= 32{\cdot}9 \times 10^{-4} = {\cdot}00329$ absolute units of force equal to the weight of ·0000517 grain. In order that this should be even approximately true the prism must be so long that the magnetisation of the middle does not interfere with that of the end. We should be able to calculate the strength of the poles of any bar short or long if we were able to find the magnetic effect produced by a series of equally magnetised

FIG. 64.

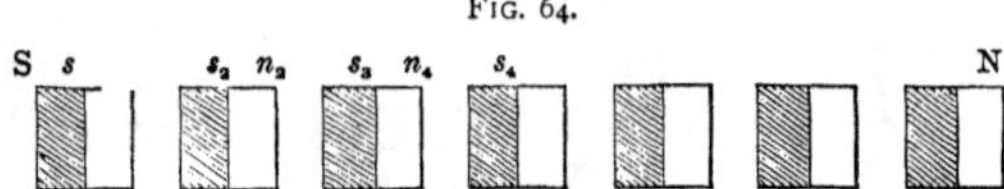

elements in a row. Let the black part of each element represent a southern pole and the white part a northern pole; then if each element were so magnetised that the black and white parts were symmetrical and if the strength of each pole were a certain multiple of the intensity of the field, then n_1 would exactly cancel s_2; n_2 would cancel s_3, and so forth, leaving S at one end and N at the other at the effective poles of the magnet; but in fact the action of each little element extends to all the others, and the summation of all these effects is so complex that we must abandon all attempt to calculate the strength of the poles from the intensity of magnetisation, except in certain very simple cases. The calculation given above applies sensibly to all long thin bars the cross section of which is small compared with one-twelfth of their length; thus our bar of one centimètre cross section would have to be at least five or six mètres long before the formula would apply.

The magnetic moment (§ 6) of a long thin bar is, $k\,\mathrm{H}\,s\,l$, where H is the intensity of the field, s the cross section of the bar, l its length, and k the coefficient of magnetic induction; the magnetic moment of a sphere in the same field will be

$$\frac{4}{3}\pi r^3 \text{H} \frac{k}{1 + \frac{4}{3}\pi k} \quad . \quad . \quad . \quad (5)$$

and from this formula the intensity of magnetisation of a given piece of steel or other metal can easily be calculated if k be known, or k may be determined from actual observation of the magnetic moment.

§ **18.** The coefficient k is constant only for low magnetic intensities, and gradually diminishes according to an unknown law as the maximum intensity for each material is approached. The maximum intensity of magnetisation for iron can be obtained from an experiment by Dr. Joule, who found that the maximum attraction he could produce between an electro-magnet and its armature was 200 lbs. per square inch of surface. Calling this maximum attraction F, the intensity i, and A the area of the surfaces between which the attraction is exerted, we have, when the distance between the surfaces is very small

$$\text{F} = 2\pi i^2 \text{A} \quad . \quad . \quad . \quad (6)$$

200 lbs. per square inch is 14061 grammes per square centimètre, or about 13,800,000 absolute units of force per square centimètre. Giving this value to F in the above equation when A is unity, we find for i the value of about 1490, as the maximum intensity of magnetisation of which iron is capable. If the value of 32·8 k were constant, a magnetic field of the intensity of about 45 would be sufficient to magnetise iron to saturation. Probably k can only be regarded as sensibly constant while the magnetisation of the iron is below one quarter of its maximum value, and from some experiments by Müller [1] we might infer that the value of k near the point of saturation is about one-third of the value given above, so that a field of magnetic intensity equal to

[1] Pogg. Ann. vol. lxxix. 1850.

about 135 would be required to give an electro-magnet the maximum possible strength.

§ **19.** The relative intensity of magnetisation in the same field for different substances has not been very fully studied; in other words, the values of k for different materials and different values of i are not well known. The following table is deduced from relative values obtained by Barlow, to which I have added nickel and cobalt, from relative values given by Plücker:

Soft wrought iron .	32·8	Soft cast steel .	23·3
Cast iron . . .	23	Hard cast steel .	16·1
Soft steel . . .	21·6	Nickel . .	15·3
Hard steel . .	17·4	Cobalt . .	32·8

These values can only be approximately true. A complete table of the values of k would require to contain a set of values for each material, and each value of i; whereas the value of i for which the above values hold good is not known. The maximum intensity of magnetisation for hard steel is less than for soft iron, and from some experiments of Plücker,[1] it appears that this difference is about 37 per cent., but a much greater intensity of field is required to produce the maximum of magnetisation.

With small values of i, the value of k for nickel was found by Weber to be five times that of iron, but with higher values of i it rapidly became smaller than for iron, reaching a maximum when i is about 30, increasing after this only about 2 per cent. when i was doubled.

§ **20.** According to experiments made by Plücker I estimate the value of k for water at

$$-10{\cdot}65 \times 10^{-6}.$$

The following values of k for different diamagnetic substances are calculated on this assumption from relative values obtained by Plücker:

Water	$-10{\cdot}65 \times 10^{-6}$
Sulphuric acid (spec. grav. 1·839) .	$-6{\cdot}8 \times 10^{-6}$

[1] Pogg. Ann. vol. xciv.

Mercury	− 33·5	$\times 10^{-6}$
Phosphorus	− 18·3	$\times 10^{-6}$
Bismuth	− 250	$\times 10^{-6}$

From an observation by Weber, the value of k for bismuth is about $- 16{\cdot}4 \times 10^{-6}$.

These figures are given to show very roughly the relative value of magnetic and diamagnetic action; they cannot be relied upon as even approximately true. Different observers give different relative values of k, differing twenty for the same substance. It must also be remembered that they are intended to indicate the value of k for equal volumes, not equal weights, of the substances.

§ **21.** It follows from equation (6) above, that the attraction between a magnet and its keeper or armature is proportional to the square of the intensity of the magnetisation, and therefore in an electro-magnet to the square of the current multiplied into k.

It also follows that where the intensity of magnetisation is the same throughout the mass of iron, the attraction will be simply proportional to the cross section of the iron. The object of increasing the length of an electro-magnet is to get a uniform field and to place the poles so that they do not interfere with one another.

By rounding or pointing the ends of a magnet, a more intense magnetisation is produced at the ends than elsewhere; hence a greater attraction per square centimètre of surface.

The attraction between a magnet and a keeper is directly proportional to the intensity of the magnetism induced in the keeper, if the keeper does not by its mass or great intensity of magnetisation react on the magnet, altering its intensity. The relative attraction of a large magnet for small volumes of different substances does therefore truly measure the relative values of k for each substance, if the volumes are the same and the intensity of the magnetic field the same throughout all the volume; but these values of k are almost useless unless the value of i in absolute measure is also determined.

CHAPTER VII.

MAGNETIC MEASUREMENTS.

§ **1.** BEFORE proceeding to study further the laws of the action of currents upon currents, it is convenient to examine the methods by which the forces exerted by magnets one upon another can be definitely measured or expressed in numbers depending solely on the centimètre, gramme, and second of time. To do this, we require to measure two things only: 1st, the intensity or strength, T, of magnetic field which a given magnet or arrangement of magnets produces at a given point. 2nd, the magnetic moment, $M = ml$, of the magnet which is acted upon by the assumed magnetic field. Knowing these two quantities, we can, in virtue of the laws already stated, calculate the couple experienced by the magnet in the field. The simplest experimental determination of the magnetic strength of a field requires that the field shall be sensibly uniform throughout the space in which the experiment is to be tried. The magnetic field due to the earth is sensibly uniform within the space occupied by the experiment, and after giving a general description of the magnetic field due to the earth's magnetism, we will proceed to examine how its intensity is to be measured.

§ **2.** The direction of the lines of force in this field is not horizontal except at some places near the equator. The earth may be (very roughly) conceived as a large bar magnet, and Fig. 58 shows that the lines of force are parallel to the axis of the magnet only at points half-way between the poles. The inclination of the lines of force at any place to the plane of the horizon is called the dip or magnetic inclination at that place. If a magnet were suspended by its centre of figure, and were free to assume any direction, it would not remain horizontal, but its axis would lie in the direction of the lines of force; in the northern hemisphere its north pole would point downwards, and the angle which this axis makes with the horizontal plane is the dip or inclination.

The lines of the earth's magnetic force do not usually lie in planes running due north and south. The vertical plane in which they lie at a given place is called the *magnetic meridian* of that place; the magnet points to the magnetic north. This magnetic north is not any one point, i.e. the magnetic meridians at different parts of the earth's surface do not cut at one point as the true meridians do.

The geographical or true meridian of a place is the plane passing through the place and containing the true axis of the earth. The angle contained by the magnetic and true meridians is called the magnetic declination at that place; the declination is said to be east if the north pole of the magnet points east of the true or geographical meridian. The declination is west if the north pole of the magnet points west. The north and south points of the mariner's compass indicate the magnetic meridian.

§ **3**. The declination and dip, or inclination, not only vary from place to place, but also at any one place from hour to hour and from day to day. There are some irregular variations, but there are others which are evidently periodic.

1. There are secular variations, the duration of which is not accurately known. In 1580, the declination at Paris was 11° 30′ E.; in 1814, this had become 22° 34′ W., and since then the needle has gradually returned towards the E.; in 1865 the declination was 18° 44′ W. In certain parts of the earth the magnetic and geographical or true meridians coincide; these points may be joined by an imaginary line, called the *agonic* line, or line of no variation.

2. There are annual oscillating variations of declination not exceeding 15′ or 18′, and varying at different epochs.

3. There are diurnal oscillating variations of declination

amounting at Paris on some days to about 25′, on others not exceeding 5′.

4. There are accidental variations or perturbations said to be due to magnetic storms. These variations occur with great rapidity, causing deflections to the right and left, comparable in their rate or period of alternation with ordinary telegraphic signalling; accidental variations of 70′ have been observed.

The dip also varies from place to place; it is greatest in the polar regions, being 90° at the magnetic pole. At a series of points near the equator there is no dip; the line joining these is called the magnetic equator. In the southern hemisphere the direction of the dip is reversed, the south pole pointing downwards. Lines connecting places where the dip is equal are called *isoclinic* lines.

§ 4. The total intensity of the earth's magnetism is the intensity measured in the direction of the lines of force at the point where the experiment is made. It is difficult to make the experiment in this way, especially as the direction varies so frequently. The *strength* of the horizontal component is therefore experimentally determined, and the *direction* of the total force. These two elements give the intensity and direction of the total force; for let H (Fig. 65) be the horizontal component, R the total intensity, and θ the dip, then

FIG. 65.

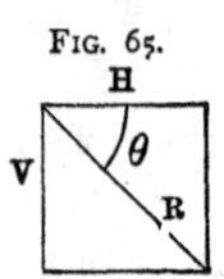

$$R = \frac{H}{\cos \theta} \quad . \quad . \quad . \quad (1)$$

§ 5. In order to determine the effect of any magnet upon another or upon an electric circuit, its moment, $M = m\,l$, must be determined. Two experiments are sufficient to determine at once the moment M and the force H. The first of these gives the value of the product M H by an observation of the directing force which the earth exerts on the magnet; the second gives the ratio $\frac{M}{H}$ by an observation of the relative strength of the magnetic fields due to the magnet and to the earth. The following are the two experiments:

1. Let the magnet be hung so as to oscillate freely in a horizontal plane round its centre of figure, being directed by the horizontal component of the earth's magnetism. Let the moment of inertia of the magnet relatively to the axis round which it oscillates be called I.[1] The quantity I is easily calculable for any regular figure, and can, moreover, be directly determined by experiment. Let the magnet now be allowed to oscillate freely, and let the number of complete or double oscillations per second be n; then

$$\mathrm{M\,H} = \frac{4\,\pi^2\,n^2\,\mathrm{I}}{g} \quad . \quad . \quad . \quad (2)$$

In Rankine's 'Applied Mechanics,' (§ 598) we have, equation (5), $\mathrm{M}_1 = \frac{4\pi^2\,n^2\,i_1\,\mathrm{I}}{g}$, where M_1 is the moment of the couple causing gyration, i_1 the semiamplitude of gyration in angular measure. Let us call F the *force* of the couple due to the magnetic field; the arm of the couple will be i_1L, where L is the distance between the poles; hence $\mathrm{M}_1 = i_1\,\mathrm{L\,F}$; but the moment of the couple due to magnetism when the magnet stands straight across the magnetic field is M H, and the arm of the couple being then L, the force must be then and always $\frac{\mathrm{M\,H}}{\mathrm{L}} = \mathrm{F}$, or $\mathrm{F\,L} = \mathrm{M\,H}$; hence $\mathrm{M}_1 = i_1\,\mathrm{M\,H} = \frac{4\pi^2\,n^2\,i_1\,\mathrm{I}}{g}$ or $\mathrm{M\,H} = \frac{4\pi^2\,n^2\,\mathrm{I}}{g}$. Q. E. D.

2. To obtain $\frac{\mathrm{M}}{\mathrm{H}}$, fix N S with its axis perpendicular to the magnetic meridian, and observe the deflection which it causes on a short magnet *n s* freely suspended so that when in the magnetic meridian the prolongation of its axis bisects N *o* S (Fig. 66). The deflection θ of *n s* will depend on the relative magnitudes of H and the field produced by N S.

[1] Rankine's 'Applied Mechanics,' § 571. I have here taken I as equal to the *weight* multiplied into the square of the radius of gyration, following Professor Rankine's example. Many writers define I as equal to the mass multiplied into the square of the radius of gyration, and if this value of I be used, the divisor g in equation 2 must be cancelled.

Let $r = o\,S = o\,N$; then $\frac{M}{H} = r^3 \tan\theta$ . . . (3)

Let m be the strength of the poles of the magnet N S; then the force which S will exert at o on a unit south pole will be $\frac{m}{r^2}$ . . . in the

FIG. 66.

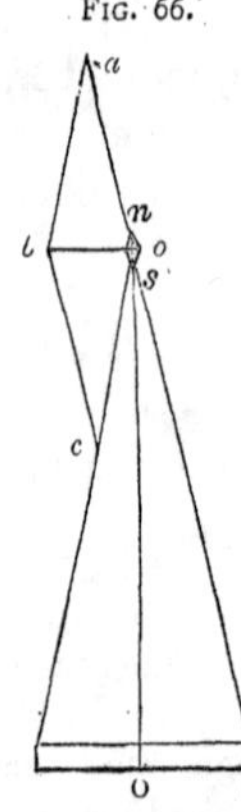

direction S o; the pole N will exert an equal force in the direction $o\,c$. Let $o\,a$ and $o\,c$ represent these forces in magnitude and direction; then $b\,o$ = T will represent the magnitude and direction of the lines of force of the magnetic field at o. We have $a\,o : o\,b = o\,S : N\,S$, or if $L = N\,S$; $\frac{m}{r^2} : T = r : L$; or $T = \frac{L\,m}{r^3} = \frac{M}{r^3}$.

Let M_1 be the moment of the little magnet, the couple due to T tending to turn it out of the magnetic meridian will be $M_1\,T\cos\theta = \frac{M\,M_1}{r^3}\cos\theta$. The couple due to H tending to bring it back will be $M_1\,H\sin\theta$; and when one balances the other $M_1\,H\sin\theta = \frac{M\,M_1}{r^3}\cos\theta$; or $\frac{M}{H} = r^3\frac{\sin\theta}{\cos\theta}$; or $\frac{M}{H} = r^3\tan\theta$. Q. E. D.

From equations (2) and (3) we have

$$H = 2\,\pi\,n\sqrt{\frac{1}{g\,r^3\tan\theta}} \quad . \; . \; . \quad (4)$$

$$\text{and } M = 2\,\pi\,n\sqrt{\frac{r^3\tan\theta}{g}} \quad . \; . \; . \quad (5)$$

§ 6. By means of the single experiment last described and illustrated by Fig. 66, the moment M of any permanent or temporary magnet can be readily determined if H be known, for from equation (3) we have $M = r^3\,H\tan\theta$; H is sufficiently constant throughout England, and from year to year, to give the value of M with sufficient accuracy for most practical purposes. This method can be used for horse-shoe magnets or magnets of any shape if care be taken to fix N S, the line joining the poles of this magnet, exactly perpendicular to the magnetic meridian To do this, suspend the magnet by its centre of figure, and let it take up its position on the magnetic meridian. Then noting this position turn the magnet through exactly 90° and fix it there.

§ **7.** In order that the values in the above formulæ should be expressed in absolute measure, consistent with that hitherto adopted, we must be careful to measure I in centimètres and grammes. As an example, the moment of inertia of a rectangular prism of steel, two centimètres long, and with a square section, each side of which measures two millimètres, and weighing 1·248 grammes is

$$\text{I} = 1{\cdot}248 \, \frac{1^2 + {\cdot}1^2}{3} = {\cdot}00416,$$

$\frac{1^2 + {\cdot}1^2}{3}$ is the square of the radius of gyration.[1]

To convert the value of H found by the above formulæ into grammes, divide by the value of g in centimètres (981·4 at Glasgow). The mean horizontal component H in England for 1862 was 0·175 (centimètres, grammes, seconds) in absolute measure. If a free unit pole weighed one gramme, it would, under the action of the horizontal component of the existing magnetism acquire a velocity of 0·175 centimètres at the end of a second. To convert this value into English absolute measure (grains, feet), we must multiply it by 21·69.

§ **8.** The value of I for a given magnet or other suspended mass of simple form can as above be calculated from measurements of its figure and its specific gravity or weight; but when the form is complex and the suspended mass of various materials, it is better to determine I experimentally by comparison with a body of known moment of inertia. To do this, first observe the time of one complete or double oscillation t of the magnet (directed by the earth's force alone), and then add some weight of simple form with a known moment of inertia I_1, and observe the time t_1 in which the compound body completes an oscillation; then, if

[1] Rankine's 'Applied Mechanics,' § 578.

n be the number of oscillations per second, $t = \frac{1}{n}$, and we have from equation (2)

$$\text{I} = \frac{\text{M H}}{4\pi^2} t^2 ; \qquad \text{I} + \text{I}_1 = \frac{\text{M H}}{4\pi^2} t_1{}^2 ;$$

$$\text{or} \quad \text{I} : \text{I} + \text{I}_1 = t^2 : t_1{}^2 ; \text{ whence}$$

$$\text{I} = \frac{\text{I}_1 t^2}{t_1{}^2 - t^2} \quad . \quad . \quad . \quad (6)$$

The method by which the value of T [or the line $o\ b$] was calculated in § 5 enables us to determine the intensity of the field at any point due to a magnet, so soon as the moment M and length l are known. The action of each pole on a unit pole at the distance r will always be equal to $\frac{m}{r^2} = \frac{\text{M}}{l\, r^2}$; and by compounding the forces due to each pole we obtain the resultant in direction and intensity.

The magnetic moments of two magnets of known moments of inertia I and I_1 can be compared by means of their times of oscillation t and t_1 in the same magnetic field; it follows from equation (2) that

$$\text{M} : \text{M}_1 = \frac{\text{I}}{t^2} : \frac{\text{I}_1}{t_1{}^2} \quad . \quad . \quad . \quad (7)$$

Similarly, the horizontal intensity of two magnetic fields can be compared by observing the times t and t_1 required for a complete oscillation of any given magnet in the two fields:

$$\text{H} : \text{H}_1 = t_1{}^2 : t^2 \quad . \quad . \quad . \quad (8)$$

In making this experiment, we must not assume the constancy of any given magnet even for two successive days.

§ **9.** In calculating the effects of a real magnet, we must never forget, that although we may experimentally determine the value of $m\ l$, we cannot really separate m from l, because we can never feel certain that the length l is equal to the length of the magnet, or to any given fraction of it. If the material were uniformly magnetised, i.e. if it would form a number of absolutely equal magnets when cut up into a number of absolutely uniform pieces, then, indeed, the length l would be the exact length of the magnet. In any actual magnet the strength of magnetisation is found to fail off near the ends, and this makes l shorter than the length of the magnet; moreover, the distribution of electricity is such that the magnetic field produced by it is different in many respects, from that which could be produced by poles.

CHAPTER VIII.

ELECTRO-MAGNETIC MEASUREMENT. ACTION OF CURRENTS ON CURRENTS AND ON MAGNETS.

§ 1. THE series of units described in Chapter V. would suffice for all electrical purposes, but they are not very well adapted for the calculation of the effect of electric currents upon one another, or upon magnets.

We obtained the set of electrostatic units from a series of equations which did not involve the forces acting between currents and magnets; starting from the measurement of these latter forces, we obtain a distinct system of units, which will be termed *electro-magnetic* units, from a series of equations which do not involve the forces of electrostatic repulsion and attraction. Electro-magnetic units are more commonly used in telegraphy than electrostatic units. In Chapter VI. § 12 a definition of the unit current was suggested, depending on the force with which a current acts on a magnetic pole. According to this definition, the unit current is such that every centimètre of its length acts with unit force on a unit magnetic pole at a distance of one centimètre from all parts of the current. To obtain this last condition the wire conveying the current must be bent in a circle, at the centre of which hangs the free magnetic pole. The force (f) exerted on the pole of a magnet in its neighbourhood is proportional to the magnetic strength (m) of the pole of the magnet, and to the strength of the current C; and if the conductor be at all points equi-distant from the pole, the force is proportional to the length of the conductor L. It is also inversely proportional to the square of the distance k of the pole from the conductor, and is affected by no other circumstances. Hence we have

$$f = \frac{\text{C L } m}{k^2} \quad . \quad . \quad . \quad (1)$$

from which $\text{C} = \frac{f\,k^2}{\text{L}\,m}$, giving the definition of the unit current stated above.

§ 2. Let us use the capital letters Q, I, R, C, and S to indicate the quantities in electro-magnetic measure which were indicated by q, i, r, c, and s in electrostatic measure; then, taking the unit of current as determined by the equation in § 1, we have, from the equations $\text{Q} = \text{C}\ t$, $\text{I} = \frac{w}{\text{Q}}$, $\text{R} = \frac{\text{I}}{\text{C}}$, and $\text{S} = \frac{\text{Q}}{\text{I}}$, a complete new series of units bearing a definite ratio to the electrostatic units; by experiment it has been found that $c = 28{,}800{,}000{,}000$ C. This numerical coefficient will be termed v.

$$\text{C} = \frac{c}{v} \quad \Big| \quad \text{Q} = \frac{q}{v} \quad \Big| \quad \text{I} = v\,i \quad \Big| \quad \text{R} = v^2\,r \quad \Big| \quad \text{S} = \frac{s}{v^2}$$

The above series of equations express the relations between the numbers expressing electrical magnitudes in the two series of units; they all follow directly from the fundamental equations. The relations of the electro-magnetic units to one another, and to the mechanical units may be summed up as follows: The unit current conveys a unit quantity of electricity per second across any section of the circuit. The unit current will be produced in a circuit of unit resistance by the unit electromotive force. The unit current in a conductor of unit resistance produces an effect equivalent to the unit of work per second. Lastly, the unit current flowing through a conductor of unit length will exert the unit force on a unit pole at a distance of one centimètre. It is this last condition which is peculiar to the *electro-magnetic* series.

§ 3. Let a very short magnet *n s* (Fig. 67), say $\frac{1}{4}$ inch in length, be freely hung at the centre of a circular coil A, of considerable relative diameter, say 18 inches, and let the plane of the coil be placed in the magnetic meridian, then the value C in electro-magnetic measure of any current passing through the coil and deflecting the magnet through the angle θ, is given by the following expression:

FIG. 67.

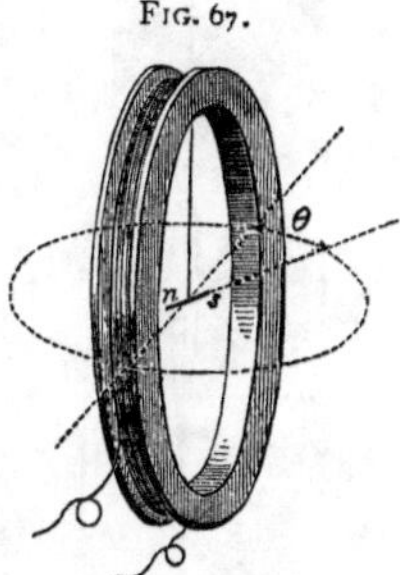

$$C = \frac{H\,k^2}{L} \tan \theta \quad . \quad . \quad . \quad (2)$$

where H is the horizontal component of the earth's magnetism and L is the length of the wire forming the coil. All dimensions must be in centimètres if H is measured in the units already adopted.

From this equation we see that the current will be proportional to the tangent of the angle of deflection, and a galvanometer of this construction is therefore called a tangent galvanometer; moreover, knowing the value of H, we shall, with tangent galvanometer, be able directly to measure currents in absolute measure, independently of any knowledge of the magnetic moment of the needle employed, and independently also of any peculiarity in the instrument used. A current so measured in Australia is therefore at once comparable with a current measured in England.

The resultant electro-magnetic force (f) exerted at the centre of a circular coil of radius k, by the current C, will by equation 1 be $f = \frac{C\,L}{k^2}$: the two poles of a short magnet hung in the centre, with its magnetic axis in the plane of the circular coil, will experience equal and opposite forces, each equal to $f\ m$, where m is the strength of each pole of the magnet. If l be the distance separating these poles or forces (equal sensibly to the length of the magnet), then the magnet experiences what is termed a couple, the moment of which is $f\ m\ l = \frac{\text{C L}\ m\ l}{k^2}$. Let N S be the plan of the magnet (Fig. 68) as it hangs in the plane of the coil of wire, and let $\text{N}_1\ \text{S}_1$, making an angle θ with N S, be any new position which it takes up under the influence of the current. Then, supposing the magnet to be small compared with the diameter of the coil, the poles remain sensibly at the centre; the force f remains the same, but the perpendicular distance N_1 C between the poles on which the equal and opposite forces are exerted is now equal to $l \cos \theta$, and hence the couple is now $\cos \theta \frac{\text{C L}\ m\ l}{k^2}$. This couple is opposed by the directing couple due to the earth's magnetism. Let us call H the horizontal component of the earth's magnetism at the place in question; then the force due to its action on each pole will be H m; the perpendicular distance S_1 C separating the two parallel forces will be $l \sin \theta$, and whole couple will therefore be $\sin \theta\ \text{H}\ m\ l$; and when the magnet is in equilibrium, under the combined forces of the directing current and the earth's magnetism, we have

FIG. 68.

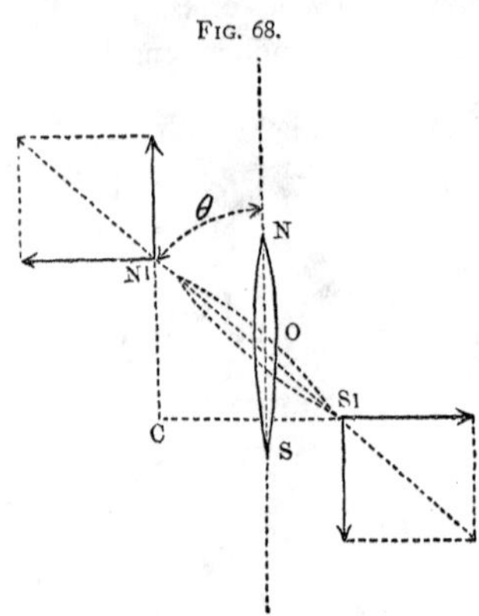

$$\cos \theta \frac{\text{C L}\ m\ l}{k^2} = \sin \theta\ \text{H}\ m\ l; \text{ whence}$$

$$\text{C} = \frac{\sin \theta}{\cos \theta} \frac{\text{H}\ k^2}{\text{L}} = \tan \theta \frac{\text{H}\ k^2}{\text{L}}$$

§ 4. All the relations between force and currents of a given form and strength may be deduced mathematically from the following theory, due to Ampère. 1. The force with which two small lengths or elements of currents act upon each other is in the direction of the line joining the centres of these elements, and this force is inversely proportional to the square of the distance between the elements.

2. Let there be two short wires $m\ n$ and $m_1\ n_1$ (Fig. 69), parallel to one

another, and perpendicular to the line d joining their centres. Let the current c flow through $m\ n$, and c_1 through $m_1\ n_1$; then the force with which these two little elements of currents attract one another if flowing in the same direction or repel one another if going in opposite directions is $\frac{c\ c_1 \times m\ n \times m_1\ n_1}{d^2}$.

FIG. 69.

3. If the two short wires be placed as in Fig. 69*a*, so as to lie in the direction of the line d joining their centres, the force acting between them is half the above: it is a repulsion if the currents flow in the same direction, an attraction if they flow in opposite directions.

FIG. 69*a*.

4. If the two short wires be placed so as to be both perpendicular to the line d, but so that $m\ n$ is also perpendicular to $m_1\ n_1$, as in Fig. 69*b*, then the currents neither attract nor repel one another.

5. If one element lies along d, and the other is perpendicular to it, the currents neither attract nor repel one another.

6. Let A B (Fig. 69*c*) be any short wire conveying any current c in any direction relatively to the short wire $A_1\ B_1$, conveying another current c_1. Let the line d join the centres of A B and $A_1\ B_1$; draw the line x_1 in the direction of d and draw y_1 perpendicular to x_1, and of such magnitude that the resultant of two forces y_1 and x_1 would be equal to the current c_1, and lie in the direction $A_1\ B_1$. On a similar plan draw y parallel to y_1, and draw x and z, rectangular components such that if y, x, and z were forces, their resultant would be equal to c, and lie in the direction A B. Then the resultant action of the current in A B on the current in $A_1\ B_1$, will be the sum of that of the three currents y, x, and z on the two currents y_1 and x_1. We may observe that this reduces itself to the sum of the action of x on x_1, which we can calculate from 3. above added to the action of y on y_1, which we can calculate from 2. above: for z is inoperative on y_1, y does not attract or repel x_1, nor does y_1 attract

FIG. 69*b*.

FIG. 69*c*.

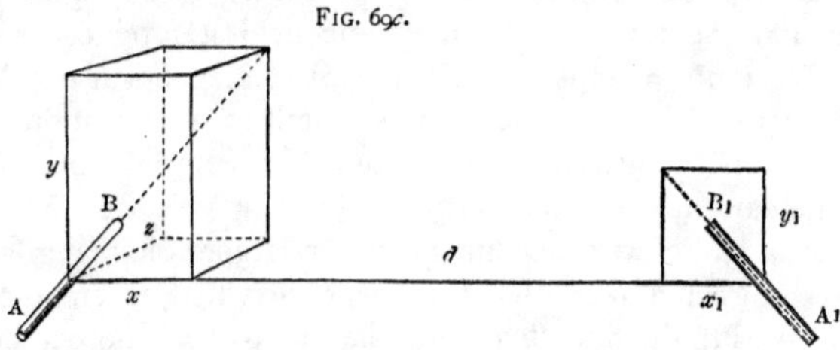

or repel *x*. In dealing with wires of any considerable length, the action of each little element of one wire on all the elements of the other must be taken into account, and the results summed. This summation or integration gives the results detailed in the following paragraphs; and these results, being confirmed by experiments on closed circuits, establish the truth of the theory as applied to closed circuits.

It follows from the above theory, that the action of a small closed circuit at a distance is the same as that of a small magnet having its axis placed perpendicularly to the plane of the current, and having a moment equal to the product of the current into the area encompassed by the circuit; thus, if the circuit be circular, the moment of the magnet will be $C \pi k^2$. Let two small circles, with radii k and k_1, be placed at a great distance D from one another, in such a manner that their planes are at right angles to each other and that the line D is in the intersection of the planes. Let an equal current C circulate in each of these conductors; forces will act between them, tending to make their planes parallel and the direction of the currents opposite; these forces will produce a couple, of which the moment will be

$$M = \frac{C \pi k^2 \times C \pi k_1^2}{D^3} \quad . \quad . \quad . \quad (3)$$

If now, M, D^3, πk^2, πk_1^2 be all made unity, this will give a value for the unit of current C, which will be the same as that founded on the action between a current and a magnet. It also follows that the unit current enclosing a circle of unit area will produce the same couple on a magnet at a distance as would be produced by a small magnet of unit moment.

§ 5. We found one means of measuring the strength of a current by comparing the magnetic field it produced with the horizontal component of the earth's magnetism H. We may determine or measure the strength of a current in the same units by measuring the action between different parts of the current itself as determined by Ampère's theory.

Let a coil of wire A be hung inside a larger coil B (Fig. 70), and so directed by means of its suspension that, when no currents pass through the two coils, the plane of A is perpendicular to that of B. When one and the same current is allowed to flow simultaneously through A and B, they experience a deviating couple proportional to C^2, and depending for its absolute value on the value of the diameters k and k_1 of A and B, and on the number of turns V and v_1 in these coils. If the plane of the coil B be so turned that, when the current is passing, the plane of A lies in the magnetic meridian, then the only couple tending to bring A back into its original position will be that due to its suspension. Then calling the deflection or angle between the planes of the coils θ, expressed in circular measure, we have

FIG. 70.

$$C = \sqrt{a \frac{\theta}{\cos \theta}} \quad . \quad . \quad . \quad (4)$$

where a is a constant, varying in different instruments, but which for any one instrument can be found experimentally or determined once for all by the maker. This method was first employed by Weber, and the instrument is called Weber's Electro-Dynamometer.

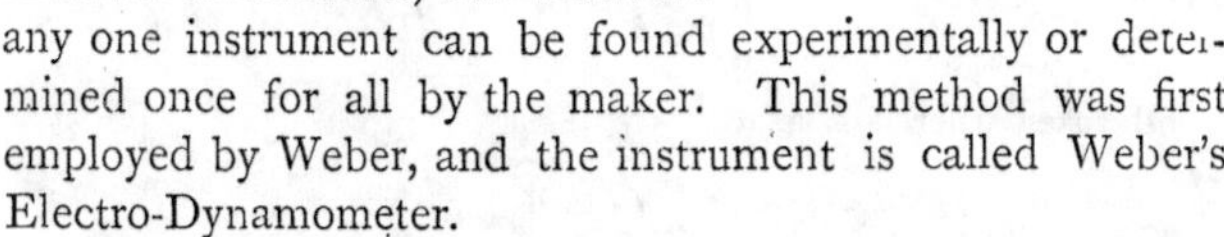

Let us call the directing couple G and the deviating couple M. When the coil A is in equilibrium, M = G. The value of G depends on the mode of suspension ; if it be by a single wire, the torsion varies simply as the angle of deflection θ, or

$$G = \mu \theta \quad . \quad . \quad . \quad (5)$$

where u stands for the expression

$$\frac{4 \pi^2 n^2 I}{g} = \frac{4 \pi^2 I}{g t^2} = \frac{4 \pi^2 I_1}{t_1^2 - t^2} \quad . \quad . \quad . \quad (6)$$

in which the several letters have the same meaning as in Chapter VII. § 8 ; I being now the moment of inertia of the suspended coil instead of the suspended magnet, and I_1 the moment of inertia of a mass of simple form added to determine experimentally the value of I.

The value of the deflecting couple is given by the equation

$$M = \beta C^2 \cos \theta \quad . \quad . \quad . \quad (7)$$

in which β is a constant determined by Ampère's theory. Let k be the radius of the large coil B, k_1 the radius of the small coil A. Let k_{11} be the distance from the centre of coil A to the periphery of coil B ; $k_{11} = k$ when the coils have a common vertical axis ; let V be the number of turns of wire in the large coil ; v the number of turns in the small coil, then

$$\beta = \frac{k_{11}^3}{2 \pi^2 V v k^2 k_1^2} \quad . \quad . \quad . \quad (8)$$

Since M = G from equations (7) and (5) we have

$$\text{C} = \sqrt{\frac{\beta}{u} \frac{\theta}{\cos\theta}} \quad . \quad . \quad . \quad (9)$$

The values of β and μ are evidently constant for any one instrument.

If the suspension is bifilar, equations (5) and (6) must be modified: we then have

$$\text{G} = u \sin\theta \quad . \quad . \quad . \quad (10)$$

and

$$u = \frac{4\,\pi^2\,\text{I}_1}{t_1^{\,2}\left(1 + \frac{w_1}{w}\right) - t^2} \quad . \quad . \quad . \quad (11)$$

where w_1 is the weight of the added mass and w the weight of the coil A.

Then from equations (10) and (7) we have

$$\text{C} = \sqrt{\frac{\beta}{\mu}\tan\theta} \quad . \quad . \quad . \quad (12)$$

for both cases, where θ is small,

$$\text{C} = \sqrt{\frac{\beta}{\mu}\,\theta}$$

θ being in circular measure.

§ 6. The following is another method, due to F. Kohlrausch, of measuring currents in absolute measure by means of a tangent galvanometer and a single coil suspended by two wires.

FIG. 71.

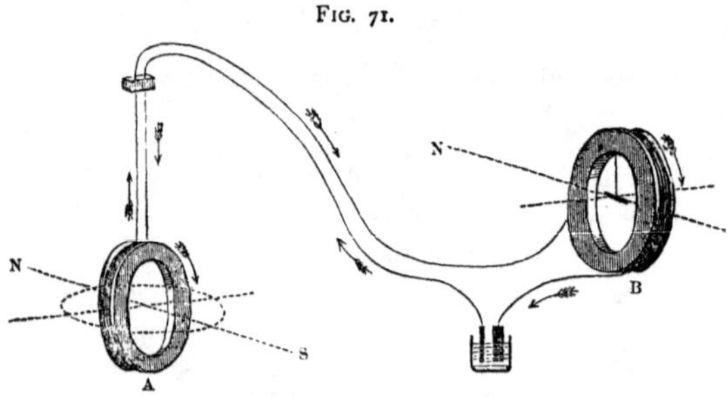

Let a coil A (Fig. 71) of k radius and n turns be hung by a bifilar suspension, with its plane perpendicular to the plane of the magnetic meridian. Observe the deflection θ produced in this coil by the current C and the simultaneous deflection θ_1 produced by the same current on the needle of a tangent galvanometer B of radius k_1, then

$$C = \sqrt{\frac{k_1^2 \mu}{L n \pi k^2} \tan\theta \,.\, \tan\theta_1} \quad . \quad . \quad . \quad (13)$$

The coil A, when the current C flows through it, is equivalent to a magnet of the moment $C\, n\, \pi\, k^2$; and calling H the horizontal component of the earth's magnetism, the couple experienced by the coil when deflected through the angle θ will be $H\, C\, n\, \pi\, k^2 \cos\theta$. The directing couple due to the bifilar suspension is $\mu \sin\theta$. Hence, when the one balances the other,

$$H\, C\, n\, \pi\, k^2 \cos\theta = \mu \sin\theta$$

$$\text{and } C = \frac{\mu}{H \,.\, n\, \pi\, k^2} \tan\theta \quad . \quad . \quad . \quad (14)$$

The value of μ can be found as by the last section. From this equation alone we might find C in terms of H; but we have also, calling θ_1 the deflection produced by the same current C passing through the tangent galvanometer of radius k_1,

$$C = \frac{H\, k_1^2}{L} \tan\theta_1$$

hence, eliminating H, we have equation (14) as given above (eliminating C, we might find H from the same equations). It should be observed that $n\, \pi\, k^2$ is more strictly the sum of the areas enclosed by the turns of different diameter of which the coil is composed.

§ 7. Let a current traverse two wires in succession, each bent so as to enclose a circle of the radius k. Let these wires be hung in parallel planes at the distance a, with their centres in the same axis. Then, if the current be sent round the wires in the same direction, they will attract one another; if in the opposite direction, they will repel one another with a force

$$F = 4\, \pi\, C^2 \frac{a}{k} \quad . \quad . \quad . \quad (15)$$

If two coils, each containing n turns, be thus hung, the force with which they attract or repel each other will be

$$F_n = 4\, \pi\, n^2\, C^2 \frac{a}{k} \quad . \quad . \quad . \quad (16)$$

hence, knowing the current, we can determine the force, or, weighing the force, can measure the current.

By placing two fixed parallel coils, A and B, opposite each other, as in Fig. 72, and passing a current round them in opposite directions, we obtain a sensibly uniform field of magnetic force between the flat coils. If a third flat coil D be hung between them it will be attracted by one and repelled by the other, and a good electro-dynamometer may be constructed on this principle. The actual value of the

FIG. 72.

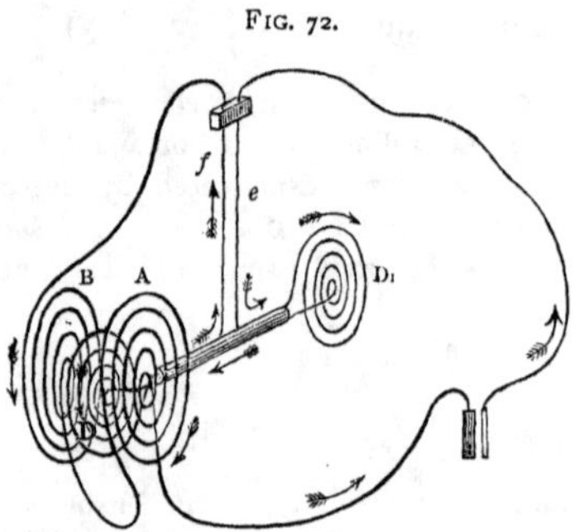

current corresponding to a given couple experienced by the suspending wires *e* and *f*, indicated by the torsion of a wire, is experimentally determined once for all by comparison with a standard instrument. A second suspended flat coil D_1 is required to make the system independent of the earth's magnetism, and this coil D_1 may advantageously be placed between two more fixed flat coils arranged so as to double the couple experienced by the suspended system.

FIG. 73.

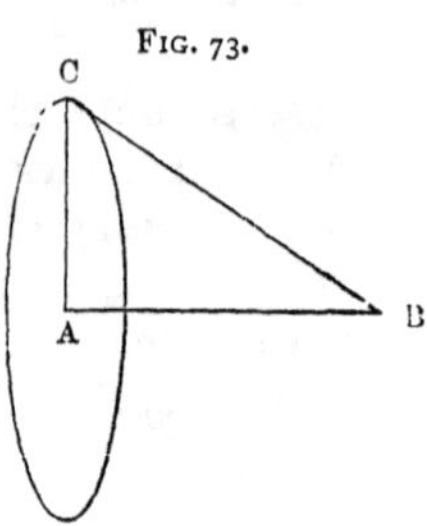

§ **8.** The intensity of the magnetic field produced by a circle at any point B on an axis perpendicular to the plane of the circle is given by the following formula :

Let A C (Fig. 73), the radius of the circular conductor, be $= k$. Let C = the current. Let A B $= x$. Let F = the intensity of the field.

$$\text{Then } F = \frac{\pi C k^2}{(k^2 + x^2)^{\frac{3}{2}}} \quad . \quad . \quad . \quad (17)$$

At A, the centre of the coil, the intensity is

$$T_1 = \frac{2 \pi C}{k} \quad . \quad . \quad . \quad (18)$$

Let an insulated wire be wound round a cylinder of the length $2\,l$, forming a spiral. Let the distance of the point M (Fig. 74) from the

FIG. 74.

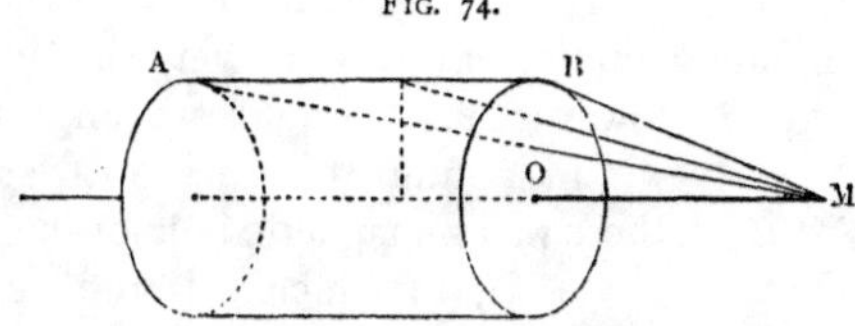

nearest end of the cylinder = M O = a. If the point were inside the spiral, a would be affected with the negative sign.

Let the line joining an element of a spiral with M = e.

Let the number of turns be n, then the intensity of the magnetic field at M is

$$T = \frac{C\,\pi\,n}{l}\left(\frac{a + 2\,l}{\sqrt{k^2 + (a + 2\,l)^2}} - \frac{a}{\sqrt{k^2 + a^2}}\right)$$

Let the angle A M O = ψ, and the angle B M O = ψ_1; then

$$T = \frac{C\,\pi\,n}{l}(\cos\psi - \psi_1).$$

This applies to inside as well as outside, remembering that $\cos\psi_1$ will be negative inside the spiral, so that we have virtually $\cos\psi + \cos\psi_1$.

The force is at a maximum in the centre.

Call the diagonal of the spiral $2\,d$; then the intensity of the magnetic field at the centre will be

$$F_m = \frac{2\,C\,\pi\,n}{d}.$$

If the length of the spiral be 40 times its diameter, the intensity of the magnetic field does not vary by one per cent. throughout $\frac{7}{8}$ of its length, and not ·1 per cent. throughout $\frac{1}{1000}$ of its length.

§ 9. A long spiral of insulated wire of small diameter relatively to its length is commonly called a solenoid, although, strictly speaking, this name applies only to a series of perfectly parallel and equal rings all perpendicular to a common axis and in all of which an equal current is flowing. The material representation of the solenoid differs experimentally little from its hypothetical type. We have seen that a current flowing round a circle or a series of circles in one plane acted upon a magnetic pole or upon an electric current at a distance as if it were a short magnet of the moment C $n \pi k^2$, where n is the number of turns.

If a solenoid beginning at A were very far prolonged towards B, it would act on all points within a finite distance of A, as if at A there was a magnetic pole of the strength C $n \pi k^2$, in which n is the number of turns in the solenoid *per centimètre.*

FIG. 75.

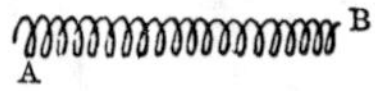

A B

An actual solenoid acts as if two such endless solenoids were superposed, having the same current flowing through them in opposite directions; one beginning at A and the other at B. Then we should have one north pole, say at A, and one south pole at B, and all the rest of the turns cancel one another; hence the magnetic moment of the solenoid is C $n \pi k^2$ L, where L is its length.

If keeping the actual number of turns constant we shorten the length L, we increase n just as we diminish L, so that the moment does not vary.

Imagine a watch hung in a solenoid in such a position that the current circulates with the hands of the watch. Then the south pole will be at the end towards which the face of the watch is turned.

§ **10.** If a magnet be hung with its north pole downwards over the centre of a vertical solenoid in which the current is circulating in the direction of the hands of a watch (looking at spiral and watch from above), then the north pole will be attracted when outside the solenoid, as if by a south pole; it will continue to be sucked into the solenoid, even after entering in it, although the force with which it is pulled down will diminish. The south pole of the magnet is repelled upwards, but with less force than the north pole is sucked downwards. When the centre of the magnet has reached the centre of the solenoid, the magnet will be in equilibrium so far as magnetic forces are concerned; if allowed to fall further, the magnetic forces will resist the motion, and if the current be powerful enough, these forces will carry the weight of the magnet and prevent it from falling further.

Feilitsch made the following experiment, showing how the force diminishes, using a magnet 10·1 centimètres long, 2·03 centimètres diameter, weighing 23·678 grammes, and a spiral or solenoid of 126 turns, 29·5 centimètres long, and 12·9 centimètres internal circumference. The following table gives the distances a of the centre of the magnet from the centre of the spiral, and g the force in milligrammes :

a.	18·7	16·7	14·7	12·7	10·7	8·7	6·7	4·7	2·7	·07	−·13
g.	190	382	493	474	313	115	32	16	11	2	−1

The poles of the magnet when in equilibrium inside the solenoid are placed relatively to the spiral, as if the spiral had magnetised a piece of soft iron of the same length. Soft iron is therefore drawn in just as the magnet would be, and the north pole of the soft iron corresponds to the north pole of the solenoid.

§ **11.** A hollow magnet does not in this respect resemble a solenoid.

If the north pole of a magnet A were introduced into the interior of a hollow magnet B at its south pole, A would be repelled from B after it had penetrated to a very short distance ; and if a rod of soft iron was placed inside a hollow steel magnet, the north pole of the magnet would induce a south pole in the end of the iron next it.

This experiment proves conclusively that we cannot regard a magnet as simply produced by a series of currents circulating round its exterior periphery; but it agrees with the hypothesis that the magnet consists of an immense number of little solenoids lying side by side. In fact, conceive a number of such solenoids, side by side, the end views of which are shown, as in Fig. 76, with the current flowing in the direction shown by the arrows, then all the elements of

FIG. 76.

each little circuit inside the ring would move in the direction followed by the hands of a watch; all the elements outside would move in the opposite direction. On a point at y the former would be most powerful; on a point at x, the latter; the radial currents counteract one another, for there are as many in one direction as in the other.

§ **12.** For general purposes, we may regard a solenoid as equivalent to a magnet, so far as regards all points outside of the cylinder; the effect of introducing soft iron into the interior of the cylinder is to make the field of force outside the cylinder, more intense. It may thus become as much as about 32·8 times more intense than before. The direction of the lines of force is very little altered. Fig. 77 shows

FIG. 77.

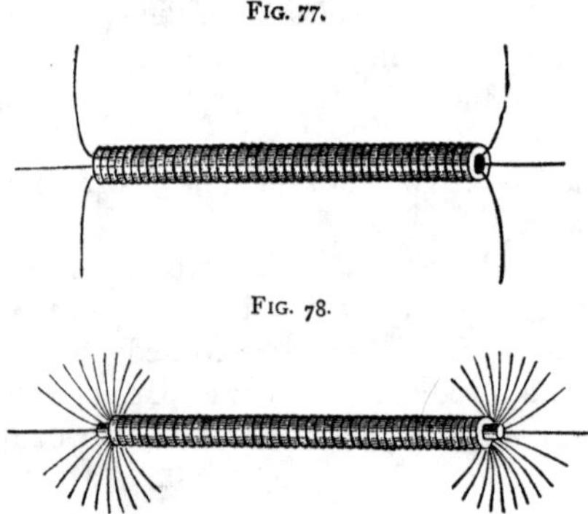

FIG. 78.

roughly the field of force due to a solenoid, Fig. 78, the field of force after a soft iron wire has been introduced. The soft iron wire concentrates the lines of force near the poles, and thus over a limited space enables the current passing through the solenoid to produce very powerful effects; its action in this respect is somewhat analogous to that of a lens used to concentrate light on a spot where illuminating action is required.

CHAPTER IX.

MEASUREMENT OF ELECTRO-MAGNETIC INDUCTION.

§ 1. A DESCRIPTION of the principal phenomena of magnetic induction has already been given, and we will now consider how to estimate *numerically* the effects produced under various circumstances.

Electro magnetic force.—When the intensity of a given magnetic field produced by a magnet or by electrical currents, has been determined, the induced current produced in a conductor moving in that field is easily determined. Every part of the conductor moving in a field and conveying a current (induced or not) is acted upon by a force perpendicular to the plane passing through its own direction and the lines of magnetic force in the field. This force is equal to the product of the length of the conductor into the strength of the current in electro-magnetic measure, the intensity of the magnetic field, and the sine of the angle between the lines of force and the direction of the current. Thus, if A B (Fig. 79) be the element of the conductor, and the lines of force be in the plane of the paper as dotted, then the direction of the force due to the field and current is perpendicular to the plane of the paper. Let the intensity of the magnetic field $= T$, the strength of the current in A B $= C$, the angle A B C $= \alpha$, and $f =$ the force.

FIG. 79.

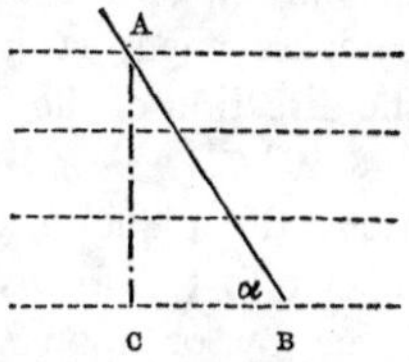

A

α

C B

Then $f = TC \times AB \sin \alpha$ . . . (1)

The force is exactly the same as if the conductor, instead of being of the length and in the direction A B, were really of the length and in the direction A C. Let A B (Fig. 80) represent a piece of the conductor in which a current C is flowing from A to B. Let D O be the direction of the lines of magnetic force so that a magnet N S would place itself in the field as shown in the figure. The force f experienced by the conductor will tend to lift it perpendicularly to the plane A O D. Let F O represent in magnitude and direction the current C and D O the magnitude and direction of the intensity of the magnetic field, then f per unit of length = T C sin α, but C sin α = the perpendicular distance from E F to O D and T = D O; hence the area of the parallelogram E F O D = f per unit of length.

FIG. 80.

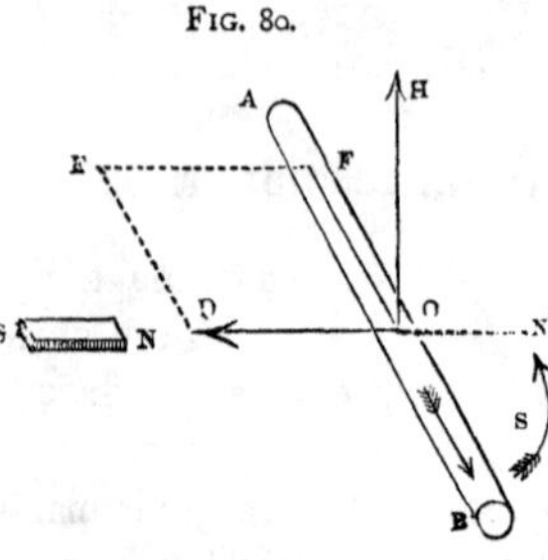

A current flowing from west to east is lifted by the earth's magnetism. The following is a rule by which to remember which way the magnetism of any field would impel any current. Place a corkscrew perpendicular to the plane E F O D and turn it, as shown by the arrow S, from the direction of the current to the direction in which the north end of the compass needle would point,[1] the screw will then move in the direction of the force.

§ 2. *Electromotive force.*—If the conductor A B is moved along the plane in which O F E D lies, its motion is perpendicular to the forces acting upon it, and no work is done either by or upon A B. When this is the case no induced current can be produced in A B, either in augmentation or diminution of the original currents, for no work is done by the motion or required to produce the motion; a current can only be increased by the exertion of energy upon it, and diminished by expending its energy.

If, however, the conductor moves in the direction O H (Fig. 80), or across the dotted lines in a direction perpen-

[1] I.e. considering O as the centre the handle would turn from the line O B to the line O N_1.

dicular to the paper (Fig. 79), the motion is either helped by the force or opposed by it. To move the conductor against the force, we must do work. The measure of this work is the product of the force into the distance moved against it. If the conductor moves obliquely across the lines of force it is resisted with a force proportional to that component of the motion which is perpendicular to the lines of force, and the work done is equal to the force multiplied into this perpendicular distance.

The work done on the conductor is found by observation to be represented by an increment or diminution in the current flowing through that conductor; now the work done by a current is by definition equal to $\text{E Q} = \text{E C} t$, where $\text{E} =$ the electromotive force acting between the ends of the conductor.

If a unit length of the conductor be moved a distance L across the lines of magnetic force in a field of intensity H, the work done will be $f\text{L} = \text{C H L}$: hence, as the work done by the current must be equal to the work expended in moving the conductor, we have $\text{E C} t = \text{C H L}$

$$\text{or E} = \frac{\text{H L}}{t} \quad . \quad . \quad . \quad (2)$$

Now $\frac{\text{L}}{t}$ is the velocity with which the conductor is moving, so that the electromotive force per unit of length is equal to the intensity of the magnetic field multiplied into the velocity of the motion.

This law still holds good if the motion be oblique to the lines of force, provided L be the component of the motion perpendicular to those lines; and if the conductor A B was also oblique to the lines of force, the unit length must be measured perpendicular to those lines of force. Thus, let the direction of the lines of force in a magnetic field be represented by O O_1; let (Fig. 81) $a\,b$ be perpendicular to O O_1 in the plane A O O_1, let A a and B b be perpendiculars let fall from A and B on the line $a\,b$, and let A B be moved to the position $\text{A}_1\,\text{B}_1$, so that P O_1, perpendicular to the plane

A O O_1, represents the distance A B has moved across the lines of force; then the E. M. F. due to the motion will be $\frac{H \times ab \times PO_1}{t}$.

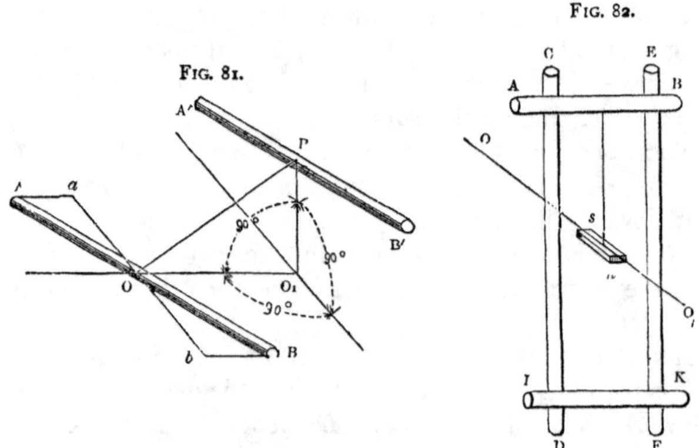

FIG. 81.

FIG. 82.

Observe that *the unit electromotive force will be produced by a rod of unit length moving with unit velocity across a field of unit intensity.*

§ **3.** Let there be two fixed rails C D and E F (Fig. 82) in a plane perpendicular to the lines of magnetic force O O_1. Let the bars A B and I K, perpendicular to the lines of magnetic force, complete a closed circuit A B I K, round which a current might circulate. Then if A B be moved downwards with the velocity V, the electromotive force due to induction will be H $\times$ A B $\times$ V; but this product is equal to the number of lines of magnetic force subtracted from the area of the closed circuit per unit of time; hence, calling this number N, we find that the E. M. F. $= \frac{N}{t}$. The direction of the current produced by this E. M. F. would be such as to oppose the motion, i.e. from B to A. If I K were moved at the same rate in the same direction there would be an equal E. M. F. in it, tending equally to produce a current from I to K, and this would balance the E. M. F. in A B, so that no current would flow. In this case the motion of I K would add just as many lines of force to those crossing the area A B C D as the motion of A B would subtract, so that the total number N added or subtracted would be nil, and the electromotive force on the whole would also be nil.

If I K moves fastest, its electromotive force would be greatest, and the difference between the E. M. F. in I K and in A B would be equal to $\frac{N_1 - N}{t}$, calling N_1 the number of lines cut by I K during its motion; the current would then run round the parallelogram from I to K B A. Similarly, if A B moved fastest there would be a resultant E. M. F. $= \frac{N - N_1}{t}$ sending a current from A to B K I. Hence in both cases the E. M. F. in the current would be equal to the number of lines of magnetic force added to or subtracted from the area per second. Now it follows from the principles developed in the previous paragraph that this is true not only of this simple case but of all cases whatever. Let the circuit be of any shape whatsoever and moved in any direction, the E. M. F. tending to send a current round the circuit due to motion in a magnetic field will be $\frac{N}{t}$.

§ 4. An apparatus for showing the phenomena or induction with a fixed pair of rails would be extremely difficult to construct; the motion could not be continued for any length of time, and the resistance in the circuit would vary at each moment, as the stationary portion was shortened or lengthened during the motion of the bar. Let a closed circuit (Fig. 83) rotate in a uniform magnetic field, and for simplicity sake let us suppose the field uniform, the circuit circular, and the axis perpendicular to the direction of the lines of magnetic

FIG. 83.

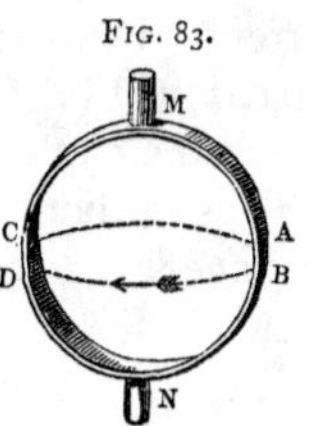

force. Let the rotation be in the direction of the hands of a watch held with its face upwards; let the direction of the lines of magnetic force be perpendicular to the plane of the paper, and such that a north pole would be impelled from the spectator down through the paper. Consider the short elements A B and C D, which are sen sibly parallel to the axis and perpendicular to the lines of magnetic force. When these are just crossing the plane of the paper they are moving in the direction of the lines of magnetic force, and a current in them would neither be assisted nor resisted; but when the circle has made a quarter of a turn they are crossing the lines of force at right angles. If the current in A B is descending, the motion of A B will be resisted by the lines of force, for a descending current in A B would impel a north pole in front of the paper from right to left, and would therefore itself be repelled from left to right. (The north pole must be in front of the paper to give lines of force which would repel a free north pole from the spectator to the paper.) Hence while A B crosses the lines of force an E. M. F. is produced in it, tending to send a current downwards. The same is true of each element in all the semicircle M A B N, the E. M. F. diminishing in each element proportionately to the sine of the angle between the element and the lines of force. Next, consider the element C D. This is simultaneously crossing the same lines of force in the opposite direction. This motion would be resisted by an upward current; hence the electromotive force in the semicircle N D C M will be from N towards M or upwards through this half of the circle.

Thus in both halves of the circle the E. M. F. tends to produce a current moving from M to A B, N, D C, and back to M.

This electromotive force will evidently be strongest at all points of the circle when this is crossing the lines of force at right angles, i.e. when the plane of the circle is in the direction of the lines of force. It will begin feebly as the circle in its rotation leaves the position sketched and advances as shown by the arrow, for at first the inclination of the direction of each element to the lines of force will be small; and again, after reaching its maximum, this inclination diminishes until it becomes nil after half a turn has been made. During the next half-turn, while M A B N is behind the paper, the E. M. F. will tend to send current up from A to M through B A; the direction of the current will therefore, during this half-turn, be reversed in the material circuit. Relating to a fixed exterior point, the current is, however, always in one direction, though varying from zero to a maximum at every half-revolution. The circuit might evidently be not a single circle but a coil of wire. The E. M. F. would increase with the length of the coil. If, however, the only resistance be that of the coil, the current will be constant whatever number of turns were taken, for the resistance will increase in the same proportion as the electromotive force. If some exterior constant resistance be connected with the coil, by sliding contacts near the axis, the current will be larger with many than with few turns.

There is no difficulty in calculating the exact electromotive force due to a coil of any given shape rotating in any magnetic field, except the mathematical difficulty of summing up the different E. M. F. in all the different elements of the coil at each moment, or, what comes to the same thing, determining the value of N during the motion.

It is now clear that the electromotive force produced by the motion of a closed circuit in a magnetic field of known intensity can be expressed in terms of that intensity and of velocity only; this measurement gives the value of the E. M. F. in absolute electromagnetic measure. We have also seen how to measure the value of any current C in the same measure, and since $R = \frac{E}{C}$ in any circuit, the resistance R of that circuit can be experimentally determined by measuring the values of E and C. When the resistance of a single circuit has been thus ascertained, a material standard coil equal to some multiple of the absolute unit can be prepared by comparison with this experimental circuit. When this has been done once for all, the resistance of other conductors can be easily determined by comparison with this standard. The following statements describe the experiments by which such a standard has been prepared.

§ **5.** Let us consider a circular coil of radius K rotating with an angular velocity A in a field of the intensity H. Then during each half-revolution the number N, equal to $\pi K^2 H$, will be alternately added and subtracted. Every addition and subtraction tends to send a current in the same direction relatively to an external point. Let n be the number of turns per second, then $n = \frac{A}{2\pi}$, and the total number of lines of force added and subtracted per second will be $4 \pi K^2 H \times \frac{A}{2\pi} = 2 A K^2 H$. The E. M. F. due to this will be $2 A K^2 H$, and the equivalent current produced $\frac{2 A K^2 H}{R}$, where R is the resistance of the circuit. If there be m turns the length of the wire in the coil $L = 2 \pi K m$, and the area enclosed $= \pi K^2 m = \frac{L K}{2}$. The number of lines added per second expressed in this manner will be $\frac{A L K H}{\pi}$ and the current $= \frac{A L K H}{\pi R}$. This current may be measured on a stationary electrodynamometer or galvanometer, and when it has been thus measured in absolute measure the only remaining unknown quantity is R.

§ **6.** The determination of R by this method requires a knowledge of the intensity of the magnetic field H, and a contemporaneous measurement of the absolute value of a current.

These two observations can be dispensed with by hanging, according to Sir William Thomson's method, a small magnet in the centre of the rotating coil and observing its deflection. The induced currents will all deflect this magnet in the direction of the rotation of the coil; the couple exerted on a magnetic needle of the moment $m\ l$, when deflected to the angle d, will be $\frac{L^2 A H}{4 K^2 R} m\ l \cos d$. The equal and opposite couple exerted by the earth's magnetism will be $\text{H}\, m\, l \sin d$; hence

$$\tan d = \frac{\text{L}^2\text{A}}{4\text{K}^2\text{R}}, \text{ or}$$

$$\text{R} = \frac{\text{L}^2\text{A}}{4\,\text{K}^2 \tan d} \quad . \quad . \quad . \quad (3).$$

This gives a simple expression for the resistance of the circuit in absolute measure in terms of known and simple magnitudes. In practically making the experiment several corrections have to be introduced, as for the inductive effects of the magnet on the coil. The experiment was carefully carried out by a committee of the British Association, and the absolute resistance of a certain standard determined in this way serves to determine the absolute resistance of any other circuit.

§ 7. When the induction takes place, not in consequence of the motion of a wire in a magnetic field, but in consequence of the sudden creation of a magnetic field, as when a neighbouring current is suddenly commenced, the effect is exactly as if the wire had been suddenly moved from an infinite distance to its actual position on the new magnetic field. The electromotive force is in this case also equal to $\frac{\text{N}}{t}$, where N is the additional number of lines of magnetic force introduced into the circuit in the time t; when the induction takes place in consequence of the cessation of a current, the electromotive force is in the opposite direction, and is equal to $\frac{\text{N}}{t}$; where N is the number of lines withdrawn. If t be made very small, the E. M. F. tending to produce an induced current may be indefinitely increased; and similarly if a current can be made to reach its full strength in a very short time, it will produce an E. M. F. in a wire close beside it much greater than that required to produce the original current. The wire in which the inducing current circulates, is often called the primary wire; the one in which the current is induced is called the secondary wire.

§ 8. In order to determine the electromotive force

produced in a secondary circuit by the commencement or cessation of a current C in a primary circuit, we require to calculate the number N of lines of force produced, cutting the surface inclosed by the secondary circuit. (Of course lines of force going in opposite directions through the surface must be reckoned positive and negative, and their addition made accordingly.) This number N divided by t gives the electromotive force. It is extremely difficult to determine t, for no current begins instantaneously, and the laws of its increase are extremely complex. The fact that the current is employed to induce a current or currents in secondary conductors, increases t. The statical induction, when sensible, increases t, and magnetisation due to currents increases t. The actual determination of the E. M. F. in any secondary circuit will not be here attempted, but the notions given serve to show how we may increase or diminish this E. M. F. in designing inductive apparatus.

§ **9**. I have now shown how, theoretically, resistance, electromotive force, and currents can all be measured in absolute electro-magnetic measure. Quantity can be measured either by observing the total current which it produces when flowing away, for which purpose a simple method will hereafter be given, depending on the use of galvanometers, or it may be measured by observing its electrostatic effects, and being then known in electrostatic measure, it may be converted into electro-magnetic measure by multiplication into the constant 28,225,000,000. Capacity is obtained by observing the quantity which the given conductor contains when electrified to a potential E. Theoretically, therefore, we may be said, while studying the laws of electro-magnetic induction, to have discovered how it is possible to measure all electrical magnitudes in this series of units. The practical methods adopted will be described hereafter.

§ **10**. The examples given of the modes of calculating induced currents in the two simple cases of a straight bar moving across a uniform field, and a circular coil rotating in such a field, serve to show how all similar problems must be attacked. The exact solution of them requires mathematical analysis of the highest kind; but correct views of the general nature of the effects to be expected are very readily obtained from the general elementary propositions now laid down. Thus it is easy to examine whether the electromotive force in some parts of the circuit is acting in a direction opposed to that in others; if so, it is easy to see that to reduce the opposing action we must reduce the velocity of those parts, and place them in the weakest portion of the magnetic field, while the efficient portions of the circuit must be placed in the strongest portions of the field, and made to move with the greatest velocity. The best direction of motion is also easily ascertained. The general effect of adding to the length of the wire or coil in which induction is taking place is also easily perceived, and the object of making the coil of materials which have but little electrical resistance. Increasing the thickness of the wire does not at all increase the electromotive force, but inasmuch as it diminishes the resistance, a thick and short wire may give a very considerable current, if outside the moving coil there be no considerable additional resistance to overcome. But if we desire a considerable or even sensible current through an external wire of great length, or of great resistance, then our inducing coil must be long in order to give great E. M. F., and in such a case its internal resistance will not greatly diminish the current, because it will not greatly increase the resistance of the whole circuit. If currents of very short duration are required, we may move our coil or wire rapidly across a magnetic field of small size but great intensity, whereas if a current of longer duration is required, the motion must be prolonged, and it will be necessary to have a large magnetic field.

CHAPTER X.

UNITS ADOPTED IN PRACTICE.

§ 1. IN the last chapter I have described the manner in which the strength of a current may be measured in electromagnetic measure. The method, although not offering any extreme difficulty, is yet too complex for continual use, and currents will certainly not be commonly expressed in this manner, until electrodynamometers are habitually sold of such construction that by simply multiplying the observed deflection into a constant number, the strength of the current is obtained.

The direct measurements of electromotive force and of resistance in the same series of units are still more complex. It is unnecessary that each electromotive force or resistance should be directly measured in absolute measure by these complicated methods. A standard of electrical resistance approximately equal to one thousand millions of absolute units of resistance (centimètre, gramme, second) has been prepared by a committee of the British Association. This standard is an actual wire of the required resistance. The measurement of any other resistance x in absolute measure consists, therefore, in a comparison of x with this standard or a copy. The process in this case is the same as that of measuring length in mètres. Theoretically the measurement of a length x in mètres means the comparison of x with a certain diameter of the earth; practically it means the comparison of x with a measure authorized by Government to be called a mètre.

§ 2. The standard of resistance has been called an *ohm*, and is now in common use.

Gauges of electromotive force ought for similar reasons to be issued, and might be of various forms. Thus the gauge might indicate a given difference of potential in virtue of the attraction which two opposed plates exert on one another, or, even more roughly, in terms of the distance at which sparks pass across air between two given balls. There can be no doubt that within a few years gauges of this kind will be issued with the same authoritative stamp as attaches to the ohm. Meanwhile electromotive force or difference of potential is often expressed in terms of the electromotive force produced by the special form of voltaic battery known as the Daniell's cell. The E. M. F. of this cell is about 100,000,000 absolute units, centimètre, gramme, second, and is fairly uniform. A much better standard of electromotive force is the cell introduced by Mr Latimer Clark, and described by him as follows, (Proceedings R. S. No. 136, 1872): 'The battery is composed of pure mercury as the negative element, the mercury being covered by a paste made by boiling mercurous sulphate in a thoroughly saturated solution of zinc sulphate, the positive element consisting of pure zinc resting on the paste.' 'Contact with the mercury may be made by means of a platinum wire.' 'The element is not intended for the production of currents, for it falls immediately in force if allowed to work on short circuit. It is intended to be used only as a standard of electromotive force with which other elements can be compared by the use of the electrometer, or condenser, or other means not requiring the use of a prolonged current.' The electromotive force of this cell is, in electro-magnetic units, $1{\cdot}457 \times 10^8$ (centimetre, gramme, second), or $1{\cdot}457 \times 10^5$ (metre, gramme, second). There is already a unit of electromotive force in practical use called a *volt*. The volt is intended to represent 10^8 absolute units, centimètre, gramme, second; the E. M. F. of Latimer Clark's cell is 1·457 volt.

The capacity of a given conductor can be determined in absolute measure with less trouble than either the electromotive force or the resistance, and condensers of the approximate capacity of $\frac{1}{10{,}000{,}000{,}000{,}000}$ or 10^{-13} absolute units, and called *microfarads*, are in common use.

§ **3.** We thus find that in ordinary electrical measurements, even when we require to calculate the relations between forces, work, or heat and electrical magnitudes, we need only compare these electrical magnitudes with known standards, these standards having been chosen with distinct reference to the units of force and work. To the ordinary electrician it is therefore much more important to know how to compare accurately one resistance with another, one current with another, and so forth, than to be able to determine resistances or currents in absolute measure. Indeed, when an electrician is said to measure a current or a resistance, it is this comparison with a recognised unit, which is in all cases understood. The unit employed is important only so far as it is widely adopted and allows a more or less ready application of the measurement in formulæ, involving other electrical magnitudes. The series of units most generally adopted in Great Britain have received distinctive names, and are all based on the absolute system. They are, however, all multiples or submultiples of the absolute units, which are themselves of inconvenient magnitudes.

§ **4.** The unit of resistance is termed an ohm and $= 10^9$ absolute units (centimètre, gramme, second).

The unit of electromotive force is termed a volt $= 10^8$ absolute units.

The unit of capacity is termed a farad $= \frac{1}{10^9}$ absolute unit.

The unit of quantity is that which will be contained in one farad when electrified to the potential of one volt: it has no distinctive name, and may be called a farad also.[1] This unit of quantity $= \frac{1}{10}$ absolute unit. The absolute units referred to throughout are those based on the centimètre, gramme, and second. There is a strong objection to the use of the words absolute unit, inasmuch as they do not indicate the series of fundamental units on which

[1] Mr. Latimer Clark calls it a *Weber*.

the derived unit is based. The volt, farad, and ohm are free from this ambiguity.

The unit of current is one farad per second; it is one-tenth of the absolute unit of current, and is frequently termed for brevity a farad, just as in speaking of velocity we often speak of a velocity of 100 feet, the words per second being understood.

§ 5. Inasmuch as the electrician deals with magnitudes differing in greatness very widely from one another, it is convenient to use multiples and submultiples of the above units, each having its appropriate name.

The megavolt = one million volts.
" megafarad = " farads.
" megohm = " ohms.

Similarly,

The microvolt = one millionth of a volt.
" microfarad = " " farad.
" microhm = " of an ohm.

The following table (p. 162) gives the value of each unit in three systems of absolute units, in which the mètre, centimètre, and millimètre, and in a fourth in which the milligramme is substituted for the gramme, are respectively made the basis or starting-point.

When we require to convert measurements expressed to absolute units based on any given system of fundamental units into absolute measurements based on some other system, it is necessary, in order to calculate the multiplier or divisor to be used for the conversion, that we should know what are called the *dimensions* of the units. In other words, we must know at what power each fundamental unit enters into the particular derived unit; thus, in the case of velocity, which is perhaps the simplest derived unit, the dimensions are said to be $\frac{L}{T}$, or a length divided by an interval of time, because the magnitude of the unit is directly pro-

		Mètre, gramme, second, series.	Centimètre, gramme, second, series.	Millimètre, gramme, second, series.	Millimètre, milligramme, second, series.
Resistance	Megohm	10^{13}	10^{15}	10^{16}	10^{16}
	Ohm	10^{7} = 10000000	10^{9} = 1000000000	10^{10} = 10000000000	10^{10} = 10000000000
	Microhm	10	10^{3}	10^{4}	10^{4}
Current	Megafarad per second	10^{4}	10^{5}	$3{\cdot}162 \times 10^{5}$	10^{7}
	Farad per second	10^{-2}	10^{-1}	$3{\cdot}162 \times 10^{-1}$	10
	Microfarad per second	10^{-8}	10^{-7}	$3{\cdot}162 \times 10^{-7}$	10^{-5}
Quantity	Megafarad per volt	10^{4}	10^{5}	$3{\cdot}162 \times 10^{5}$	10^{7}
	Farad per volt	10^{-2}	10^{-1}	$3{\cdot}162 \times 10^{-1}$	10
	Microfarad per volt	10^{-8}	10^{-7}	$3{\cdot}162 \times 10^{-7}$	10^{-5}
E. M. F.	Megavolt	10^{11}	10^{14}	$3{\cdot}162 \times 10^{15}$	10^{17}
	Volt	10^{5}	10^{8}	$3{\cdot}162 \times 10^{9}$	10^{11}
	Microvolt	10^{-1}	10^{2}	3.162×10^{3}	10^{4}
Capacity	Farad	10^{-7}	10^{-9}	10^{-10}	10^{-10}
	Microfarad	10^{-13}	10^{-15}	10^{-16}	10^{-16}

Some idea of the actual magnitude of each of these units may be obtained from the following considerations:—

The ohm is about equal in resistance to 48·5 mètres of pure copper wire, one millimètre diameter at 0° Centigrade. The volt is from five to 10 per cent. less than the E. M. F. of one Daniell's cell. About one farad per second would flow if one Daniell's cell were employed to produce a current in a circuit the total resistance of which is one ohm. Few Daniell's cells can actually produce such a current because they have themselves a larger resistance than an ohm. A single Daniell's cell would send about 125 microfarads through the Atlantic Cable.

The capacity of most submarine cables is about one-third of a microfarad per knot.

portional to the magnitude of the unit used to measure length, and inversely proportional to that of the unit used to measure time. Similarly the absolute unit of force is directly proportional to the unit of length and the unit of mass employed; it is inversely proportional to the square of the unit of time used; hence the dimensions of the unit of force are $\frac{LM}{T^2}$.

When we wish to convert a measurement expressed in absolute units based on the units L, M, T, (say foot, grain, second) into an absolute measurement based on some other system of units *l*, *m*, *t*, (say metre, gramme, second), we require to know the ratios $\frac{L}{l}$, $\frac{M}{m}$, $\frac{T}{t}$, of the actual magnitudes of each pair of units. Thus in the example chosen $\frac{L}{l} = 0{\cdot}3048$, $\frac{M}{m} = {\cdot}0648$, $\frac{T}{t} = 1$; then to effect the conversion from English to French measure we must multiply the number expressing the measurement in English measure by each ratio raised to the power at which the corresponding letter appears in the expression for the dimensions of the unit. If the power is negative, we divide by the ratio instead of multiplying; thus to convert a velocity expressed in English measure into a velocity in French measure, we multiply by 0·3048, and divide by 1: to convert a measure of force (foot, grain, second) into French measure we multiply by $\frac{{\cdot}3048 \times {\cdot}0648}{(1)^2} = {\cdot}01975$.

The following table of dimensions and constants is taken from the British Association Report on Electrical Standards 1863.

Fundamental Units.

Length = L. Time = T. Mass = M.

Derived Mechanical Units.

Work = W = $\frac{L^2M}{T^2}$. Force = F = $\frac{LM}{T^2}$. Velocity = V = $\frac{L}{T}$.

Derived Magnetic Units.

Strength of the pole of a magnet . . . . $m = L^{\frac{3}{2}}\ T^{-1}\ M^{\frac{1}{2}}$
Moment of a magnet . . . . . . $ml = L^{\frac{5}{2}}\ T^{-1}\ M^{\frac{1}{2}}$
Intensity of magnetic field . . . . $H = L^{-\frac{1}{2}}\ T^{-1}\ M^{\frac{1}{2}}$

Electro-magnetic System of Units.

Quantity of electricity . . . . $Q = L^{\frac{1}{2}}\ M^{\frac{1}{2}}$
Strength of electric current . . . . . . $C = L^{\frac{1}{2}}\ T^{-1}\ M^{\frac{1}{2}}$
Electromotive force . . . . . . $E = L^{\frac{3}{2}}\ T^{-2}\ M^{\frac{1}{2}}$
Resistance of conductor . . . . . . $R = L\ T^{-1}$

Electrostatic System of Units.

Quantity of electricity . . . . $q = L^{\frac{3}{2}}\ T^{-1}\ M^{\frac{1}{2}}$
Strength of electric currents . . . . . . $c = L^{\frac{3}{2}}\ T^{-2}\ M^{\frac{1}{2}}$
Electromotive force . . . . $e = L^{\frac{1}{2}}\ T^{-1}\ M^{\frac{1}{2}}$
Resistance of conductor . . . . . . $r = L^{-1}\ T$

Table for the conversion of British (foot grain second) system to centimetrical (centimetre gramme second) system.

	Number of centimetrical units contained in a British unit.	Log.	Log.	Number of British units contained in a centimetrical unit.
1. For M . .	0·0647989	$\bar{2}$·8115678	1·1884321	15·43235
2. For L, v, R, $\frac{1}{r}$ & V	30·47945	1·4840071	$\bar{2}$·5159929	·03280899
3. For F (also for foot grains and metregrammes)	1·97504	0·2955749	$\bar{1}$·7044250	·506320
4. For W . .	60·198	1·7795820	$\bar{2}$·2204179	·01661185
5. For H and electro-chemical equivalents	·0461085	$\bar{2}$·6637804	1·3362196	21·6880
6. For Q, C and e .	1·40536	0·1477874	$\bar{1}$·8522125	·711561
7. For E m q and c	42·8346	1·6317949	$\bar{2}$·3682051	·0233456
8. For heat . .	0·0359994	$\bar{2}$·5562953	1·4437046	27·7782

British system.—Relation between absolute and other units.

Let v be the ratio of the electro-magnetic to the electrostatic unit of quantity $= 28{\cdot}8 \times 10^9$ centimetres per second approximately, and we have

$$q = v\,\mathrm{Q} \;\Big|\; c = v\,\mathrm{C} \;\Big|\; e = \frac{1}{v}\mathrm{E} \;\Big|\; r = \frac{1}{v^2}\mathrm{R} \;\Big|\; s = v^2\,\mathrm{S}$$

One absolute unit of $\left\{\begin{matrix}\text{force}\\ \text{work}\end{matrix}\right\} = 0{\cdot}0310666 \left\{\begin{matrix}\text{weight of a grain}\\ \text{foot grains}\end{matrix}\right\}$ in London.

In London $\left\{\begin{matrix}\text{weight of a grain}\\ \text{one foot grain}\end{matrix}\right\} = 32{\cdot}1889$ absolute units of $\left\{\begin{matrix}\text{force.}\\ \text{work.}\end{matrix}\right.$

One absolute unit of $\left\{\begin{matrix}\text{force}\\ \text{work}\end{matrix}\right\} = \frac{1}{g}\left\{\begin{matrix}\text{unit weight}\\ \text{unit weight and unit length}\end{matrix}\right\}$ everywhere.

g in British system $= 32{\cdot}088\,(1 + 0{\cdot}005133 \sin^2\lambda)$, where $\lambda =$ the latitude of the place at which the observation is made.

Heat. The unit of heat is the quantity required to raise the temperature of one grain of water at its maximum density 1° Fahrenheit.

Absolute mechanical equivalent of unit of heat = 24861 = 772 foot grains at Manchester.

Thermal equivalent of an absolute unit of work = ·000040224.

Thermal equivalent of a foot grain at Manchester = ·0012953.

Electro-chemical equivalent of water = ·02 nearly.

Metrical system. Relation between absolute and other units. (Centimetre gramme second.)

One absolute unit of $\left\{\begin{matrix}\text{force}\\ \text{work}\end{matrix}\right\} = {\cdot}0010195 \left\{\begin{matrix}\text{weight of a gramme}\\ \text{centimetre gramme}\end{matrix}\right\}$ at Paris.

At Paris $\left\{\begin{matrix}\text{the weight of a gramme}\\ \text{or centimetre gramme}\end{matrix}\right\} = 980{\cdot}868 \left\{\begin{matrix}\text{absolute}\\ \text{units of}\end{matrix}\right\} \begin{matrix}\text{force.}\\ \text{work.}\end{matrix}$

One absolute unit of $\left\{\begin{matrix}\text{force}\\ \text{work}\end{matrix}\right\} = \frac{1}{g}\left\{\begin{matrix}\text{unit weight}\\ \text{unit weight} \times \text{unit length}\end{matrix}\right\}$ everywhere.

g in metrical system $= 978{\cdot}024\,(1 + 0{\cdot}005133 \sin^2\lambda)$, where $\lambda =$ the latitude of the place where the experiment is made.

Heat. The unit of heat is the quantity required to raise one gramme of water at its maximum density 1° centigrade.

Absolute mechanical equivalent of the unit of heat = 41572500 = 42354·2 centimetre grammes at *Manchester.*

Thermal equivalent of an absolute unit of work = ·000000024054.

Thermal equivalent of a centimetre gramme at Manchester = ·0000236154.

Electro-chemical equivalent of water = ·00092 nearly.

CHAPTER XI.

CHEMICAL THEORY OF ELECTROMOTIVE FORCE.

§ 1 IN Chapter III. § 15, the phenomenon of electrolysis was described and water was shown to be an electrolyte; the decomposition of water is much facilitated by the addition of a little acid, which has the effect of diminishing the resistance of the liquid and of allowing a larger current to pass from a given battery than would traverse pure water. The acid is not decomposed, or, if it is, the elements recombine so as never to appear at the *electrodes*, as the metal terminals plunged in the liquid are called. Platinum or gold electrodes are used to show the decomposition of water; otherwise the oxygen carried to the positive electrode would not be set free, but would oxidise the metal instead of appearing in the test tube (Fig. 41). Three or four galvanic cells are usually employed to decompose water. The electromotive force of one of the usual Daniell's cells is insufficient for the purpose, and this we shall be able to prove from a consideration of the chemical affinity of the materials employed, and of the work required to be done, measured in absolute measure. When the tubes are graduated so that the volume of the gases can be measured, the apparatus shown in Fig. 41 is called a voltameter. Owing to the absorption of gas by the water, neither the true relative nor absolute volumes of the gases appear in the test tubes.

With very few exceptions, electrolysis occurs only in liquids. Fused saline bodies are electrolytes, and probably many fused oxides are electrolytes, but the reoxidation takes place so readily that this is not easily verified. Conduction through electrolytes is subject to Ohm's law.

so far as is known. Electrolytes apparently conduct very small currents without being decomposed.

§ **2**. Electrolytes are not necessarily decomposed into simple or elementary substances. Many electrolytes are decomposed into two groups of components; each group, or each simple element, is called by Faraday an *ion*; with any given electrolyte, the same group, or ion, always appears at the same electrode, so that ions may be classed as electropositive or electronegative; the electropositive ion appears at the negative electrode, and the electronegative ion at the positive electrode.

When the electrolyte is changed, an ion may change its electrode, and ions can be classed in a list such that each is electropositive to all which follow; so that an ion such as sulphur, which is electronegative towards hydrogen, is electropositive towards oxygen.

Hydrogen and metals are electropositive relatively to acids and oxygen: oxygen is the most electronegative, and potassium the most electropositive element.

§ **3**. The bases of salts may practically be classed as electropositive ions. When we decompose salts composed of two or of three elements, we find the base at the negative electrode and the acid at the positive electrode; but this classification is not strictly scientific, for chemists do not consider the decomposition of sulphate of potassium, for instance, as consisting in the separation of the base potash from the sulphuric acid, but rather as the separation of potassium from the other constituents of sulphate of potash. When, however, the potassium appears at the negative pole, it decomposes water and combines with oxygen to form potash, while at the other pole sulphuric acid and one element of oxygen appear. When the decomposition goes on rapidly, oxygen and hydrogen in small quantities do appear at each electrode; otherwise they recombine and form water. The practical result is that the base behaves as an electropositive and the acid as an electronegative ion.

§ **4**. The following table is an electro-chemical series, in which the most electropositive materials come last :—

Oxygen	Chromium	Silver	Manganese
Sulphur	Boron	Copper	Aluminium
Nitrogen	Carbon	Bismuth	Magnesium
Fluorine	Antimony	Tin	Calcium
Chlorine	Silicon	Lead	Barium
Bromine	Hydrogen	Cobalt	Lithium
Iodine	Gold	Nickel	Sodium
Phosphorus	Platinum	Iron	Potassium
Arsenicum	Mercury	Zinc	

§ **5**. The quantity of any electrolyte decomposed by a current is proportional to the strength of the current and to its duration ; in other words, to the whole quantity of electricity which during decomposition passes through the electrolyte.

The weights of different electrolytes decomposed by a constant current are in direct proportion to their combining numbers. Tables of these numbers are given in all works on chemistry.

It follows from the above propositions that if we know the weight of any electrolyte which has been decomposed by any known current in a known time, we can calculate the weight of any other electrolyte which in a given time will be decomposed by any given current. It does not follow that a given battery will decompose two electrolytes at such rates that the quantities decomposed in a given time are simply proportional to the combining numbers ; the resistance of one electrolyte may be so different from that of the other, that in order to obtain the same current very different batteries may be required in the two cases.

The quantity of each electrolyte decomposed by the unit current in a second is perfectly definite and constant ; we shall denote this quantity by the symbol ϵ, and call it the *electro-chemical equivalent* of the substance. Since the weights of the electrolytes decomposed by the unit current are proportional to the combining numbers of the compounds, the weights of the ions appearing at each electrode will be proportional to these numbers, and hence, knowing the weight of any one ion produced at either electrode by the unit current in a given time we can calculate the weights of all the others; in other words, we can calculate the electro-chemical equivalent of each ion, and therefore of all simple bodies. The following is a table of the electro-chemical equivalents of some bodies expressed in grammes and calculated from that of water experimentally determined to be ·00092; that is to say, the table is calculated on the assumption that one absolute electro-magnetic unit of current (centimètre gramme second) will in one second decompose ·00092 gramme of water.

Aluminium	·00141	Iron	·00186
Antimony	·00624	Lead	·01058
Arsenicum	·00383	Magnesium	·00123
Barium	·00700	Manganese	·00280
Bismuth	·01073	Mercury	·01022
Boron	·00056	Nickel	·00301
Bromine	·00409	Nitrogen	·00072
Calcium	·00204	Oxygen	·00082
Carbon	·00061	Phosphorus	·00158
Chlorine	·00181	Platinum	·01007
Chromium	·00268	Potassium	·00199
Cobalt	·00301	Silicon	·00143
Copper	·00324	Silver	·00552
Fluorine	·00097	Sodium	·00118
Gold	·01007	Sulphur	·00164
Hydrogen	·00010	Tin	·00604
Iodine	·00649	Zinc	·00342

§ 6. When a current is passed from metal electrodes through an electrolyte and decomposes it, the current performs an action equivalent to the performance of work or expenditure of energy—an action which may be measured in the units employed to measure energy. Let I be the electromotive force between the two electrodes, and Q the quantity of electricity passing, then the work done by the electricity is, as we know, necessarily equal to I Q; and if this energy is wholly spent in decomposing the electrolyte, this product measures the energy which must be expended on the electrolyte to overcome the chemical affinity of the ions. In expending work in this manner on the electrolyte, we may be said to add intrinsic energy to the ions: after being decomposed they possess a potential energy in virtue of which they can recombine, and during the recombination they must manifest in some form the energy given them when they were decomposed. They may manifest this energy in the form of heat, and if allowed to do so, the total amount of this heat of combination must be equivalent to the energy expended in decomposing them. Thus, calling θ the heat produced by the combination of a unit of weight of one ion with the other, and ϵ the electro-chemical equivalent of the first ion, then $\theta\epsilon$ will be the heat produced during the combination of as much of that ion as would be decomposed by the unit quantity of electricity, and $J\theta\epsilon$ will be the mechanical equivalent of that heat where J is 41572500, being Joule's coefficient, or the number of absolute units of work equivalent to the heat which will raise one gramme of water one degree centigrade. Thus the equation expressing the equivalence between the heat resulting from the combination of two ions, and the work done in decomposing them, will be —

$$I Q = Q J \theta\epsilon,$$

$$\text{or } I = J\theta\epsilon \quad . \; . \; . \; . \; . \; . \; . \; . \quad 1^{\circ}$$

This equation gives the value of the electromotive force which is absolutely necessary to effect the decomposition. If we have less electromotive force than this, I Q can never equal $Q J \theta\epsilon$; or the work done by the current, no matter what the resistance may be, can never be sufficient to separate the weight $Q\epsilon$ of the ion from its electrolyte. If a greater electromotive force than this be maintained between the electrodes, the decomposition will proceed very rapidly, but since I Q will be greater than $Q J \theta \varepsilon$, some of the energy of the current will be spent otherwise than in decomposing the electrolyte.

§ 7. If we look on the work done in separating two ions as a product of two factors, one factor being the weight of one ion M, and the other factor the chemical affinity E, then $M E = I Q$, or $E = I \frac{M}{Q}$.

But the ratio $\frac{M}{Q}$ is equal to ε; hence $E = I \varepsilon$, or $\frac{E}{\varepsilon} = I$, so that the chemical affinity of the ions per electro-chemical equivalent is equal to the electromotive force required to just decompose the electrolyte.

§ 8. The ions which by their combination form an electrolyte may generate a current instead of producing heat. If the whole energy due to chemical affinity is so employed, the value of the energy will, as before, for each electro-chemical equivalent ε be the product $J \theta \varepsilon$. The mechanical equivalent of the current produced is $I_1 Q_1$, where I_1 and Q_1 are the electromotive force and quantity of electricity produced by the combination of the ions; but the electromotive force just required to decompose the ions is exactly balanced by the E. M. F. which the combination of the ions can produce. In other words, $I_1 = I$, and therefore $Q_1 = Q$. Hence the electromotive force due to the combination of any pair of ions is equal to $J \theta \varepsilon$ or the mechanical equivalent of as much of the chemical action as goes on with the unit of the current in the unit of time.

ε may be taken for either ion. $\theta \varepsilon$ is constant, whichever is taken.

A table giving the values of θ is required before we can calculate from the table of electro-chemical equivalents the E. M. F. which any given combination will produce.

§ 9. When a series of chemical actions take place in a circuit, some of these may tend to produce an E. M. F., the others to resist it. We express this fact by saying that the respective values of I for the several reactions may be positive or negative. The resultant value or actual electromotive force tending to produce a current, or to resist decomposition, is the algebraic sum of all the values of I. Thus, in the galvanic cell known as Daniell's cell, the electrodes are copper and zinc; next the copper there is a saturated solution of sulphate of copper, and next the zinc a solution of sulphate of zinc. The chemical action is as follows: 1. The zinc electrode combines with oxygen. 2. The oxide thus formed combines with sulphuric acid and forms sulphate of zinc. 3. Oxide of copper is separated from the sulphate. 4. The copper in this oxide is separated from the oxygen.

The oxygen of the water is separated at the zinc electrode from the hydrogen, and at the other electrode this hydrogen recombines with the oxygen from the oxide of copper, but this alternate decomposition and recombination of the elements of water can neither increase nor decrease the E. M. F. of the cell, the actions being opposite and equal.

1. The heat evolved by the combination of one gramme of zinc with oxygen is 1,301 units.

2. The heat evolved by the combination of the 1·246 gramme of oxide thus formed with dilute sulphuric acid is 369 units.

3. The heat evolved by the combination of the equivalent quantity ·9727 of a gramme of copper with oxygen is 588·6 units.

4. The heat evolved by the combination of 1·221 gramme of the oxide thus formed with dilute sulphuric acid is 293 units.

The thermal equivalent of the whole chemical action due to one gramme of zinc is therefore $1301 + 369 - (588{\cdot}6 + 293) = 788{\cdot}4$; but we require the thermal equivalent of a weight of zinc equal to ε, and this we obtain by multiplying 788·4 into ·00342, giving for $\theta\,\varepsilon$ the value 2·696; next, to obtain the value of I, this product is multiplied by J or 41572500, and we then obtain for the electromotive force of a Daniell's cell about 112,000,000 units, a value which agrees closely with the result of direct experiment. This theory and example are taken from Sir W. Thomson's paper in the 'Philosophical Magazine' for 1851.

§ **10.** The separation of substances into ions which appear separately at the two electrodes is a fact made useful in many ways. The elements or elementary groups gather at the electrodes in a state of great purity, and hence the process of electrolysation is made use of to obtain pure chemicals. Metals may be deposited in this way on an electrode of any form which it is desired to copy. The metal copy thus formed is called an electrotype. The nobler metals are often deposited on electrodes of baser materials for the sake of ornament. These electrodes are then said to be electro-plated with the nobler metals. Some substances can only be decomposed by electrolysis, and some ions can only be maintained in a state of separation while the current is passing.

§ **11.** The passage of an ion from the place where it is first decomposed to the electrode appears to take place by a series of combinations and decompositions. Thus, when a molecule of water half-way between the electrodes is decomposed, neither the hydrogen nor oxygen cross the water as

FIG. 84.

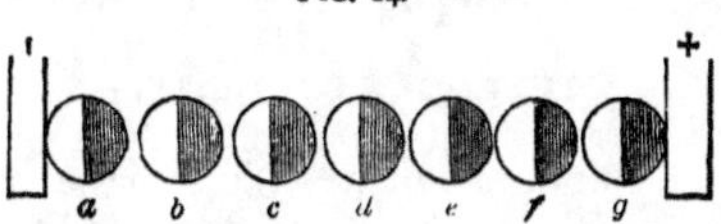

free gases, but the hydrogen of *d*, shown by the white half of the molecule, Fig. 84, combines with the oxygen of *c*, shown by the black half of that molecule. This sets the hydrogen of *c* free to combine with the oxygen of *b*, and finally the hydrogen of *b* combines with the oxygen of *a*, leaving the hydrogen of *a* free at the negative electrode. A similar series of compositions and decompositions leaves the oxygen of *g* free at the positive electrode. This is shown by the fact that ions can be transmitted through materials for which they have a strong chemical affinity without combining with them.

FIG. 85.

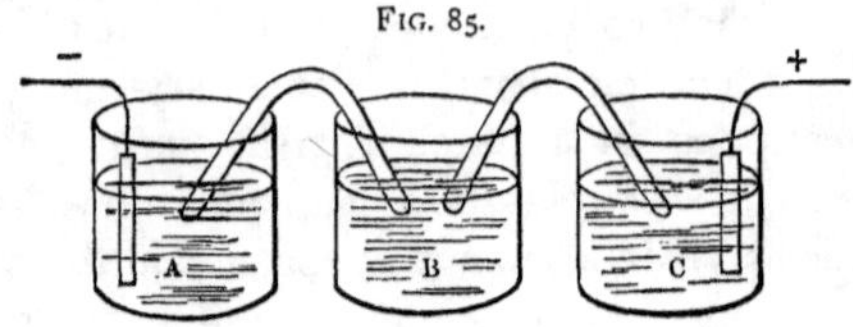

Thus, put a solution of sulphate of sodium into A, Fig. 85; dilute syrup of violets into B, and pure water into C; pass a current from an electrode in C to an electrode in A. The sulphate in the vessel A will be decomposed. Soda will be found in A, and sulphuric acid, which must have come from A, will be found in C. Nevertheless, the colour of the solution in B will not have been altered; whereas the addition of a very small quantity of free acid to B will produce a distinct red colour.

CHAPTER XII.

THERMO-ELECTRICITY.

§ 1. WHEN the junctions of a circuit made of two metals are at different temperatures, a current of electricity generally flows through the circuit. The electromotive force producing this current depends, 1, on the metals employed; 2, on the difference of temperature between the junctions; and, 3, on the mean temperature of the junctions.

When the mean temperature of the junctions is kept the same for circuits containing pairs of metals in various combinations, and when the difference of temperatures between the junctions is small and constant, the electromotive force of each circuit depends only on the metals employed. Let us call ϕ (A B) the numerical factor by which the difference of temperature τ between the junctions must be multiplied to give the E. M. F. of a circuit composed of two metals A and B at the mean temperature t, and let us call the value of this numerical factor, when τ is equal to unity, the *thermo-electric power* of the circuit A B at the temperature t. Then, calling ϕ (A C) and ϕ (B C) the thermo-electric powers of the pair A and C and of the pair B and C, we find experimentally that ϕ (B C) $=$ ϕ (A C)$-\phi$(A B). This equation expresses the fact that the thermo-electric power of any pair of metals is equal to the difference between the thermo-electric powers of those metals relatively to some one standard metal A. In order therefore to calculate the thermo-electric power of any pair of metals it is sufficient that we determine experimentally the thermo-electric power of all metals relatively to some one metal used as a standard. In what follows *lead* will be taken as the standard metal.

§ 2. We call a metal thermo-electrically positive to another, when the E. M. F. in a circuit of these two metals sends a current from the first to the second across the hot junction; the difference of temperatures τ being supposed small. It follows from § 1 that the metals may *for any one mean temperature t* be arranged in a series such that each will be positive relatively to that beneath it; it follows, moreover, that a number may be assigned to each metal proportional to its thermo-electric power relatively, say, to lead, and such that the algebraic difference between these numbers for any two metals will express in any arbitrary units the E. M. F. of a circuit of those two metals when the junctions are at the mean temperature t, but differ by a small constant difference τ or, say, by unity. The thermo-electric series printed in most books give approximately numbers of this kind, but the experiments on which they are based have generally been conducted without reference to the condition that the mean temperature t

should be constant, and this temperature is seldom given. The thermo-electric series differs entirely at different temperatures. The following is compiled from Dr. Matthiessen's experiments, and is such that approximately the thermo-electric power relatively to lead is expressed in microvolts per degree Centigrade.

Substance	Power	Substance	Power
Bismuth pressed commercial wire	+97	Pressed Antimony wire	− 2·8
Bismuth pure pressed wire	89	Silver pure hard	− 3
Bismuth crystal axial	65	Zinc pure pressed	− 3·7
Bismuth crystal equatorial	45	Copper galvanoplastically precipitated	− 3·8
Cobalt	22	Antimony commercial pressed wire	− 6
Argentine	11·75	Arsenic	− 13·56
Quicksilver	·418	Iron pianoforte wire	− 17·5
Lead	0	Antimony axial	− 22·6
Tin	− ·1	Antimony equatorial	− 26·4
Copper of commerce	− ·1	Red Phosphorus	− 29 7
Platinum	− ·9	Tellurium	− 502
Gold	− 1·2	Selenium	− 807

The mean temperature for which these numbers are approximately true may be taken at from 19° to 20° Centigrade.

§ **3.** Any two metals joined by a third metal so as to form a circuit have an E. M. F. equal to that which they would have had if directly joined, provided both junctions with the third metal are at one temperature; thus in Fig. 86 the three circuits all have the same E. M. F.—that due to zinc and antimony alone. The copper wire might be replaced by any complex arrangement of substances without interfering with the E. M. F. of the circuit, provided the junctions were all at one temperature, except those intended to be effective. Thus the E. M. F. of a thermo-electric pair—such as zinc and antimony—may be tested by observing the current flowing through a complex circuit composed, for instance, of the copper wire of a galvanometer having brass terminals, and of German silver resistance coils. We must, however, in such cases test the equality of the temperatures at the other junctions by observing whether any current is produced when the thermo-electric element is removed, and the copper, brass, and German silver connections joined so as to make an independent circuit exactly similar to that previously used except as regards the removal of the zinc and antimony, or other thermo-electric pair.

FIG. 86.

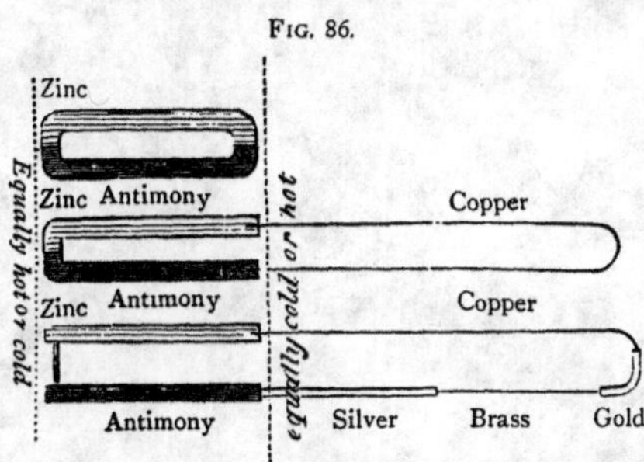

§ 4. The thermo-electric powers of different combinations not only change with a change of mean temperature, but they change in very different proportions. Thus the thermo-electric power of copper-silver differs little for temperatures between 0° and 100°, but the thermo-electric power of iron-copper varies rapidly; so rapidly, indeed, as to fall to zero at about 230°, and then again to increase, but with the opposite sign; so that whereas copper is positive to iron below 230°, it is negative to iron above that temperature. It follows that, if we are to possess accurate knowledge as to the thermo-electric relations of metals over a considerable range of temperatures, we must have sufficient knowledge to construct such a diagram as is shown in Fig. 87, where the vertical ordinates indicate temperatures in degrees Centigrade, and the horizontal ordinates the thermo-electric powers in microvolts of the metals relatively to lead.

Fig. 87.

Degrees Centigrade
50
100
150
200
250
300
350
400
450
500
550
600
+15
+10
+5
0
-5
-10
-15
Microvolts
PALLADIUM
PLATINUM
IRON
TIN
ALUMINIUM
LEAD
BRASS
COPPER
SILVER
CADMIUM
ZINC

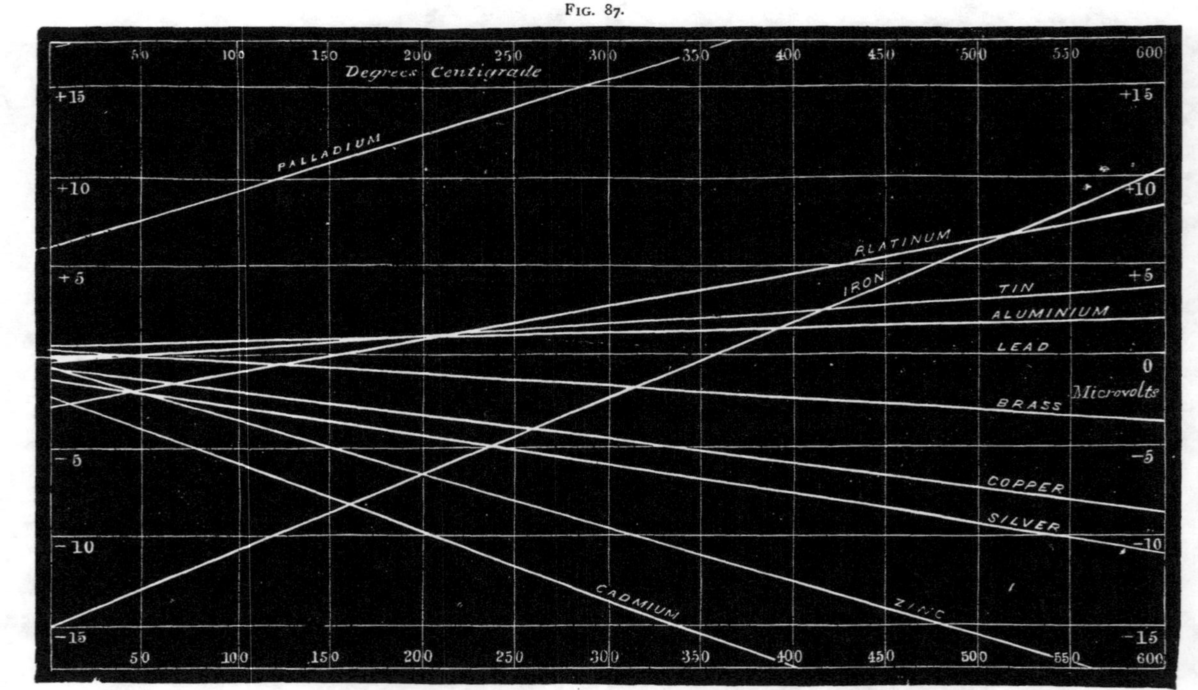

This diagram may be looked upon as simply one mode of tabulating the thermo-electric powers of metals relatively to one another at different temperatures, the horizontal scale being so arranged that the distance between the two lines of any given metals at any temperature gives the thermo-electric power of the two metals at that temperature.*

Thus the thermo-electric power of copper and iron at 50° is nearly 11·4, and at 260° is zero, and at 400° it is −7·6. I here call the thermo-electric power + when the current is from the first-named to the second-named of a thermo-electric pair across the hot junction.

§ 5. For any very small differences of temperature the electromotive force of a pair is equal to the product of the difference of temperature between the junctions into the thermo-electric power, so that the area of a narrow strip (approximately a parallelogram) represents this E. M. F. on the diagram. When the breadth of this strip is unity, or the difference of temperature 1°, the electromotive force is simply equal to the ordinate or to the thermo-electric power. When the difference of temperatures is considerable—say 50° —the electromotive force is the same as if we had 5 pairs of junctions arranged as in Fig. 88; thus if while $a\ a_1$ were

FIG. 88.

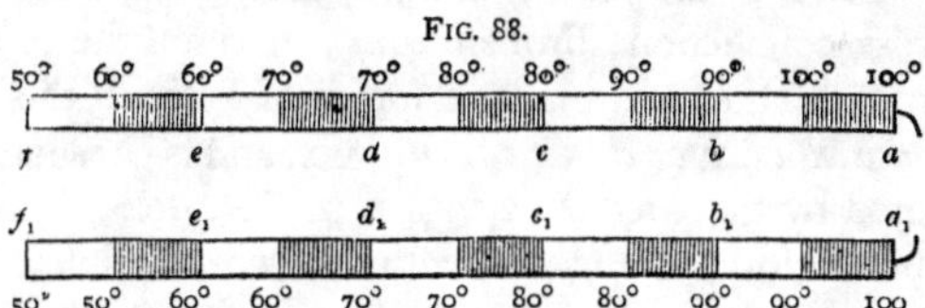

joined we were to complete a circuit by joining the junctions $b\ b_1$, we should in this circuit have an electromotive force equal to the parallelogram $a\ a_1$ in Fig. 89, where M N and O P represent the thermo-electric lines for copper and iron.

* The first diagram of this kind was given by Sir William Thomson in the Bakerian Lecture on the electro-dynamic qualities of metals, Phil. Trans. 1856, p. 708.

If we now were to break the circuit at $a\ a_1$, Fig. 88, and leaving $b\ b_1$ joined were to join $c\ c_1$, we should have a circuit $b\ b_1\ c_1\ c$, in which the E. M. F. would be represented in Fig. 89 by the parallelogram $b\ b_1$. Similarly in the circuit $d\ d_1\ c\ c_1$, Fig. 88, the E. M. F. would be represented by the area $c\ c_1$ in Fig. 89, &c.

FIG. 89.

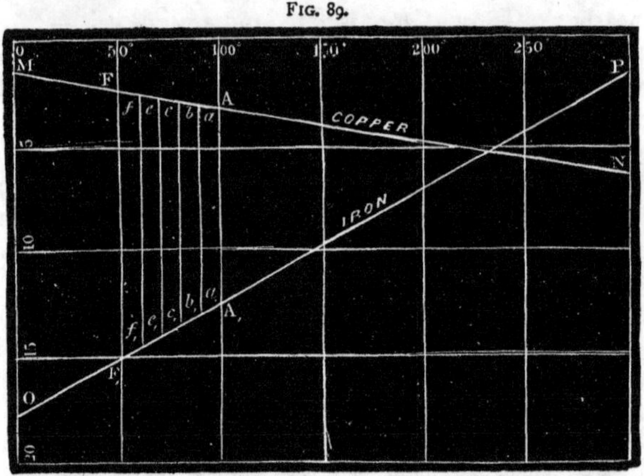

Now when $a\ a_1$ are joined, and $f\ f_1$ are joined, and all the other cross connections broken, the E. M. F. of the series is the sum of all the electromotive forces of each of the little circuits $a\ a_1\ b\ b_1$, $b\ b_1\ c\ c_1$, $c\ c_1\ d\ d_1$, &c., and is consequently represented by the area A A_1 F_1 F in Fig. 89. Thus the electromotive force of any pair with the two junctions at any two temperatures can be calculated by calculating the area enclosed between the two thermo-electric lines of those metals, and the ordinates corresponding to the two extreme temperatures.

§ **6.** In taking out this area we must, however, observe that if the areas to the left of any point where two lines cut are called positive, those to the right must be termed negative, for they represent an E. M. F. tending to send the current in the reverse direction. If, therefore, the two junctions are at such temperatures that the areas are equal, no E. M. F. will be produced in the circuit.

The points where the two lines for any metals cut are called the *neutral points* for those metals, because at that temperature the metals are neither positive nor negative relatively to one another, their thermo-electric powers being equal. When the lower junction is so far from the neutral point that the triangular area intercepted by the ordinate of its temperature is greater than the triangular area cut off by the ordinate of the higher temperature, the current will go from the metal highest on the scale below the neutral point to the other through the hot junction. The direction of the current will be the opposite if the triangular area above the neutral point is the greatest.

§ **7.** So far, we have been following Sir William Thomson. Professor Tait, led by theoretical considerations, has experimentally proved that the thermo-electric lines are in most cases approximately straight between 0° and 300° Centigrade, and probably at much higher temperatures. This greatly facilitates the calculation of E. M. F., because the areas to be dealt with are simply triangles, or trapezes. Let m be the distance separating the lines of the two metals forming the pair at the mean temperature of the junctions; let $t_1 - t_2$ be the difference of temperatures: then $m\ (t_1 - t_2)$ is the E. M. F. of the pair under those conditions, being the area of the trapeze, or triangle, above described. It follows from the above, that when the mean temperature of the two junctions is that of the neutral point, no current will flow through the circuit. This gives a means of determining the neutral points of metals with great accuracy. Professor Tait has also established the curious fact that the thermo-electric line of iron, whether pure or commercial, when prolonged towards red heat, is a sinuous or broken straight line, so that there may be two or more neutral points in one circuit when iron or steel is one of the two metals.

The E. M. F. of any pair may be calculated in microvolts

from the diagram (Fig. 87), taking the measurement of the mean distance between the lines of the metals by the horizontal scale, and the vertical measurements in degrees Centigrade; but it is obviously more convenient to calculate than to measure the length of the mean distance between the lines, and for this purpose the following table is given, containing the tangents of the angles at which the lines are inclined. Let k_1 and k_2 be the tangents for two given

Prof. Tait's Thermo-electric Table (converted to give E.M.F. in microvolts).

Metals.	Neutral Point with Lead. Degrees Centigrade. n	Tangent of Angle with Lead Line. k
Cadmium	−69	−·0364
Zinc	−32	−·0289
Silver	−115	−·0146
Copper	−68	−·0124
Brass	+27	−·0056
Lead	—	—
Aluminium . . .	−113	+·0026
Tin	+45	+·0067
German silver . .	−314	+·0251
Palladium . . .	−181	+·0311
Iron	+357	+·0420

Note.—The straightness of the thermo-electric lines has not been verified below 0°; hence the table must only be used to calculate E. M. F. for couples between 0° and 400° or 500° Centigrade.

The metals used were not chemically pure.

This table is calculated from the iron series in Prof. Tait's table, p. 599. Proc. R.S.E. 1871–72, taking the E.M.F. of a Grove's cell as 1·93 volts.

metals. Let n_1 and n_2 be the temperatures of their neutral points with lead. Let t_m be the mean temperature of the junctions; then the mean ordinate or m is given by the formula

$$m = k_1 (n_1 - t_m) - k_2 (n_2 - t_m)$$

Thus, let the mean temperature of a pair of copper-iron junctions be 50°, and the difference of the temperatures of the junctions 100°; then (50 + 68) (−·0124) = − 1·46 is one portion of the mean ordinate (for copper), and (50 − 357) (·042) = − 12·9 is the other (for iron). Their difference is 11·43, and this multiplied into 100° gives 1143 as the E. M. F. of the copper-iron pair in microvolts. When the thermo-electric lines of two metals are nearly parallel, the E. M. F. produced by a pair of those metals will be nearly proportional to the difference of temperatures maintained between their junctions. For metals or alloys, the lines of which diverge, no such law even approximately holds good, and it is necessary, before the E. M. F. can be calculated, that we should know not only the difference of temperatures, but the actual temperatures of the junctions.

§ 8. A number of thermo-electric pairs, or elements, may be joined in series, so as to give an E. M. F. which is the sum of the electromotive forces of all the elements. To do this it is only necessary to join the metals, as shown in Fig. 90, and keep all the junctions on one side, as at A, warm while the other side is cold. Batteries of this kind are easily made with exceedingly small resistance, so that when the other resistances in the circuit are also small, considerable currents will be produced—greater currents than could be obtained under similar circumstances from a Daniell's cell of moderate size. A bismuth-antimony pair may be prepared having, say, an E. M. F. of 100,000 microvolts, or about $\frac{1}{10}$ the E. M. F. of a Daniell's cell, while the resistance might be reduced to almost any desired extent by increasing the section of each element. Thus, if each element were about 2 centimètres in length, and a tenth of a square centimètre in section, the resistance of the pair would be about 3,370 microhms, and the resistance of 100 such pairs would be 337,000 microhms, or ·337 ohm, so that through a short circuit they would give

FIG. 90.

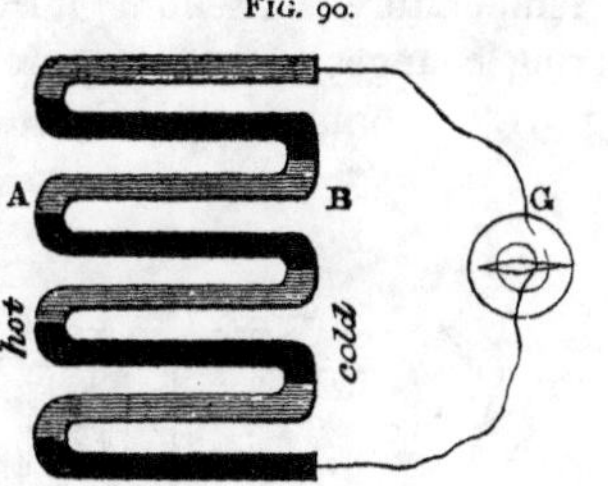

a greater current than any except the largest sized Daniell's cell. There are thermo-electric pairs which give a much greater E. M. F. than the above, but generally the increase in E. M. F. is to a great extent counterbalanced by an increase in the internal resistance of the pair.

§ **9.** Thermo-electric currents are produced by non-metallic substances. Metals and fusible salts form powerful pairs, which are generally held to be thermo-electric, and Becquerel has constructed a battery of the artificial sulphuret of copper and German silver, in which the salt is used without being fused.

Thermo-electric currents are also produced in circuits of metals and liquids, and probably in simple liquid circuits.

§ **10.** The chief practical use to which thermo-electric batteries have been put is the measurement of small differences of temperature. Melloni introduced this method of observing changes of temperature. A thermo-electric battery, Fig. 91, is connected by the terminals t t_1 with a galvano-

FIG. 91.

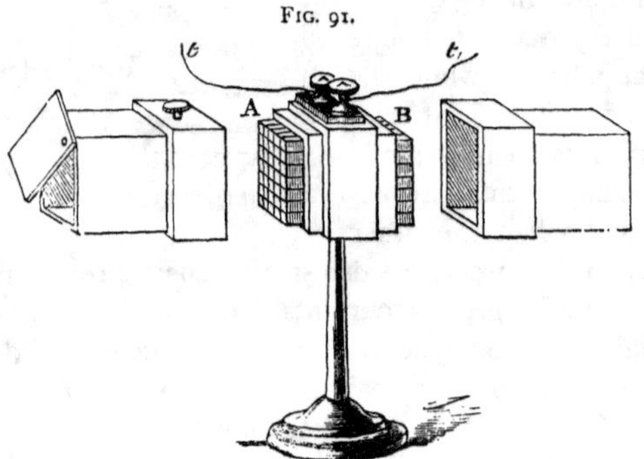

meter having a very small resistance; one series of junctions B is maintained at one temperature as nearly as possible, being enclosed in a metal case; the other series of junctions A is exposed to radiation from the objects the temperatures of which are to be compared. The junctions are screened by tubes from the radiation of other objects; these tubes are shown removed from the battery in Fig. 91.

When any substance warmer than the space opposite B is allowed to radiate heat upon the junctions A, the galvanometer is immediately deflected. When the junctions A radiate heat to a colder substance than B, so as to become colder than B, a deflection to the opposite side is produced; for small differences of temperature the currents produced are proportional to the differences of temperature. This arrangement is so sensitive, that by its aid the heat radiated by the fixed stars has been detected.

§ **11.** In accordance with the doctrine of the conservation of energy, heat is transformed into electricity in the thermo-electric circuit; the work done by the current is precisely the equivalent of the heat so transformed. If the whole work of the current consists in heating the conductors, the effect is merely a transference of heat by means of electricity from one part of the circuit to another; so that, in accordance with the law of dissipation of energy, the parts of the circuit are, on the whole, more nearly at one temperature than if no current had been produced, and heat had merely been conducted along the wires. If the current is employed to do mechanical work, an equivalent amount of heat is abstracted from the circuit, and reappears in the bearings of the working machine and the materials it works upon; similarly a portion of the work done may be electro-chemical. In whatever form the work is done, in the whole circuit this work will be equal to I Q § 2, Chap. VIII.

The heat is transformed into electricity at the hot junction, and also at unequally-heated portions of one or both metals. Peltier discovered that a current flowing through a circuit of two metals heated one junction and cooled the other. Now, the current which flows in a thermo-electric circuit flows in such a direction in general as to heat the cold junction and cool the hot one; so that for some time it was considered that the heat producing the current was wholly absorbed at the hot junction, and given out at the cold junction diminished by radiation, and by an amount equivalent to the work done in the rest of the circuit.

Sir William Thomson pointed out that this explanation was incomplete, for when a junction is at the neutral point no Peltier effect can occur; the two metals are then thermo-electrically identical; nevertheless when the hot junction is at the neutral point and the other junction at a lower temperature, a current is observed, increasing as the temperature of the lower junction is diminished, and the direction of the current is such as to heat the cold junction. Heat must therefore be absorbed at other parts of the circuit than at either junction.

§ **12**. We may, perhaps, best conceive of the manner in which this heat is absorbed by considering what would occur if a current were passed through a series of metal pieces, arranged as in Fig. 92, where each is in succession more positive than that which precedes it, *a* being the least and *k* the most positive. If a current is passed from *a* to *k*, it will flow in the direction opposed to that in which a current would

FIG. 92.

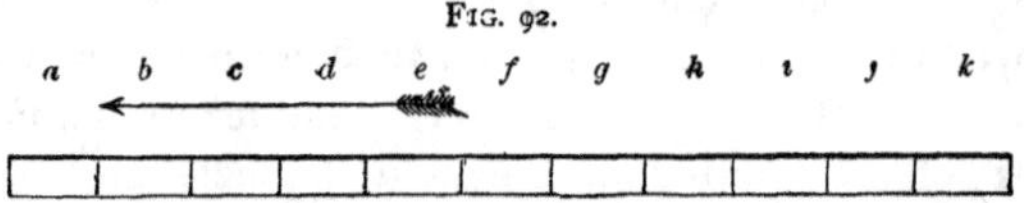

flow across any of the junctions, if that were the hot junction of a circuit made of those two metals, and therefore every junction would be heated; whereas if the current were passed in the other direction, as shown by the arrow, every junction would be cooled. If the Peltier effect at every junction were the same, the bar would be heated and cooled uniformly; but if the Peltier effect increased from *a* towards *k*, then the bar would be unequally heated or cooled by the passage of the current. The current in the direction of the arrow would cool the bar most near *k* so as apparently to heap up heat towards *a*, whereas a current in the opposite direction would heap up heat towards *k*; in other words, in such a bar as this, positive electricity might be said to carry heat with it. Now, a copper bar, or wire, with the end *k* cooler than the end *a*, behaves as if it were composed of an infinite number of such little elements ; a current from hot to cold heats it and carries heat with it ; whereas an iron bar behaves as if when the end *k* were the hotter it were the more positive, so that a current from cold to hot heats iron. The heaping up of heat in iron goes in the direction opposed to that of the current. We see that a current from hot to cold in iron absorbs heat, and one from cold to hot absorbs heat in copper ; and hence, when a pair is formed of copper and iron with its hotter junction at the neutral point, the current goes from cold to hot in the copper and hot to cold in the iron. Hence the copper and iron both absorb heat, and the electromotive forces of the two are added. With most pairs of metals the E. M. F. in the one unequally heated metal is opposed to that in the other. In this case the stronger E. M. F. overcomes the weaker, and the resultant current is due to the difference of electromotive forces. The discovery of the absorption or evolution of heat due to the unequal temperatures of metals and its convection were predicted from theoretical considerations by Sir William Thomson, who afterwards verified his conclusions by experiment.

CHAPTER XIII.

GALVANOMETERS.

§ 1. A GALVANOMETER is an instrument intended to detect the presence of a current and measure its magnitude ; all forms of the instrument consist of a coil of insulated wire and a magnet freely hung or pivoted so as to be easily deflected by the passage of a current through the coil. The wire forming the coil is so wound that each turn lies in a plane approximately perpendicular to the axis of the undeflected magnet. The current, in passing through the coil, or bobbin, of insulated wire, produces a magnetic field in the space in which the magnet hangs, and the couple tending to deflect the magnet is directly proportional to the strength of this field and to the moment of the magnet. The opposing couple tending to bring back the magnet to its undeflected position may be due to various causes.

In one class of galvanometers the magnet is suspended or supported in a horizontal plane, and the opposing couple is simply due to the earth's magnetism. In instruments of this class, no increase in the moment of the suspended magnet will increase the sensibility of the instrument—that is to say, it will not increase the deflection due to a given current—for by just as much as the deflecting couple is increased, by so much is the opposing couple also increased. The complete magnetisation of the needle therefore is not of much consequence, and a change in the magnetisation of the needle does not alter the sensibility. A small, light magnet will also in this class of instruments be deflected through the same angle as a large, heavy one, and will have the following advantages: 1st. That the small magnet will require only a small coil to surround it, and that this small coil will for the same number of turns produce a more intense magnetic field (§ 8, Chap. VIII.) than the large one, and offer much less resistance than the large coil, if made of the same wire. 2nd. That the inertia of the small magnet being less relatively to the magnetic moment, it will reach its maximum deflection more quickly, and will come to rest more rapidly than the large magnet. It will also indicate transient currents which do not last long enough to deflect the large magnet.

§ **2**. In a second class of galvanometers, the couple opposing the deflection is due not to magnetism, but to weight. The magnet is pivoted in a vertical plane, and has one end slightly weighted, so as to hang upright when undeflected. In these instruments any increase in the magnetic moment of the magnet increases the sensibility, assuming the counter-balance or directing weight to remain constant. Hence in these instruments, to ensure the greatest sensibility the needles should be magnetised to saturation, but, in order to ensure constant sensibility, the magnetism of the needle must remain constant, and these two conditions can rarely be realized together. The vertical component of the earth's magnetism exerts a certain directing force on the needles, but its effect is usually nearly insensible in comparison with that of the weight. These instruments are not generally intended for the indication of such small currents as those described in § 1. With very small magnets it is difficult to diminish the friction of the pivots and the counter-balance proportionately to the diminution of the magnetic moment. Hence in some forms of the second class it may be disadvantageous to diminish the size of the needle.

§ **3.** In choosing a galvanometer for any special purpose, we must first consider the character of the circuit into which it is to be introduced. The introduction of the coil of the galvanometer into the circuit will in all cases increase the resistance of the circuit, and therefore diminish the current. If the coil has a small resistance relatively to that of the other portions of the circuit, the diminution of the current will be small, and may in some cases be altogether neglected; but if the resistance of the original circuit be small, the mere introduction of the galvanometer intended to measure or indicate the current may reduce that current a thousandfold or more. In all cases there is some advantage in using a galvanometer coil of small resistance, but in order that a small current may produce a sensible magnetic field, it is desirable that it be led round the coil as often as possible, a condition antagonistic to the former. We can readily see that for circuits of small resistance the galvanometer giving the largest deflection will be an instrument having a coil with few turns of thick wire; but for circuits of large resistance, galvanometers having thousands of turns of thin wire will be on the whole most advantageous. In some writings these two classes of instruments are spoken of as adapted to two different classes of currents instead of to two different classes of *circuits*. The instrument with numerous turns of fine wire is said to indicate *intensity* currents, the other class to indicate *quantity* currents. These two old names survive, although the fallacious theory which assumed that there were two kinds of currents is extinct; the term 'intensity galvanometer' is used to signify an instrument with thousands of turns of thin wire in its coil, and 'quantity galvanometer' an instrument with few turns of thick wire. I shall name the two varieties 'long coil' and 'short coil' galvanometers.

§ **4.** The student must clearly understand that equal deflections on the same galvanometer always indicate equal currents. These currents may be flowing through very different circuits, and any given change may produce very different effects in the two circuits; but so long as the currents produce the same deflection in the same or equal galvanometers, the currents are equal, though the circuits may be very different. Thus, using a short coil galvanometer having a resistance of, say, 0·1 ohm, and no other external resistance in circuit, a thousand voltaic cells in series will produce about the same deflection as one cell of the same kind. The thousand cells produce 1,000 times the electromotive force that one cell does, but the resistance of each cell, which we may assume as 4 ohms, is much greater than that of the short coil galvanometer. Hence, the resistance of the thousand cells added to that of the galvanometer will be about 1,000 times greater than that of one cell added to the galvanometer, being 4000·1 in one case, and 4·1 in the other. The resistance varies in nearly the same proportion as the electromotive force, and therefore the galvanometer shows nearly the same deflection, indicating nearly the same current in the two cases. In the example taken above, the thousand cells would give a deflection greater than that of the single cell in the proportion of 41 to 40 nearly. When a long coil galvanometer, having a resistance of, say, 8,000 ohms, is employed, very different results follow. With one cell perhaps no deflection is observable, whereas with one thousand cells the needle is violently thrown against the stops limiting its deflection. The cause is simple. With one cell the resistance of the whole circuit, which will be 8004, including the long thin wire of the galvanometer, was so great that the E. M. F. of one cell did not give current enough to deflect the needle; but when a thousand cells were employed, the electromotive force was a thousandfold greater, and the whole resistance of the circuit was 8000 + 4000, or 12000 ohms. Hence if the E. M. F. of each cell be taken as one volt, the current in the first case will be $\frac{1}{8004}$ or nearly 0·000125 farads per second; whereas in the second case it will be $\frac{1000}{12000}$ or 0·0833, or about 666 fold greater. The couple deflecting the magnet of the galvanometer will also be 666 fold greater in the second than in the first case. Remark, however, that neither current will be so strong as that produced when the short coil galvanometer was used; for in that case, with a single cell the current would be $\frac{1}{4\cdot 1}$ = 0·244 farad per second, or roughly three times that due to the thousand cells as above; nevertheless the couple exerted on the magnet of the long coil galvanometer would be far greater with 0·0833 farad than that exerted on the short coil galvanometer by 0·244 farad simply because to produce the same couple the long coil galvanometer would only require about three times as many turns as the short coil galvanometer, whereas in practice it would have several hundred times more turns. The greatest deflection with any given circuit is obtained by using a galvanometer, the coils of which have a resistance equal to that of the other parts of that particular circuit.

§ 5. The sensibility of any galvanometer the needle of which is directed by a magnetic field may be increased by diminishing the intensity of the magnetic field. The opposing couple is due to the intensity of this field, and by its diminution the deflection due to a feeble current may be indefinitely increased. This diminution of the intensity of the original magnetic field is most easily brought about by laying a powerful magnet near the galvanometer, in such a position as to counteract the earth's magnetism, i.e. in the magnetic meridian, with its north pole pointing north. This magnet, often called a compensating magnet, is best placed in the same meridian as the suspended magnet. As the intensity of the field diminishes under the influence of this magnet, the rate of oscillation of the suspended magnet diminishes, and by observing this rate we can determine the increase of sensibility. The period of oscillation is inversely proportional to the square root of the intensity of the field, and as the directing couple is directly proportional to this intensity, and the sensibility inversely proportional to the directing couple, we have the sensibility directly proportional to the squares of the periods of oscillation. So long as the magnetism in the needle of a galvanometer remains unaltered, its relative sensibility with the compensating magnet at different distances can be roughly computed in this manner; I say roughly, because the number of swings which can be counted is small when the sensibility is great, owing to the resistance of the air, and this resistance would also necessitate a correction in the above series of proportions. This method of obtaining a sensitive galvanometer has the following defect: Inasmuch as the directing field is due to a difference between two nearly equal magnetic fields, a very small change in the direction or intensity of either produces a great change in the difference; and as the direction and intensity of the earth's magnetism is perpetually varying, it is nearly impossible to keep the needle pointing at a constant fiducial mark or zero, or with a constant sensibility.

The zero should be adjusted by a much smaller magnet called an adjusting magnet, placed across the lines of force of the magnetic field or pointing east and west, and fixed so as to be capable of adjustment by turning in a plane perpendicular to the magnetic meridian, and with its centre in the meridian of the suspended magnet. This adjusting magnet does not, when turned, alter the intensity of the field near the suspended magnet, but only alters the direction of the lines of force.

The suspended magnet in very sensitive instruments should be hung by a single silk fibre, such as can be obtained from the silk threads in a common silk ribbon. The viscosity and torsional elasticity of this fibre put a limit to the possible diminution of directing force as described above.

§ **6.** The most sensitive instruments employed are those known as *Astatic galvanometers.* In these instruments two magnets joined as in Fig. 93, with the north pole of one

FIG. 93. FIG. 94

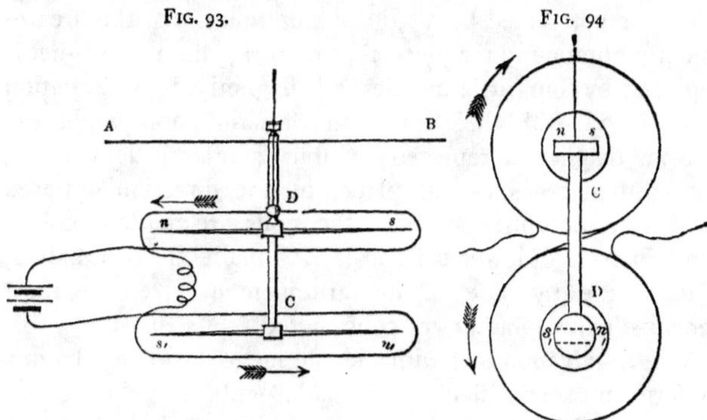

over the south pole of the other, form one suspended system. If the two magnets had exactly equal moments with axes precisely parallel, they would hang in equilibrium in any direction in any uniform magnetic field. The moment of one magnet always slightly exceeds that of the other, and by this excess directs the system. A single galvanometer coil may surround one needle, or, as is obviously better, each needle may have its own coil, the two coils being so joined that the current must circulate in opposite directions round the two so as to deflect both magnets similarly. In one common form of the astatic galvanometer, needles about a couple of inches long are used, and their deflection is observed by means of a pointer or glass needle, A B, Fig. 93, rigidly connected with the astatic system by a prolongation of the brass rod C D. This pointer oscillates over a graduated circle, and its position is observed by a microscope or simple magnifying glass. The coils are made flat, of the shape indicated in Fig. 93. To allow the introduction of the needle, the top and bottom coils are made in two halves, placed side by side, with just sufficient space between them to allow the rod C D to hang freely.

In Thomson's mirror astatic galvanometer, Fig. 94, the magnets are much reduced in size, being only about $\frac{1}{4}$ in. long. They are connected by a strip of aluminium C D, and are frequently compound magnets, that is to say, the top magnet is replaced by four little needles, all magnetised to saturation and placed with their poles in one direction while the bottom magnet is replaced by four similar little needles, having their poles also all placed in one direction opposed to that of the upper system ; the coils are made circular ; the upper and lower coils are each made in two halves, placed side by side. This arrangement gives the most sensitive galvanometer yet constructed.

§ 7. A galvanometer with a single magnet directed by any uniform magnetic field, and made with a coil large in diameter relatively to the length of the magnet hung in the axis of the coil, is called a *tangent galvanometer*, because the tangents of the angles to which the needle is deflected by the currents are proportional to the currents causing the deflections. This law has been proved above, § 3, Chap. VIII. The best form of tangent galvanometer is that in which there are two coils in parallel planes, Fig. 95, separated by a distance equal to one-half their diameter. The magnet, which should be short, is hung in the common axis of the coils half-way between them.

The object of this arrangement is to do away with the error due to the sensible length of the magnet, and to any small deviation from a truly central position.

The deflection is observed by means of a light glass pointer oscillating over a graduated limb.

§ **8.** A galvanometer, whether astatic or not, with magnets directed by any uniform magnetic field, and having the coils constructed so as to be capable of turning on the axis round which the magnet turns, is called a *sine* galvanometer, because, if the coils be turned by hand so as to lie in a vertical plane parallel to that passing through

FIG. 95.

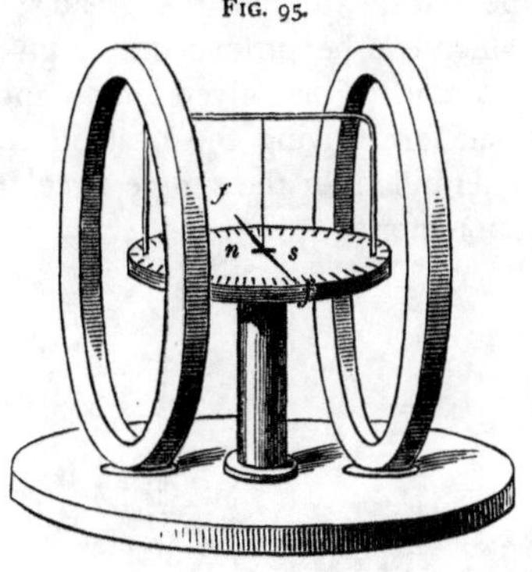

the magnet when deflected by a current, then currents deflecting the magnet to angles θ and θ_1 will be to one another in the ratio of sin θ and sin θ_1 : this follows from the considerations explained in § 3, Chap. VIII. Sine galvanometers can be easily made much more sensitive than tangent galvanometers, because they may be astatic, and because the coils may closely surround the magnets. They are inconvenient for many purposes, because an observation with them occupies a longer time than with any other galvanometer: each adjustment of the coils moves the magnet also, and many trials are necessary before perfect parallelism of the planes is arrived at. This parallelism is attained by bringing a fiducial mark attached to the coils vertically under a pointer attached to the magnet. A vernier is attached to the coils, and the angle through which they are turned from the position indicated by the fiducial mark when no current was passing to that indicated by the fiducial mark when the current flows is read off on a graduated circle. This can be done with great accuracy. The coils are generally moved by a tangent screw.

§ **9**. The form of the coil in a galvanometer is not a matter of indifference. The coil may be too broad and flat, or it may be too narrow, to give the greatest intensity of magnetic field which can be produced by a given length of wire wound into a coil. For a given length and size of wire there is always one form giving the best effect. This form has only been determined for the simple circular coil used in the mirror galvanometer.

FIG. 96.

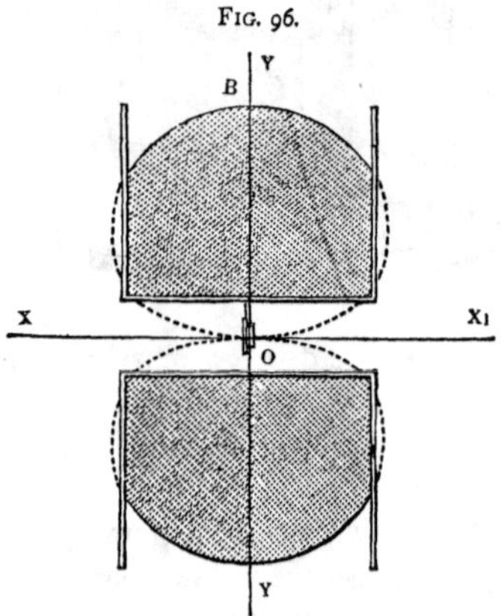

The form of the curve, bounding the best section of the

coils, is given by the following equation, due to Sir William Thomson:

$$x^2 = (a^2 y)^{\frac{2}{3}} - y^2$$

where x is the ordinate measured in a direction parallel to the axis of the coil, y the ordinate perpendicular to that axis and a the distance O B. The origin of the co-ordinates is at centre of the coil, where the magnet hangs. Fig. 96 shows the theoretical curve and a longitudinal section of a practicable coil. A portion of the area enclosed by the curve near the magnet is necessarily omitted to give room for the magnet to move; a practical approximation is made to the best form by winding the wire on a bobbin of the proportions shown, and filling with wire that portion which is cross-hatched.

FIG. 97.

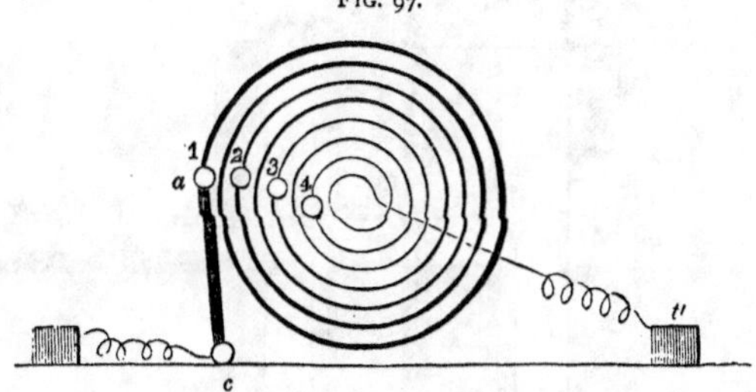

To get the best result the wire should not be all of one gauge, but should increase with the diameter of the coil, so that the cross section of the wire may be directly proportional to the diameter of the coil at each point: the resistance of every turn of the coil will then be equal. It is practically impossible to follow this plan rigidly, but three or four sizes of wire may very properly and easily be employed in winding a galvanometer coil.

§ **10.** Sir William Thomson has given the name of *graded* galvanometer to an instrument constructed as above, and having also a moveable arm or lever by which one of the two terminals *t*, Fig. 97, can be connected by an arm *a c*, hinged at *c*, with the several stops, 1, 2, 3, 4, so as to include in the galvanometer circuit either the whole of the wire, or $\frac{3}{4}$, or $\frac{1}{2}$, or $\frac{1}{4}$, but in all cases so as to use the most efficient part of the wire for the degree of sensibility required. The relative sensibility of each grade is easily determined by experiment, and is constant.

§ **11.** Sir William Thomson has given the name of *dead beat* galvanometer to a mirror galvanometer having the following peculiarities:—1. very light mirror; 2. four small magnets at the back instead of one of equal weight; 3. the cell in which the mirror moves only just large enough in diameter to allow the mirror to deflect; 4. the front

Fig. 98.

Sectional Elevation.

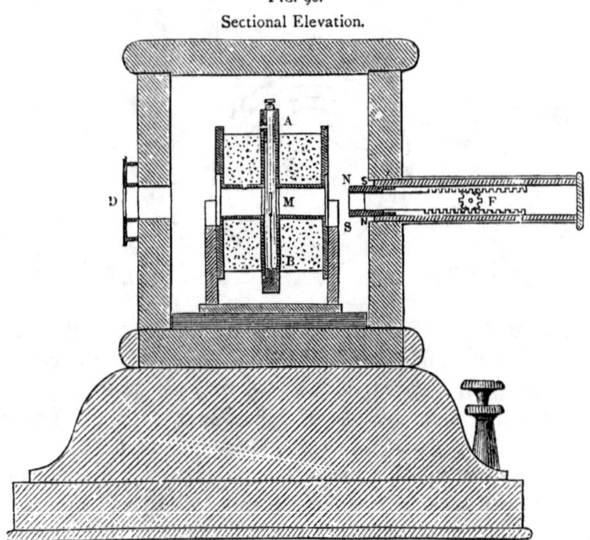

and back of the cell so close as each separately to act as a stop, preventing deflection of the mirror beyond the angle required to bring the spot of light to the end of the scale. The mirror does not strike the stops in actual use.

With instruments so made the spot of light moves to the final deflection without oscillation being checked by the viscosity of the air. The same end is much less perfectly attained in some instruments by a vane of light material hanging from the magnet. This vane sometimes dips in water, and Mr. Varley has made galvanometers in which the cell containing the magnet and mirror is full of water.

§ 12. The *Marine galvanometer* is a galvanometer adapted for use at sea. It must be so constructed that neither the motion of the ship nor the change of direction produces

Sectional Plan.

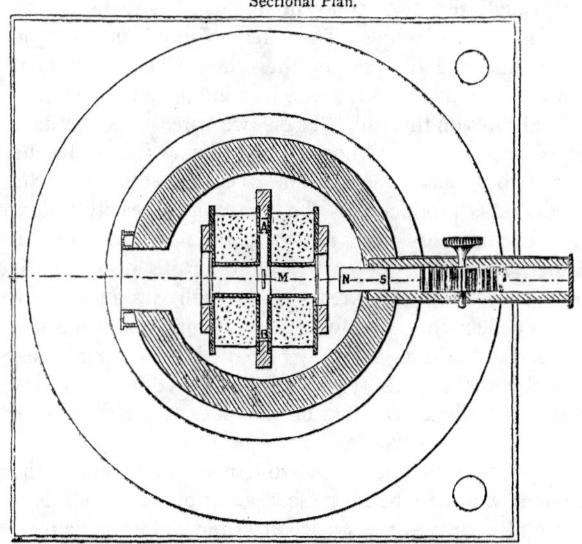

sensible deflections. This result has been obtained by Sir William Thomson in the following way: The magnet and mirror of a mirror galvanometer are strung on a bundle of straight silk fibres, stretched between A and B, Fig. 98. The suspended system is balanced so that the axis of the fibres passes through its centre of gravity. A powerful directing horse-shoe magnet, not shown in the drawing, embraces the coils, and serves to overpower the directive force of the earth's magnetism, the effect of which on the suspended magnet is moreover much weakened by a massive soft iron case, enclosing the whole system everywhere except at the little window D, by which the rays of light reflected by the mirror enter and return. An adjusting magnet N S is worked by a ratchet and pinion F.

§. **13.** The *differential galvanometer* has two equal coils, so arranged that when the same current or equal currents pass through the two coils in opposite directions, the magnet is not deflected. The effect of one coil is completely neutralised by that of the other. The differential galvanometer is most easily made by winding simultaneously two equal wires on the coil. These two wires are sometimes arranged in a sort of ribbon or plait, being joined by the silk used to insulate them. The accurate equality of the magnetic fields produced by the two coils is easily tested, for if a current pass from the battery first round one coil and then round the other in the opposite direction, it should, no matter how great its strength, produce absolutely no deflection. In most cases a small deflection will be observed, but this is easily remedied by adding a few turns to the weaker coil. If after this has been done the resistance of one coil exceeds that of the other, a length of wire can be added to the coil of least resistance, and placed in such a position as not to tend to deflect the magnet; the instrument will then be in perfect adjustment. This is a very useful instrument, as we shall see in a future chapter, for the purpose of comparing resistances. The coils are sometimes made of German silver instead of copper. German silver has a much greater resistance than copper, but its resistance varies much less with changes of temperature. In differential galvanometers intended to be used in circuits otherwise of great resistance, the total resistance of the coils is of small importance, but the equality of the resistance of the two coils is very important.

§ **14.** The sensibility of a galvanometer may be varied in a very simple manner by the use of what is termed a *shunt*. A shunt is a resistance coil, or coil of fine wire used to divert some definite portion of a current, taking it past a galvanometer instead of through its coils. Thus let G, Fig. 99, represent the galvanometer coils, and let S represent the shunt. Let the resistance of the shunt be $\frac{1}{9}$th that of the galvanometer; then, of a total current passing from C to D, 9 parts go through the shunt and do not deflect the needle, while 1 part goes through the galvanometer: only $\frac{1}{10}$th of the whole current is therefore effective in deflecting the needle, and the deflection (supposing a mirror galvanometer be used) is only $\frac{1}{10}$th of what it would have been had no shunt been used. Similarly by making the shunt equal in resistance to $\frac{1}{99}$th of the galvanometric coil, we reduce the sensibility of the instrument to the $\frac{1}{100}$th part of its original sensibility. Most galvanometers used for measuring currents are now sold with shunts = $\frac{1}{9}$th, $\frac{1}{99}$th, and $\frac{1}{999}$th, of the galvanometer coil: by these the sensibility of the instrument can be varied 1000fold. The shunts must be made of the same metal as is used for the coils, and should be placed so as to be as nearly as possible at the temperature of the coils. Calling S the resistance of the shunt, and G the resistance of galvanometer coil; calling d the deflection without the shunt, and d_1 the deflection with the shunt, we have quite generally, with a given constant current and assuming that the deflections shown by the instrument are proportional to the currents:

FIG. 99.

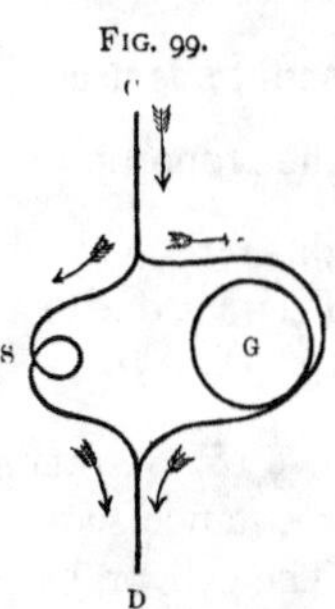

$$d : d_1 = G + S : S.$$

It must be remembered that adding the shunt will in all

cases diminish the resistance of the circuit, so that unless this resistance is so great that the resistance of the galvanometer forms no sensible part of it, the *deflections* will not be altered in the above proportion. Let R be the resistance of all parts of the circuit except the galvanometer. Then, if the E. M. F. remain constant, we have $R + G$ as the total resistance when no shunt is used, and $R + \frac{G\,S}{G + S}$ when the shunt S is used. The currents C and C_1 will therefore be in the proportion of $R + \frac{G\,S}{G + S}$ to $R + G$; and compounding this ratio with that given above, we have for d and d_1 deflections due to a constant E.M.F. with and without the shunts

$$d : d_1 = R\,(G + S) + G\,S : (R + G)\,S.$$

§ **15.** Galvanometers intended for circuits of extremely small resistance sometimes consist of a single thick ring of copper. The cell or battery used with such a galvanometer as this must be of such construction as to have very small internal resistance, or no deflection will be observed. A Grove's cell (*vide infra*, Chap. XIV. § 14) with large plates will give a current which can be observed with a single ring galvanometer. Galvanometers intended for thermo-electric experiments must have very small resistance, and are frequently made with twenty or thirty turns of No. 20 wire Birmingham wire gauge, the diameter of which is nearly 0·09 centimètres. The resistance of these galvanometers may be less than a quarter of an ohm. Galvanometers intended for use in circuits of great resistance are frequently made with wire of No. 30 or No. 36 B.W.G., corresponding to the diameters 0·0305 and 0·0106 centimètres, and the resistance of these galvanometers is frequently as much as 8,000 ohms. About half a yard of the No. 36 gauge copper may have a resistance of one ohm, so that the above resistance would require 4,000 yards of copper wire. The resistance in itself is a defect, but it is impossible to get a large number of turns into a small space without great resistance. It is very important that every coil of the galvanometer should be perfectly insulated from its neighbour: if any two coils touch or are connected through the silk, they are, in technical language, said to be short-circuited; the current does not then flow round any of the intermediate turns, and the effect of these is lost. When there is no actual metallic contact there may be imperfect and uncertain insulation, and this is the worst defect a galvanometer can have: its resistance becomes uncertain and variable; the shunts can no longer be depended upon as equal to definite fractions of the resistance, and the instrument is useless for accurate observations. The insulated wire should not only be thoroughly covered with silk, but should also be baked so as to be very dry before being wound on; and after a few layers have been coiled, the bobbin should be baked again and dipped in pure melted paraffin. When the coiling has been completed the whole coil should be again baked, and its resistance compared with the calculated resistance of the wire wound on.

Contact between coils of a differential galvanometer is obviously a radical defect; and when two or more distinct coils are wound on the same bobbin, as is sometimes done, these coils must be very carefully insulated. Serious errors in testing have arisen from bad insulation between different coils and different parts of the same coil.

CHAPTER XIV.

ELECTROMETERS.

§ 1. ELECTROMETERS indicate the presence of a statical charge of electricity by showing the force of attraction or repulsion between two conducting bodies placed near together. This force, depending in the first place on the quantity of electricity with which the conducting bodies are charged, ultimately depends on the difference of potential between them; an electrometer is therefore strictly an instrument for *measuring* difference of potential. It is used often simply to indicate the presence of electricity, but it does not measure quantity, and when used to compare quantities it can do this only because under given circumstances the differences of potential produced between the two conductors are proportional to the quantities on the bodies by which one of the conductors of the electrometer is successively charged.

The usual repulsion electroscopes have already been described. They are known as the pith-ball or Canton's electroscope; the gold leaf or Bennet's electroscope and the Peltier electroscope. Bohnenberger's electroscope, which consists of a single gold leaf hanging between two symmetrically disposed knobs maintained one at a positive potential, and the other at an equal negative potential, belongs to a different class, called by Sir William Thomson heterostatic electroscopes—or instruments in which, besides the electrification to be tested, another electrification, maintained independently of it, is taken advantage of. In Bohnenberger's instrument the independent electrification maintaining the two knobs at a constant difference of potential is produced by a kind of galvanic battery called a dry pile, consisting of thin plates of two metals soldered together, and separated by paper which remains very slightly moist in consequence of containing some deliquescent material. Sometimes the metal plates are replaced by metals in powder adhering to the paper. So long as the gold leaf is neither positive nor negative, it is neither attracted to the right nor left; positive electrification deflects it to the negative knob, and *vice versâ*.

A modification of Bohnenberger's electroscope, Fig. 100, may be made, in which the heterostatic charge may with advantage be given to the gold leaf, instead of to the two symmetrically disposed bodies A and B. Any difference of potentials between A and B will be indicated by the attraction of the gold leaf to one side. The higher the potential of the gold leaf the more sensitive the instrument. The high potential is most easily maintained by connection with a Leyden jar.

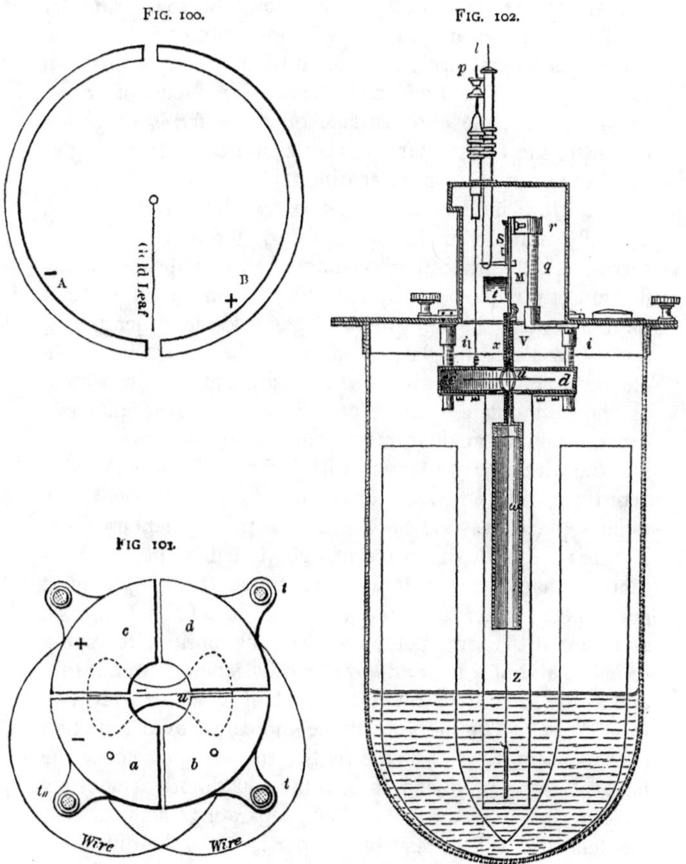

FIG. 100.

FIG. 101.

FIG. 102.

§ **2.** The most perfect form of heterostatic electrometer yet constructed is Sir William Thomson's quadrant electrometer. In this instrument the Bohnenberger's gold leaf is replaced by a very thin flat aluminium needle, *u*, shown in plan, Fig. 101, and (to a smaller scale) in elevation, Fig. 102. This flat needle spreads out into two wings, shown dotted in the plan, and is hung by a wire S from an insulated stem *q* inside a Leyden jar. This Leyden jar contains a cupful of strong sulphuric acid, the outer surface of which forms the inner coating of the Leyden jar. A wire Z, stretched by a weight, connects *u* with this inner coating.

A mirror, hidden in Fig. 102 by the metal cover *t*, is rigidly attached to the needle *u* by a rod. The mirror serves, as in the reflecting galvanometer, to indicate the deflection of the needle *u* by reflecting the image of a flame on to a scale. The needle *u* hangs inside four quadrants, *a b c d*, insulated by glass stems, $i\ i_1$; the quadrant *a* is in electrical connection with *d*, and *c* is in connection with *b*, as shown in plan. Above and below the quadrants two tubes, V and u^1, at the same potential as *u*, serve to screen *u* and the wires in connection with it from all induction except that produced by the quadrants *a b c d*. These quadrants replace the bodies A and B in the elementary form, Fig. 100. Let us suppose *u* charged to a high negative potential—then, if the quadrants are symmetrically placed, it will deflect neither to the right nor to the left, so long as *a* and *c* are at the same potential. If *c* be positive relatively to *a*, the end of *u* under *c* and *a* will be repelled from *a* to *c*, and at the same time the other end of *u* will be repelled from *d* to *b*. The motion will be indicated by the motion of the spot of light reflected by the mirror. Moreover the field of force produced inside the quadrants is sensibly uniform just over the narrow slit separating them, so that the deflection will be sensibly proportional to the difference of potential between *a* and *c*. The number of divisions which the spot of light traverses on the scale will therefore in an arbitrary unit measure the difference of potential between *a* and *c*. This instrument is therefore an electrometer, and not a mere electroscope. Two terminals *p*, of which only one is shown in the drawing, serve to charge *a* and *c*: they can be lifted up out of contact with *a* and *c* after charging them. A third terminal, *l*, serves to charge the Leyden jar. It is usually disconnected from the inner coating by being turned back, so that the tongue M is disconnected from the metal rod behind S.

With good glass, carefully washed in distilled water and dried before the fire, before being filled with sulphuric acid, the Leyden jar can be made to insulate so well as not to lose a quarter per cent. of its charge per diem. Sir William Thomson adds a little inductive electrical machine inside the jar (§ 1, Chapter XIX.), by which the charge can be increased or diminished at will, and also a gauge by which the constancy of the charge can be measured. An instrument of this class may be made so sensitive as to give a deflection of 100 divisions for the difference of potential between zinc and copper.

§ **3**. The essential parts of Sir William Thomson's *portable* electrometer are shown in Fig. 103. *g* is a flat insulated disc to which the charge to be measured may be communicated. *h* is a second insulated disc, having an opening at the centre filled by a very light aluminium plate *f*, supported by a stretched wire *i i*, and carrying an index arm below the plate *h*. This plate and wire are shown in Fig. 57, p. 100. If now *g* and *h* are at the same potential, there will be no charge on the opposed faces, and *f* will neither be attracted nor repelled by *g*. If a charge of electricity be communicated to *g* or *h*, so that the potentials differ, *f* will be attracted or repelled by *g*, and the consequent motion can be read by observing at *l* the position of a little hair, fixed to the index arm. Unless, however, the charges on *g* and *h* are very great, the forces will be very small, and this arrangement would offer little advantage: its sensibility is enormously increased by the following device:—A considerable permanent charge is given to *h*, which is maintained in permanent connection with a highly charged perfectly insulated Leyden jar; then if *g* be in connection with the earth, a charge will be induced on *g*, and *f* will be attracted by that charge with a very sensible force. Let the torsion of the wire *i i* be adjusted so as to depress *f* or elevate the hair near *l*, then there will for a given potential of *h* be one distance

FIG. 103. FIG. 103 *a*.

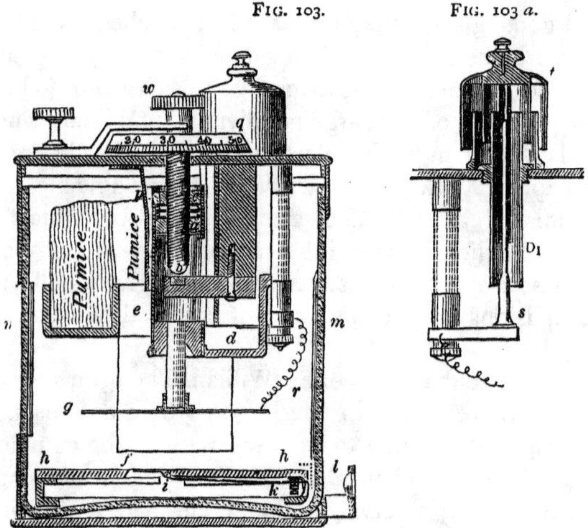

between *g* and *h*, at which the electrical attraction will just balance the torsion of the wire. The distance of the plate *g* from the plate *h* can in the instrument be adjusted by a fine screw, and this position is read off by a divided scale and vernier. Let *g* next be disconnected from the earth and connected with the body the potential A of which is to be tested, i.e. compared with that of the earth—a new charge will be induced on *g* proportional to the difference between the potential of *h* and A; if A be positive, assuming the potential of *h* to be positive also, the charge will be less than that due to the earth, and plate *g* must be lowered. If, on the contrary, A be negative, the charge will be greater than that due to the earth, and to bring the hair at *l* back to its fiducial mark *g* will have to be raised—the difference of potential between A and the earth will be proportional to the distance through which g is moved; for, from § 7, Chapter V., we have $f = \frac{V^2 M}{8 \pi a^2}$; where V is the difference of potential between two plates at a distance a. When l is at the fiducial mark, f determined by the torsion of the wire is constant, and the quotient $\frac{V^2}{a^2} = \frac{V}{a}$ must also be constant, so that the difference of potential V must vary in direct proportion to the distance a between the plates, in order to balance this constant force.

Each 100th of an inch corresponds therefore with a given potential of the plate h to a perfectly definite and constant difference of potential, so that if with one body A the disc g requires to be raised 0·01 above the position when the earth reading was taken, and with a second body B the same plate requires to be raised 0·1 above the same position, we know that the potential of B is ten times that of A, both potentials being above or below that of the earth. By making the potential of h in all cases large, the distance a may also be large for a constant force f, and a great range of measurement is thus combined with great sensibility.

The plate h h forms part of the inner armature of a Leyden jar, the glass of which is lettered m m; the micrometer screw b serves to raise and lower the insulated plate g by means of a slide which need not be specially described here. The position of g is read off by a vertical scale not shown, still further subdivided by the divided ring at q; the plate g is connected with a terminal s, shown in Fig. 103a, projecting outside the Leyden jar through an opening in the case. This rod t serves to charge the plate g, and is usually covered with a cap, t, of special form, intended to prevent the influx or efflux of air. When the instrument is not in use, the cap t is pushed down, closing the Leyden jar entirely. When the instrument is in use, the cap t is raised, and being then wholly insulated it serves as the terminal by which to charge *g*. A lead case for pumice stone and sulphuric acid is placed inside the Leyden jar to dry the air. The Leyden jar can be charged by an insulated rod, introduced temporarily through a little opening provided for the purpose in the top of the case. When the jar is once charged this hole is closed by a screw. When proper glass is chosen for the jar, well washed with distilled water, and dried by evaporation before the fire before being finally closed, the Leyden jar will not lose $\frac{1}{4}$ per cent. of its contents per diem. Care must be taken to remove the pumice stone once a month and bake it, otherwise the sulphuric acid diluted with water attracted from the atmosphere will overflow and spoil the instrument. The difference of potential produced by the contact of zinc and copper may be detected on this instrument, and the electromotive force of 20 or 30 Daniell's cells can be measured with considerable accuracy. The value of each division of the instrument alters as the charge in the Leyden jar varies. The instrument is not an absolute electrometer, but is used to compare potentials as galvanometers are used to compare currents. It is specially adapted for experiments on the potential of the atmosphere. If a burning match be attached to the terminal *s*, the plate *g* is rapidly brought to the potential of the air at the point where the match burns. The instrument is held in one hand, the position of the hair at *l* relatively to the fiducial mark observed through the magnifying glass, and the plate *g* adjusted by moving the screw head *w*. In the manufacture of the instrument so much torsion should be given to the wire as will just leave the plate *f* in stable equilibrium when *l* is at the fiducial mark. When very little initial torsion is given, the directing force of the wire varies very rapidly with the increased angle through which it is turned by the attraction or repulsion of plate *f*, and the equilibrium is then very stable. As more initial torsion is given, the change of directing force due to a deflection from the fiducial point is less, and the equilibrium may easily be made quite unstable. The torsion used should be a little less than that giving instability for the lowest position in which *g* will be used.

§ 4. The absolute electrometer is an instrument much like the portable, but on a larger scale, and so arranged that the actual force on the moveable disc can be measured. Then, calling V and V_1 the two differences of potentials which give the same force F with the two distances D and D_1 between the parallel plates, and calling A the area of the moveable plate, we have

$$V_1 - V = (D_1 - D)\sqrt{\frac{8\pi F}{A}},$$

by which equation the difference of potential $V - V_1$ is given in absolute electrostatic units : from measurements of this kind we can determine the constant multipliers required to convert the indications of a quadrant or portable electrometer into absolute measure.

CHAPTER XV.

GALVANIC BATTERIES.

§ 1. THE simplest form of galvanic cell practically in use consists of a plate of zinc and a plate of copper, immersed in water slightly acidulated by the addition of a little sulphuric acid. The zincs and coppers are generally soldered together in pairs, and placed in a long stoneware or glass trough, divided into separate cells by partitions as shown in Fig. 104. This battery is made more portable by filling the cells with sand, which supports the plates and prevents the liquid from splashing about when the trough is moved. In this form it is called the common sand battery. The copper is advantageously replaced by platinum or platinized silver; this battery without sand is then known as *Smee's* battery. The rough surface of the deposited platinum seems to have the effect of diminishing polarisation. Fig. 105 shows a common form of one cell of

FIG. 104.

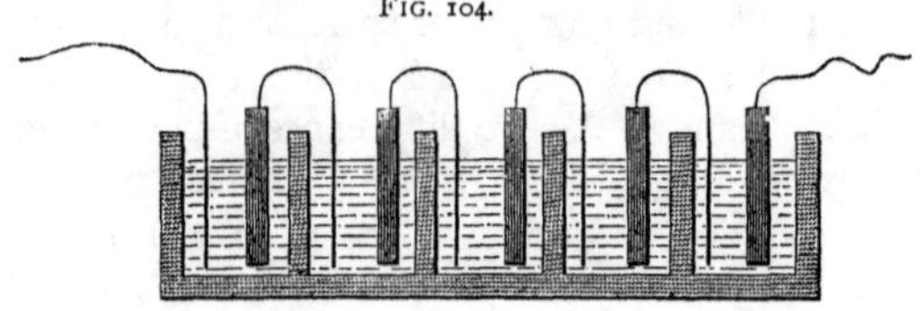

Smee's battery; the plate of platinized silver hangs from a wooden bar between two plates of zinc amalgamated with

FIG. 105. FIG. 105A.

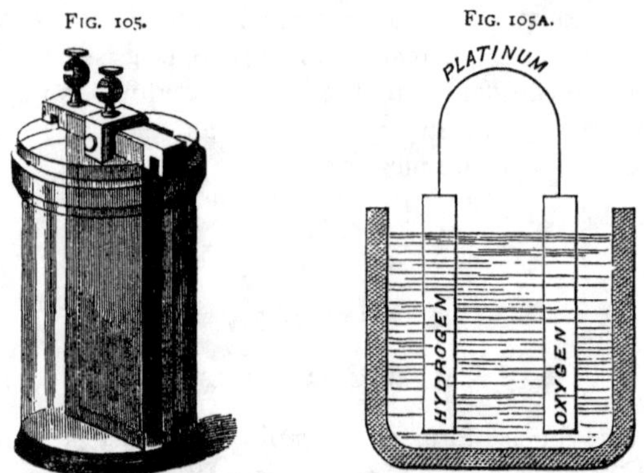

mercury; the brass terminals serve to hold the three plates together.

In *Walker's* battery the copper is replaced by graphite.

§ 2. The following are the chief merits of a galvanic cell:

1. It should produce a high electromotive force.

2. It should have small and constant internal resistance.

3. Its electromotive force should be constant whether it be employed in producing a large or small current.

4. The materials it consumes should be cheap.

5. No materials should be consumed except when the battery is employed to produce a current.

6. The form should be such that the condition of the cells can easily be seen, and fresh materials added when required.

No one battery combines all these advantages in the highest degree, and the special requirements of each case should guide us in the choice of the design to be preferred for any given purpose.

§ **3.** No single-fluid cell can give a constant electromotive force because of the polarization of the plates, § 9, Chapter IV. The electromotive force due to the metals in the batteries above described diminishes with extraordinary rapidity as soon as the poles are joined, especially when the current flowing is considerable. This diminution is due to an opposed E. M. F. consequent chiefly on the presence of free hydrogen on the copper or platinum plate. The effect of gases in setting up an electromotive force is easily shown by the voltameter, Fig. 41, p. 67. Let the wires A and B be joined by a wire, part of which is the coil of a galvanometer. A current will be perceived opposed in direction to that which decomposed the water; it will come from the hydrogen, through the water to the oxygen. This current is accompanied by the recombination of oxygen and hydrogen forming water. The direction of the current from this gas cell is such as would be produced if hydrogen were a negative metal electrode, and oxygen a positive electrode, as shown in Fig. 105*a*. Provided the oxygen and hydrogen have no chemical affinity for the metal employed to join them, this metal will have no effect on the E. M. F. of the gas cell; the hydrogen plays the part of the zinc plate, being oxidised by the water, and the hydrogen set free appears at the positive electrode (oxygen) and combines with it. The fact that hydrogen and oxygen joined by a metal conductor will recombine, whereas when simply in presence of one another they will not recombine, is probably due to the electromotive force set up at the junction between the metals and the gases Thus the junction between the platinum and hydrogen makes the hydrogen positive; the oxygen is either less positive or negative: thus the difference of potentials produced by the contacts tends to produce a current from the hydrogen electrode to the oxygen electrode through the water, and this would decompose the water, sending hydrogen to the oxygen electrode, and oxygen to the hydrogen electrode. The result is, that the decomposition of the water is balanced by the recomposition at the electrodes, and the gas gradually absorbed. The whole of the gas cannot be thus absorbed consistently with the theory of dissipation of energy. The above illustration of the action of the gases certainly is not a complete or accurate hypothesis. If it were, the electromotive force of the gas cell or polarized platinum plates would be constant, whereas it is much increased if the decomposition of the water has been effected by a high E. M. F., and gradually diminishes as the recombination of the gases occurs, as we should expect from the theory of dissipation of energy.

The electromotive force called up by the deposition of gases on electrodes is within limits nearly proportional to the E. M. F. employed in producing the deposition. This is most clearly seen when the electrodes are so formed that the gases cannot easily escape—when, for instance, the electrodes are small surfaces of metal, surrounded by an insulator, such as are produced by boring a hole so as to lay bare a small portion of the copper of a guttapercha-covered wire. We may, perhaps, conceive the high E. M. F. produced in reaction against a great decomposing E. M. F. as due to the decompositions of a row of molecules forming a number of gas cells in series imperfectly insulated from one another.

§ 4. The sand battery is the worst of all batteries as regards constancy of electromotive force, the polarization being greater in this battery than in any other because the gas cannot readily escape. The common copper and zinc cell is the next in order of demerit. Its electromotive force can at any time while it is producing a current be greatly increased by mechanically brushing the gases off the metals, or even by shaking the battery. The Smee battery is better than the copper zinc battery because it is found that hydrogen does not stick to the finely divided platinum on the surface of the plates so much as to the copper. The carbon or graphite plate in Walker's battery performs the same function of facilitating the liberation of the free hydrogen.

When any of these single fluid batteries are left with the electrodes free or insulated so that no current passes, the full electromotive force is gradually restored, partly by the liberation of the hydrogen, partly by its recombination with oxygen. The process of restoration may be assisted by passing a current through the cells against their E. M. F.

For some purposes a constant current is not required; —for instance, where batteries are employed to ring bells in houses or on railway lines they have long intervals of repose; for such purposes single fluid batteries are still employed on account of their simplicity.

§ **5.** If the voltaic theory of the cell were absolutely correct, the electromotive force of the cell would depend wholly on the electrodes or plates in the electrolyte, and not at all on the solution or electrolyte employed to connect them. We might then form a list or *potential series* (§ 22, Chapter II.) in which the difference of potential between each successive material being known, the E. M. F. produced by a cell made with any two materials would simply be equal to the difference of potential obtained by summing up the differences due to all the intermediate materials. These differences of potential have not been determined by experiment, though potential series have been determined so far as to give the order in which the materials come; but it has been found that this order is slightly changed by the solution employed to join the plates or electrodes; in order to account for this fact it is necessary to treat the voltaic theory as incomplete. The following are potential series for some of the more important solutions:—

OBSERVER .	Fechner.	Faraday.	Poggendorff.	Faraday.	Faraday.	Poggendorff.	Faraday.
SOLUTION .	Water.	Dilute sulphuric acid.		Dilute nitric acid.	Hydrochloric acid.	Sol. of Sal Ammoniac.	Sol. Sulphide of potassium (yellow).
PROPORTIONS		1 vol. water. 1 vol. oil of vitriol.		1 vol. strong acid. 7 vol. water.	1 vol. acid. 1 vol. water.		
			Amalg. Zinc				
	Zinc	Zinc	Zinc	Zinc	Zinc	Zin	Zinc
			Cadmium	Cadmium	Cadmium	Cadmium	Copper
	Lead	Tin	Lead	Lead	Tin	Lead	Cadmium
	Tin	Lead	Tin	Tin	Lead	Tin	Tin
	Iron	Iron	Iron	Iron	Iron	Iron	Silver
			Aluminium		Copper		Antimony
		Nickel	Nickel	Nickel	Bismuth		Lead
	Antimony	Bismuth	Antimony	Bismuth	Nickel	Bismuth	Bismuth
	Bismuth	Antimony	Bismuth	Antimony	Silver	Antimony	Nickel
	Copper	Copper	Copper	Copper			Iron
				Mercury (Avogrado)		Silver	
	Silver	Silver	Silver	Silver		Mercury	
	Gold		Gold (Davy)	Gold (Avogrado)	Gold	Gold	
			Platinum	Platinum (Avogrado)	Platinum	Platinum	
					Graphite	Graphite	
					Superoxide of Manganese		

The positions occupied by the several materials with the first five solutions do not differ so much as to invalidate the theory of potential series. We know from thermo-electric experiments that not only the chemical condition but the molecular condition of the metals (what may be called their tempering) affects their position in the potential series : remembering this fact, and also the effect of polarization, no surprise need be felt at the change in relative position of cadmium and tin, or bismuth and antimony. We perceive also a general law that the materials having the greatest affinity for one constituent of the solution are most electro-positive, and this agrees with the chemical theory of electromotive force already stated. The last potential series obtained with a yellow solution of sulphide of potassium is quite anomalous and inconsistent with the simple potential series theory.

§ **6.** The true absolute values of the electromotive force produced by unpolarized single fluid elements are not accurately known, and owing to the polarization produced by any current cannot be determined by galvanometric observations. This is of less consequence, because, not being constant, the value of this electromotive force could not be used in any formulæ depending on Ohm's law. The available electromotive force in a Smee's element is about 0·47 of a volt.

The solution employed has little effect on the electromotive force, but has a great effect on the resistance. Pure water has a much higher resistance than any of the solutions employed in batteries : hence a cell with pure water or nearly pure water will give only a very feeble current through an external circuit of small resistance ; when salt, or sulphuric or nitric acid are added, the current is increased at once. This is due merely to the change in the total resistance of the circuit, not to any increase of electromotive force. A solution of sulphuric acid and water containing thirty per cent. of sulphuric acid has a smaller resistance than a solution with either less or more sulphuric acid; but, when used to charge a battery, it gives rise to useless oxidation of the zinc—useless because it produces no current outside the cell. Much weaker solutions, of about one part in twelve, are therefore commonly employed; solutions of common salt and of sulphate of zinc are also employed to charge the battery; the first because of its small resistance and the second because the action of the cells causes no change in the constituents of the solution.

§ 7. Some useless oxidation of the zinc or other electrode which is consumed in the cell almost always occurs, and is due to what is called local action. This local action arises from inequalities in the condition of the zinc exposed to the liquid. These inequalities cause certain points of the zinc to be electro-negative to certain other points. These points being in metallic connexion through the mass of the zinc constitute with the fluid a galvanic cell of small E. M. F., but also of very small resistance, and a current is produced in a local circuit as indicated by arrows in Fig. 106: that portion of the zinc which is most electro-positive is eaten away, and the current produced is confined to the cell, and cannot be utilized. This local action is very much increased by diminishing the resistance of the fluid. It is much diminished by amalgamating the surface of the zinc. This is done by cleaning the surface of the zinc plates with dilute sulphuric or hydrochloric acid, and then rubbing a little mercury over the surface with a brush. The surface being then composed of a uniform material not susceptible of those differences of temper described by the words 'hard' and 'annealed' is not attacked

FIG. 106.

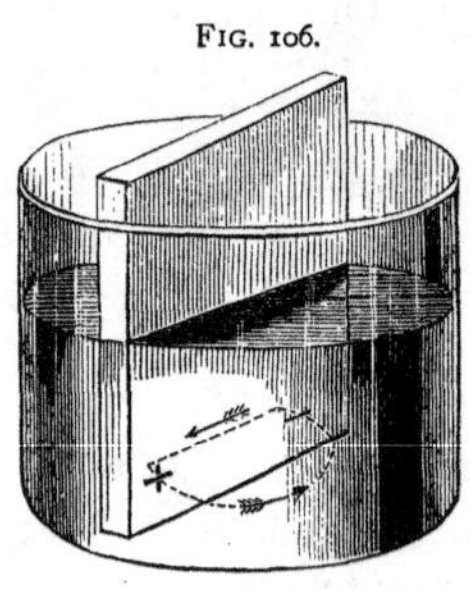

by the solution until the external circuit is closed : no zinc is consumed except in producing useful currents. Several forms of battery are in use in which the zinc plate is kept permanently in contact with a small supply of mercury.

§ **8.** Single fluid batteries are subject to another inconvenience besides that of polarization; the solution usually employed cannot by any convenient means be kept in uniform condition. For instance, the sulphuric acid used in most forms of the cell is gradually used as well as the zinc, so that the resistance of the battery is perpetually increasing, and the cell requires from time to time to be refreshed, as it is termed, by the addition of sulphuric acid. Single fluid batteries are subject, therefore, to three defects : their electromotive force is enfeebled by polarization; it is not constant; and their resistance is not constant.

§ **9.** All these defects are remedied in the two fluid batteries, of which the *Daniell's cell* was the first invented, and is a good typical example. In the most constant form of this cell, the zinc is plunged in a semi-saturated solution of sulphate of zinc, the copper in a saturated solution of sulphate of copper, and these two solutions are separated

FIG. 107.

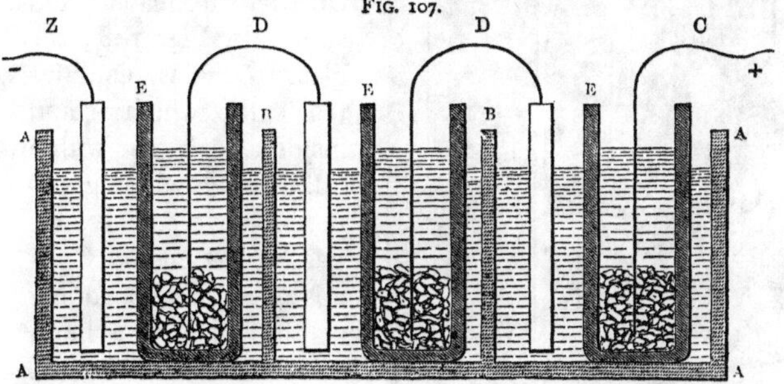

either by a porous earthenware barrier or by taking advantage of the different specific gravities of the two solutions. Fig. 107 shows three Daniell's arranged with porous cells, as used in telegraphy. The glass trough A A has glass partitions B B, which separate it into distinct cells, insulated from one another. In these cells stand the porous earthen-ware pots E E E, containing a saturated solution of sulphate of copper, and surrounded by a semi-saturated solution of sulphate of zinc. A thick plate of zinc is joined by a connecting strap to a thin plate of copper at D D; the coppers stand in the porous cells, the zincs in the sulphate of zinc. The terminal plate of copper C forms the positive pole of the battery, and the terminal zinc Z has a copper wire soldered to it, which forms the negative pole. In one common form, called Muirheads, and shown in Fig. 108, the glass trough A A contains ten cells, which stand inside a teak case with a lid, through which gutta-percha-covered wires pass at the ends. Crystals of sulphate of copper of the size of a hazel nut are placed in the porous cells to maintain the solution in a saturated condition. The copper connecting strap is cast in the zinc, having been tinned to ensure adhesion. The plates

FIG. 108.

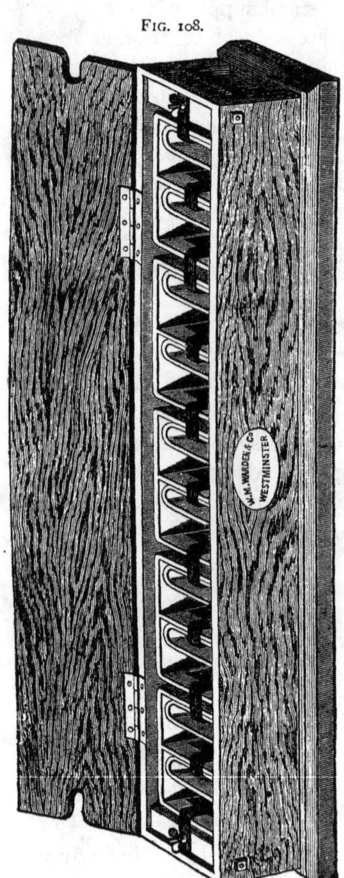

may be four inches long, and two inches wide, and the copper plates about four square inches. The zinc should hang on the upper part of the cell, and not reach to the bottom.

§ **10.** The chemical action in the Daniell's cell when in perfect working order has already been described, chap. xi. § 9; the result of the series of actions there described is that the sulphuric acid and oxygen of the sulphate of zinc are transmitted to the zinc, combine with it, and form fresh sulphate of zinc; the sulphuric acid and oxygen of the sulphate of copper are transmitted to the zinc, set free by the above process, and reconvert it into sulphate of zinc; the copper of the sulphate of copper is transmitted to the copper electrode, and remains adhering to it. The whole result is therefore the substitution of a certain quantity of sulphate of zinc for an equivalent quantity of sulphate of copper. together with a deposition of copper on the copper or negative electrode. Sulphuric acid and oxygen have a stronger affinity for zinc than for copper, otherwise there would be no source of power in the substitution.

The result differs in two material respects from that given by single fluid batteries. 1. No free hydrogen appears at the copper electrode. It is impossible to say whether water is or is not decomposed at some stage of the process, but if it is, the oxygen and hydrogen recombine without becoming visible. In the single fluid batteries described, the oxygen of the decomposed water combines with the zinc or other electropositive metal, leaving the equivalent of hydrogen free. In the Daniell's cell no oxygen is required from the water, the supply coming from the sulphate of copper. Consequently no free hydrogen appears. 2. It is comparatively easy to keep the solutions in a sensibly constant condition. The sulphate of copper solution is maintained by the presence of crystals of sulphate of copper. The sulphate of zinc solution, if it be saturated in the first instance, simply deposits the sulphate of zinc which is formed. Practically it is found better to work with semi-saturated solution of zinc, because a crust of sulphate of zinc crystals forms at the edge of the saturated solution and this impairs the action of the battery if it touches the zinc, and injures the insulation of the battery by forming a conducting film all round the edges of the cell, and on the copper junction straps.

§ **11.** The Daniell's cell will give a constant electromotive force, and retain a nearly constant resistance, for weeks together. To ensure this result, the following precautions must be taken :

The solutions must be inspected daily and kept in the proper condition by the addition of crystals of sulphate of copper and the removal of sulphate of zinc solution, water being added to replace the liquid withdrawn. No sulphate of zinc or dirt must be allowed to collect at the lips of the cells. The zinc plate must not touch the porous cell, or copper will be deposited upon it, which will set up local action. The sulphate of copper must be free from iron. To detect iron, add liquid ammonia to the solution ; both copper and iron will be at first precipitated, making the solution appear cloudy; but as more ammonia is added the copper will be redissolved, forming a bright blue solution, and leaving the iron as a brown powder. No acid should be used to set the battery in action ; it should be charged with sulphate of zinc from the first (unless a very low resistance, not constancy, be the object in view). The plates should be clean. Copper plates, if dirty, may be cleaned by being made red hot, and dipped in weak ammonia. The card used in cotton factories is a good brush for batteries. Porous cells must be examined to see that they are not cracked ; if set aside for a time after being used, they must be kept moist, or the crystallization of the sulphate of zinc they contain will crack them. The solution of sulphate of copper must be watched to see that it does not rise in the porous cells so high as to overflow the edges. This action by which liquid is drawn from one side of the porous diaphragm to the other is called osmose. The resistance of the cell described above with very porous Wedgwood pots may perhaps not exceed 4 ohms; 6 or 10 ohms is a much more common resistance.

§ **12.** The various constructions of Daniell's cell are very numerous. When the cells are large, a separate glass or earthenware jar is generally used for each cell. The porous cells are cylindrical, and the zincs and coppers are likewise parts of cylinders. Sometimes the zincs and sometimes the coppers are placed inside the porous cell; but the zinc should always be in the largest receptacle. Sometimes

FIG. 109.

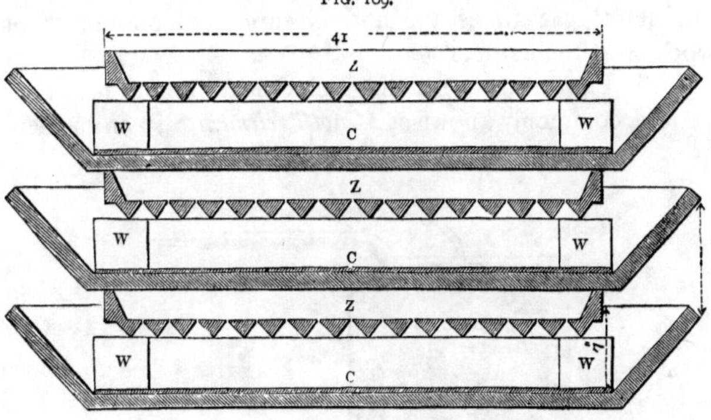

the copper electrode is made the jar to hold the sulphate of copper, the zinc being then inside the porous cell. This form of cell cannot be recommended, as the copper is frequently eaten away at the corners and allows the liquids to run out.

A more distinct form of the Daniell's cell is that in which the porous cell is replaced by *sawdust;* the copper lies at the bottom of the cell covered by crystals of sulphate of copper; on this sawdust is placed, moistened with the copper solution at the lower part of the cell and with the zinc solution near the top of the cell. On the top of all lies the zinc plate. This form of battery was first used by Sir William Thomson, who made the lower coppers in the form of trays, which rested directly on the zinc of the cell beneath. This form would be very convenient for plates of large size, if the copper were not occasionally eaten through. This defect he has remedied by making the trays of wood covered with lead, electrotyped with copper at the bottom. Fig. 109 shows three of these square trays, in which the zincs are forty-one centimètres long and broad. The trays are seven centimètres deep inside. The resistance of one of these cells is about 0·2 ohm.

The zinc is made in the form of a grating to allow bubbles of gas to escape, and is supported on blocks of wood W at the corners.

§ 13. Fig. 110 shows a slight modification of the sawdust battery, commonly known as *Menotti's element.** In an earthen-

FIG. 110.

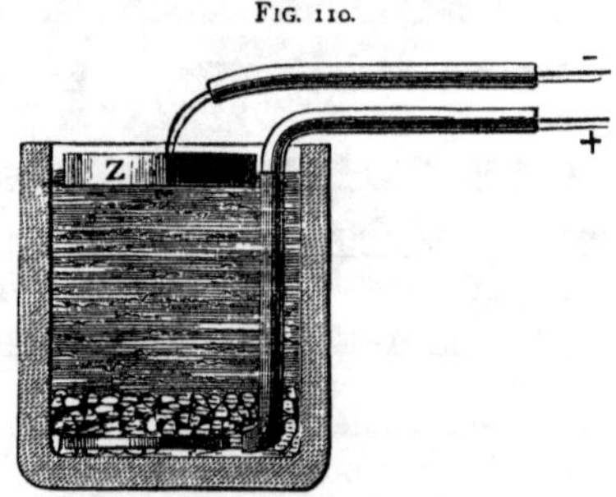

ware or glass cell, a flat circular plate of copper C is laid, with a piece of guttapercha-covered wire soldered to it; this wire comes out of the cell and forms the positive pole. The copper is covered with crystals of sulphate of copper and sawdust as above described, and the zinc lies on the top. A little oil is sometimes added to prevent evaporation.

* This element differs in no respect from one introduced for testing the Atlantic cable, by Sir William Thomson, in 1858.

The cells are usually about 10 centimètres diameter inside and 12 centimètres high. The metal plates are then made about $8\frac{1}{4}$ centimètres diameter This form of battery is portable, and has a constant E. M. F. Its resistance is high, being usually about 20 ohms when in fair condition. It is chiefly used for purposes connected with testing. The sawdust cells are well adapted for use at sea, where the wash of the solution tends to disturb the electromotive force and to produce variable polarization; for even in a Daniell's cell there is practically always some polarization.

Gravitation batteries are like the Minotti's with the sawdust removed. They must be kept perfectly still, and are found difficult to manage.

§ **14**. The following double fluid batteries are in practical use :—1. *Marie Davy's element*, which consists of a carbon electrode in a paste of proto-sulphate of mercury (Hg_2SO_4,) and water contained in a porous pot, and a zinc electrode in dilute sulphuric acid, or in sulphate of zinc. The chemical action is similar to that of the Daniell's cell; sulphate of zinc is formed, and mercury deposited at the carbon electrode.

The sulphate of mercury is apt to rise by capillary action to the junction of the carbon and copper; it then attacks the copper and destroys the continuity of the circuit. This is prevented by filling the pores of the charcoal at the top with melted paraffin; the sulphate of mercury is expensive, but very little mercury need be wasted, and it is easily re-converted into proto-sulphate. This material is poisonous. The E. M. F. of this element is about 1·5 volts, but its resistance is greater than that of a Daniell's cell.

2. *Grove's cell.*—This well-known and very useful element consists of a platinum electrode plunged in nitric acid, more or less diluted, and a zinc electrode plunged in sulphuric acid diluted with about twelve parts of water: the two solutions are separated by a porous cell. The zinc is converted into sulphate of zinc, the oxygen required being

obtained from the water; the hydrogen is prevented from remaining free at the platinum pole by forming, with the nitric acid, water and hyponitrous acid gas. This gas is in part dissolved, and in part appears as nitrous fumes. These fumes are not only disagreeable, but poisonous. The electromotive force of this battery varies from nearly two volts, when the nitric acid is concentrated and the sulphuric acid solution has the specific gravity 1·136 (20 parts sulphuric acid in 100 by weight), to 1·63 volt, when the nitric acid solution has the specific gravity 1·19 (26·3 parts N_2O_5 in 100 solution), and the sulphuric acid the sp. gr. 1·06 (9 parts in 100 by weight).

With the zinc in sulphate of zinc, and the nitric acid solution sp. gr. 1·33, the E. M. F. is 1·67.

With the zinc in solution of common salt, and nitric acid sp. gr. 1·33 (45 parts in 100), the E. M. F. is 1·9 volt.

The E. M. F. of this cell is very high, but its great merit is its low resistance which may with moderate-sized cells be reduced to $\frac{1}{4}$ of an ohm. The resistance of a cell constructed as follows was ·212 ohm; area of zinc plate 27·3 sq. in.: area of platinum plate 13·8 sq. in.; sp. gr. of sulphuric acid 1·06; nitric acid 26·3 parts by weight in 100 of solution. The double E. M. F. is easily got by doubling the number of Daniell's elements, but the size of these elements must be immensely increased to reduce the resistance to that of a small Grove's cell.

3. *Bunsen's cell.*—This element is exactly similar to Grove's, except that the platinum is replaced by porous carbon. In both Bunsen's and Grove's cells the zinc must always be amalgamated, or the local action causes intolerable fumes and waste of zinc. The electromotive force of Bunsen's cell is rather greater than that of Grove; but the resistance is also greater, and there is occasionally difficulty in securing a good contact between the carbon electrode and the metallic strap or wire used to connect it with the next zinc or with the terminal of the battery. The carbons are

specially prepared for all carbon batteries, and vary much in quality. The upper part of the carbon should be impregnated with stearine to prevent the junction from being corroded.

Faure puts the nitric acid inside the carbon pole, which is made in the form of a bottle closed by a carbon stopper. The carbon performs the double part of porous pot and electrode. The nitrous fumes rise inside the bottle, and by their pressure assist in forcing the nitric acid through the porous carbon.

The resistance of an ordinary Bunsen's element 12 centimètres high with the carbon outside the zinc is given by Blavier as equal to from 2 to 3 ohms when partially charged, but to double this amount after a few hours.

4. The *Chromate of potassium* element is thus described by Mr. Latimer Clark: 'Prepare two solutions, 'the first to be made by dissolving 2 ounces of bichromate of 'potash in 20 ounces of hot water, and when cold add 10 'ounces of sulphuric acid. As this addition will cause the 'solution to become warm, it must be allowed to cool before 'being used. The second is a saturated solution of common 'salt. To charge the battery with these solutions the 'bichromate solution must be poured into the porous jar 'containing the carbon, until it reaches about half an inch 'from the top; then pour the salt solution into the outer 'vessel containing the zinc until it reaches the same level.'

The electromotive force is said to be 2 volts.

The chlorine of the common salt unites with the zinc, forming chloride of zinc, while at the carbon electrode the sodium replaces hydrogen in sulphuric acid, forming sulphate of sodium. The nascent hydrogen reduces chromic acid (produced by the action of sulphuric acid on the bichromate of potash), so that sulphate of chromium is produced. In chemical notation,

$$3Zn;\ 6NaCl;\ 6H_2SO_4;\ 2CrO_3$$

gives $$3ZnCl_2;\ 6H_2O;\ 3Na_2SO_4;\ Cr_2(SO_4)_3.$$

Q 2

5. The *Leclanché battery*; a zinc carbon element. The zinc is plunged in a solution of ordinary commercial sal ammoniac, and the carbon is tightly packed in a porous pot, with a mixture of peroxide of manganese and carbon, in the form of a coarse powder. Its E. M. F. is about 1·48 volt. The zinc unites with chlorine, forming chloride of zinc; ammonia is set free at the negative electrode, while the nascent hydrogen from the ammonium reduces the peroxide of manganese to sesquioxide. The chemical notation of the change is that Zn; $2NH_4Cl$; $2MnO_2$ is changed into $ZnCl_2$; H_2O; $2NH_3$; Mn_2O_3.

6. Mr. Latimer Clark's cell of constant electromotive force; this element has already been described, Chap. X. § 2.

§ **15.** With all batteries it is of the utmost importance that during any delicate experiments the whole battery should be perfectly insulated, and each cell perfectly insulated from its neighbour. For telegraphic purposes this is less essential, but it is always desirable. When a battery gives no current or a much feebler current than was expected, the following are defects which should be looked for: 1, solutions exhausted; for instance, sulphate of copper in the Daniell's cell entirely or nearly gone, leaving a colourless solution; 2, terminals or connections between the cells corroded, so that instead of metallic contact we have oxides of almost insulating resistance intervening in the circuit; 3, cells empty or nearly empty; 4, filaments of deposited metals stretching from electrode to electrode. Intermittent currents are sometimes produced by loose wires or a broken electrode which alternately makes and breaks contact when shaken. Inconstant currents are also produced when batteries are shaken, unless they are in first-rate condition: the motion shakes the gases off the electrodes, increasing temporarily the E. M. F.

CHAPTER XVI.

MEASUREMENT OF RESISTANCE.

§ **1.** IN order to measure a resistance we must compare it with a standard recognised as the *unit* of resistance. In telegraphy the measurement of resistance plays a very important part, regulating the choice of materials and enabling the electrician to test the quality of goods supplied. The *ohm* (Chap. X. § 4) is the unit of resistance almost universally adopted in this country. Multiples and submultiples of the ohm are so arranged in boxes of resistance coils that any given resistance from one ohm to 10,000 or 100,000 ohms can be readily obtained for comparison with any other resistance. The general arrangement of these boxes is shown in the diagram, Fig. 111.

FIG. 111.

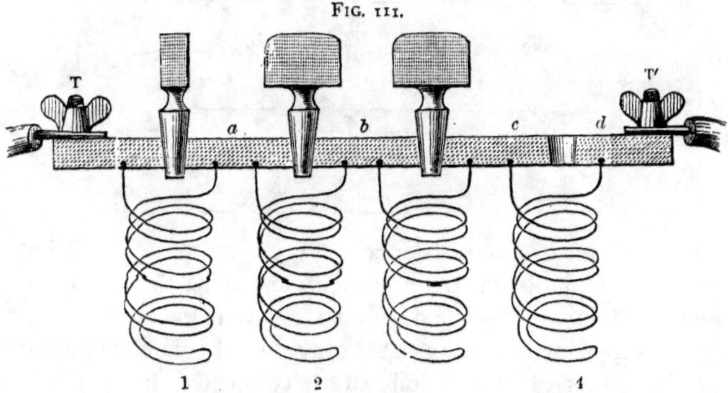

Between two terminal binding screws T and T_1 secured on a vulcanite slab, are fixed a series of brass junction pieces, *a*, *b*, *c*, *d*; each of these is connected by a resistance coil to its neighbour, as shown at 1, 2, 3, and 4. A number of brass conical plugs with insulating handles of vulcanite are provided, which can be inserted between any two successive junction pieces, as between T and *a*, or *a* and *b*. Conical holes are bored for this purpose at the opening between the junction pieces. When the plugs are withdrawn, no electrical connection exists between the junction pieces except through the coils.

Let us assume that the resistance of the first coil is one ohm, that of the second two ohms, that of the third three ohms, and that of the fourth four ohms. Then if the plugs are arranged as in the figure the whole resistance between T and T_1 will be 4 ohms, because the resistance of the large metallic junction pieces directly connected by plugs would be insensible between *c* and T. If all the plugs

Fig. 112.

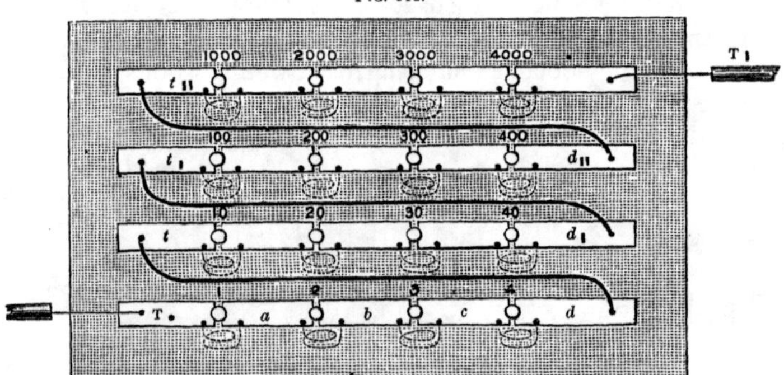

were withdrawn, the resistance between T and *d* would be 10 ohms, and obviously by properly arranging the plugs we could obtain any resistance from 1 to 10 between T and *d*. Now suppose that *d*, instead of being the final terminal of the set of resistance coils, were connected by a thick copper bar to *t* as in Fig. 112, showing a plan of the lid of the box containing the coils; and that a similar series of junction pieces were used to connect coils of 10, 20, 30, and 40 ohms, precisely as *a, b, c,* and *d* connected the coils 1, 2, 3, and 4; then between *t* and d_1, if all the plugs were out, we should have a resistance of 100 units, but by inserting the proper plugs we could at will have 10, 20, 30, 40, 50, 60, 70, 80, or 90 units. Thus for 80 units, withdraw the 1st, 3rd, and 4th plug, giving 10 + 30 + 40 or 80 units. Now between d_1 and T we can obviously by proper plugging obtain any number of units between 1 and 110; d_1 is connected by a thick bar with t_1, the last of five junction pieces joining coils of 100, 200, 300, and 400 units, by means of which, between d_{11} and T, we can get with the twelve plugs any number of units from 1 to 1110; similarly with four more junction pieces and four more coils we have between T_1 and T, the final terminals of the box, a series of sixteen coils and sixteen plugs, by the proper arrangement of which we can between T and T_1 obtain any number of units of resistance from 1 to 11110; when all the plugs are in their places the resistance between T and T_1 ought to be very small relatively to the resistance of one ohm; and, if this is not the case, the plugs and holes must be well cleaned, as any resistance observed when all the plugs are in, can only be due to imperfect metallic contact between the holes and plugs.

§ 2. Many other arrangements of resistance coils may be adopted. Thus, instead of the 1, 2, 3, 4 series, we might have had ten equal coils in each row of junction pieces, but this would have required 40 plugs instead of 16. We might also have arranged ten coils in a circle, and joined them to 11 equidistant junction pieces, as in Fig. 113. Then the resistance between the wires t and t_1 would be 2 if the arm A was on the second stop, or 5 if on the fifth stop. The end of the arm A may be so arranged that, before leaving one junction piece, it makes contact with the next, so that the circuit between t and t_1 is never wholly broken.

In all boxes of resistance coils the following precautions should be observed during the manufacture. Large gauges of wire should be used for the smaller coils instead of short pieces of fine wire. Better adjustment and less liability to derangement by a powerful current is thus obtained. The metal used for the wire must be such that its resistance varies little with changes of temperature. German silver is a good material. The wires should be insulated with

FIG. 113.

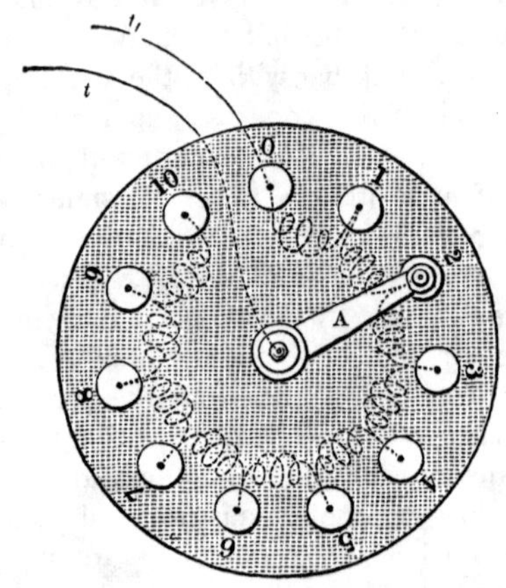

two coatings of silk saturated with solid paraffin or other suitable insulating mixture. No solderings should be permitted inside the coils—above all, no solderings in making which acid is used. The wire should be wound double, so that the current makes as many turns from left to right as from right to left. There is no self-induction (Chap. III. § 21) in a coil so wound, nor does the current affect galvanometers in the neighbourhood. The junction pieces must be firmly fixed, well insulated, and so formed that the vulcanite on which they stand can be easily cleaned. It is a good plan to make the bobbins hollow, and rather of large than small diameter, to promote uniformity of temperature. All the bobbins should be in one box.

FIG. 114.

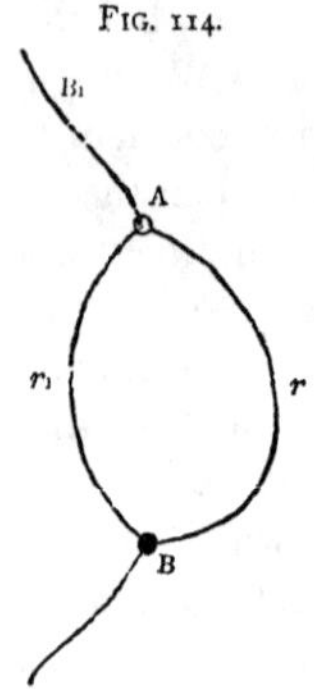

§ **3.** Let two points A and B, Fig. 114, be joined by two conductors having resistances r and r_1, these conductors are said to be joined in *multiple arc*; with a difference of potentials 1 between A and B.

The current C through r will be equal to $\frac{1}{r}$, and similarly the current through r_1 will be $\frac{1}{r_1}$; the whole current between A and B will be $\frac{1}{r} + \frac{1}{r_1}$, or $\frac{1\,(r_1 + r)}{r r_1}$; this current will be the same as if A and B had been joined by a single resistance equal to $\frac{r r_1}{r + r_1}$, which is therefore the resistance of the two conductors joined in multiple arc. With three wires r, r_1, r_{11}, connecting the same points by a multiple arc, the resistance between A and B will be $\frac{r\, r_1\, r_{11}}{r\, r_1 + r_1\, r_{11} + r\, r_{11}}$

If a galvanometer with the resistance G be shunted by a shunt of the resistance s, the resistance of the shunted galvanometer will be $\frac{G\,s}{G + s}$. Let $u = \frac{G + s}{s}$, then the sensibility of the shunted galvanometer will be to that of the unshunted galvanometer as 1 to u; then calling C the current flowing in other parts of the circuit, $\frac{C}{u}$ will flow through the galvanometer, and $\frac{G}{u\,s}$C will flow through the shunt; the resistance of the shunted galvanometer will be $\frac{G}{u}$.

Example.—We have a galvanometer with a resistance of 8,000 ohms, and wish to find the shunt which will reduce its sensibility 100 fold, $u = 100 = \frac{8{,}000 + s}{s}$, or $s = \frac{8{,}000}{99} =$ 80·8.

The resistance of the galvanometer when shunted will be $80 = \frac{8{,}000}{u}$.

§ **4.** *Definition.*—The *conductivity* of a given wire or conductor is the reciprocal of its resistance.

That is to say, if a be the resistance of the wire, $\frac{1}{a}$ is its conductivity; if the resistance of a conductor is 10 ohms, its conductivity is 0·1.

The conductivity of a number of wires joining two points in multiple arc is the sum of the conductivities of the several wires. For the current in each wire with a unit difference of potential between the ends is $\frac{1}{r}$.

The sum of all the currents is

$$\frac{1}{r} + \frac{1}{r_1} + \frac{1}{r_{11}} + \ldots \frac{1}{r_n},$$

which is the same current as if a single conductor joined the two points with a conductivity of $\left(\frac{1}{r} + \frac{1}{r_1} + \frac{1}{r_{11}} \ldots + \frac{1}{r_n}\right)$ The resistance of the wires in multiple arc is the reciprocal of the conductivity of the multiple arc. This rule gives the same expression for the resistance as is given in § 3.

Example.—Let two points be joined by wires in multiple arc with resistances of 2, 18, 27, and 64 ohms respectively.

The conductivities are 0·5, 0·05555, ·03704, ·01562. The sum of the conductivities is 0·6082; and the resistance of the four wires in multiple arc $= \frac{1}{\cdot 6082} = 1\cdot 644$ ohms.

§ **5.** We may compare one resistance with another by comparing the deflections produced by a given battery through the same galvanometer, but with the different resistances in circuit. Thus, let G be the galvanometer resistance, B the battery resistance, R a resistance chosen at pleasure from those at our disposal in the box of resistance coils, and x the unknown resistance which we wish to measure or compare with R. Let us first observe the deflection d obtained with a circuit containing G, B, and R only, arranged in any order, and next the deflection d_1, obtained with G, B, and x only in circuit: then, if the galvanometer be a mirror galvanometer, the deflections of which are proportional to the currents flowing through it, we have, by Ohm's law, the proportion

$$G + B + R : G + B + x = d_1 : d;$$

for the E. M. F. being the same in both cases, the currents and therefore the deflections must be inversely proportional to the total resistances. From the above we find

$$x = \frac{d}{d_1}(G + B + R) - (G + B) \ . \ . \ . \ 1^\circ.$$

When G and B are so small that they can be neglected relatively to R, we have approximately $x = \frac{d}{d_1}R$. This case seldom arises; but frequently, as, for instance, when x is the resistance of some insulating substance, we may neglect G + B as insensible relatively to x, and then we have

$$x = \frac{d\,(G + B + R)}{d_1} \ . \ . \ . \ 2^\circ.$$

The number $d\,(G+B+R)$ is in telegraphy called the *constant* of the instrument with the given battery. If $d_1=1$, we shall have the whole resistance of the circuit $x = d(G+B+R)$; hence the constant is often defined as the resistance of the circuit with which the given battery would give the deflection 1. Obviously when a tangent galvanometer is used, we must write tan d and tan d_1 in the above formulæ instead of d and d_1; and if a sine galvanometer is used, we must write sin d and sin d_1.

§ **6.** By the use of shunts the application of this method is greatly extended; calling the resistance of the shunt S, the resistance of the shunted galvanometer becomes $\frac{G\,S}{G+S}$; hence if the shunt be used when both d and d_1 are observed, we must in equation 1 substitute $\frac{G\,S}{G+S}$ for G, the only effect being to diminish the resistance of the galvanometer; but when d is observed with the shunt in, and d_1 without the shunt, the sensibility is different in the two cases. Then let the ratio $\frac{G+S}{S}$ be called u as before; we have by Ohm's law:—

$$\frac{G}{u} + B + R : G + B + x = d_1 : u\,d.$$

$$\text{or } x = u \frac{d}{d_1}\left(\frac{G}{u} + B + R\right) - (G + B) \quad \ldots \quad 3^\circ.$$

The constant of the unshunted galvanometer or resistance for which $d_1 = 1$, is $u\ d\ (\frac{G}{u} + B + R)$.

Thus with a shunt reducing the sensibility 100-fold, a deflection of 90 divisions, with $G = 8000$, $B = 20$, and $R = 4000$, the constant will be 36,900,000; this will be the whole resistance of the circuit including G and B with which the battery used would give the deflection 1 on the galvanometer used without a shunt. In practice R is chosen so that $\frac{G}{u} + B + R$ may be some whole convenient number; thus in the above case an experienced observer would have made $R = 3900$ when the constant would have been 36,000,000. A series of shunts are usually sold with each galvanometer of such resistance that u may by them be made 10, or 100, or 1000 at pleasure. The constant is determined at the beginning of the experiment when the galvanometer is not shunted, and the value of the resistance in circuit giving a deflection d_1 is obtained by simply dividing u times the constant by d_1. To get x, the resistance of $G + B$ must be subtracted from the whole circuit, but when the sum of $G + B$ is small, this subtraction is often omitted.

This mode of measuring a resistance is much used in testing insulating materials, such as gutta-percha. The battery and wire covered with gutta-percha are arranged as in Fig. 115.

The negative current flows from Z through the shunted galvanometer to the copper wire inside the gutta-percha; then through the gutta-percha to the water in the tub T, and from T to the copper pole of the battery. The resistance x is the resistance of the gutta-percha.

FIG. 115.

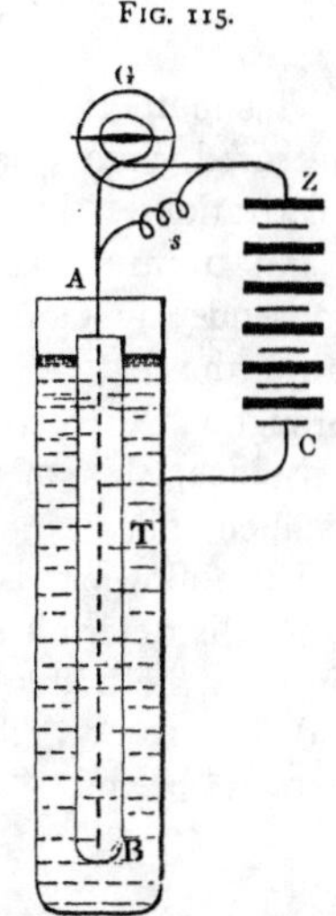

§ 7. The value of R in the above equations is always known, and the value of G or of $\frac{G}{u}$ is also generally known, and can always be directly determined by experiment; for instance, it may be measured as any other resistance would be measured, a second galvanometer being used for the purpose. The value of B should be determined at least once a day, since the resistance of any battery is found to vary considerably from day to day. There are several methods of determining the value of B. The following is the most common:—

Make a circuit consisting of the battery B, the galvanometer G, and a set of resistance coils R; shunt the galvanometer with a piece of short thick wire connecting the terminals; put all the plugs of the resistance coils in their places so as to reduce R sensibly to zero; let the wire shunt be so short and thick as to have no sensible resistance relatively to the battery, but adjust it of such length that a sensible deflection D is shown by the galvanometer; the greater part of the current is shunted, but enough goes through the sensitive galvanometer to give the deflection D; under these circumstances the whole resistance of the circuit is B, that of the battery, for R is reduced to nothing, and the resistance of the shunt is insensible; now increase the resistance in the box to R by taking out plugs until a deflection D_1 is obtained; then

$$R + B : B = D : D_1, \text{ or } B = \frac{R\,D_1}{D - D_1} \quad . \quad . \quad . \quad . \quad . \quad 4^\circ$$

$$\text{If } D_1 = \frac{D}{2} \text{ we have } B = R$$

This method has the defect that the battery resistance is measured when a powerful current is passing, increasing the polarization. Moreover the current is very different when D and D_1 are taken, and the polarization very different. Consequently Ohm's law is seldom strictly applicable, because the E. M. F. of the battery is not strictly constant throughout the experiment. With a battery of very small resistance, this method would be liable to injure the resistance coils.

The following is a second method by which the sum $B + G$ is determined. Observe two deflections D and D_1 given by the battery when the two circuits are $B + G + R$ and $B + G + R_1$; then we have $G + B + R : G + B + R_1 = D_1 : D$, and

$$G + B = \frac{R_1\,D_1 - R\,D}{D - D_1} \quad . \quad . \quad . \quad . \quad 5^\circ$$

Mr. Varley recommends that three deflections D, D_1, and $D_{1,}$, be taken with additional resistance R, R_1, and R_{11} for the purpose of testing whether polarization interferes much with the experiment; if there be no polarization. adjusting the values of R, R_1, and R_{11} so that $D_{11} = 4\,D$ and $D_1 = 2\,D$, we should have $R = 3\,R_1 - 2\,R_{11}$.

The following is a third method. Arrange the connections as in Fig. 116; let D be the deflection when the circuit is $B + R + G$; next insert the shunt of known resistance S, by making contact at *a*; reduce R to R_1 until the deflection is the same as before, then

$$B = S\frac{R - R_1}{G + R_1} \quad . \quad . \quad . \quad . \quad 6^\circ$$

or, fourthly, leaving R unaltered, let D_1 be the deflection observed when contact is made at *a*; then

$$B = S \frac{D - D_1 \frac{G}{G + S}}{D_1 - D \frac{S}{G}} \quad . \quad . \quad . \quad 7^\circ$$

or approximately

$$B = S \frac{D - D_1}{D_1} \quad . \quad . \quad . \quad . \quad 8^\circ$$

This method is especially applicable to batteries of very small resistance.

§ **8.** The accuracy with which a resistance can be measured by any of the above methods is limited by the accuracy with which a deflection can be observed. If we cannot make certain that any deflection is correct within one per cent., still less can we feel confident that the resistance calculated from the deflection is correct within one per cent.

FIG. 116. FIG. 117.

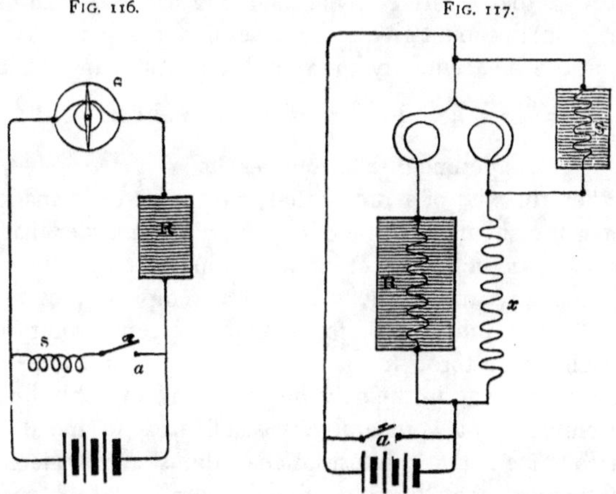

The following methods, which may all be termed differential methods, admit of much greater accuracy. The simplest differential method has already been described (Chap. IV. § 3), and the arrangement of the connections is shown in Fig. 117. With a sensitive galvanometer it admits of extreme accuracy, for by increasing the battery power we may increase at pleasure the deflection which the difference between the currents in the two branches produces. We may also shunt either branch of the galvanometer so as to reduce its resistance and sensibility u times. Calling the resistance of each branch of the instrument G, we then have, when the galvanometer is undeflected on completing the circuit, assuming that the known resistance is connected with the shunted branch of the galvanometer,

$$R + \frac{G}{u} : x + G = 1 : u$$

$$\text{or } x = u\,R \quad . \quad . \quad . \quad . \quad 9^{\circ}$$

If, as in the figure, x is connected with the shunted branch we have $x = \frac{R}{u}$.

Resistances one thousand times greater or one thousand times less than R, are easily measured in this way. In order that the plan should give accurate results, it is necessary that the ratio u be accurately known and that it remain constant. Now $u = \frac{G + S}{S}$; and if the resistance of either G or S varies during the experiment fallacious results are given.

When the wire of a differential galvanometer is made of copper the shunts must be of copper also, in order that the ratio u may be constant at all temperatures; but even with this precaution, the very current employed in testing disturbs the value of u, for a much larger current flows through S than through G, and hence more heat is generated in the shunt than in the galvanometer coil, and this heat is concentrated in a comparatively small mass of metal; the consequence is that the resistance of the shunt is increased relatively to that of the galvanometer by every current which passes, and this seriously impairs the value of the method. Differential galvanometers made of German silver give much more accurate results than copper wire instruments, because their resistance and that of their shunts are less affected by temperature. The circuit should be completed for the shortest possible time by making contact with a key at *a*, and breaking it as soon as a deflection to right or left has been observed. This may, however, lead to error if the unknown resistance *x* is so formed that any self-induction can take place, or if *x* has any sensible electrostatic capacity like a gutta-percha-covered wire in water. In either of these cases the currents in the two branches will not increase at the same rate when contact is first made. Assuming the coils in R to be properly wound, while *x* is a simple bobbin of wire not wound double, the current in *x* will lag behind, and hence a momentary contact at *a* will always show *x* as greater than R when it is really equal to it. The first jerk of the galvanometer needle must, in this case, be neglected, and *x* measured by means of the permanent deflection arrived at after the currents in the various branches have become constant.

§ 9. When a steady current C through a resistance R is due to a difference of potentials I between the ends of the conductor, then the difference of potentials *i* between any two intermediate points separated by a resistance *r* must be equal, by Ohm's law, to *r* C; the smaller the resistance between the two points the less the difference of potential between them, and if one end of the conductor be at zero potential or uninsulated, the potential of any point in the conductor will be proportional to the resistance *r* between the earth and the point in question, and equal to *r* C. In the diagram Fig. 118, if the line A E represents to any scale the

FIG. 118.

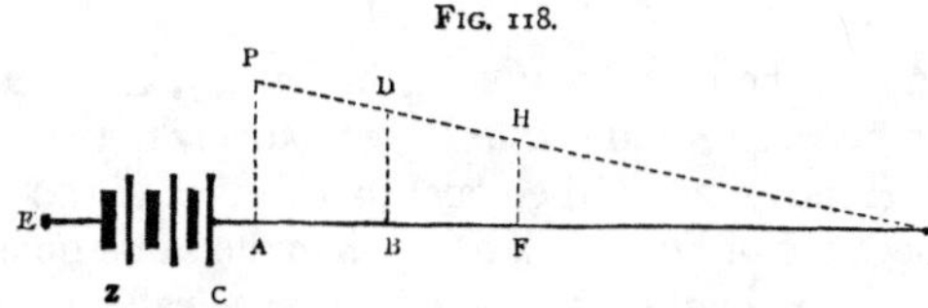

length of a uniform conductor separating the battery C Z from the earth at E, and if the line or ordinate P A represent the E. M. F. of the battery to any scale: then, joining P E by a straight line, the ordinate F H will represent the potential of the conductor at the point F. If, for instance, P A is equal to 12 volts, and F is half-way between A and E, F H will be equal to 6, and 6 volts will be the potential of the conductor at that point.

If A were separated from E by several conductors of different resistances, we must draw A E so as to represent the *total resistance* instead of the mere length of the conductor. Then as before, F H will represent the potential at the point F, separated from E by a *resistance* equal to F E; if F is so placed that the resistance of F E is equal to that of A F, the potential at F will be half that at A, in whatever manner the resistances A F and F E are made up.

The difference of potential between B and F is equal to the difference of the length of the lines B D and F H.

Let us now suppose that the two ends of the resistance A D B are joined to the two poles of an insulated battery, and that at the middle of the resistance at D the conductor is connected with earth, Fig. 119. The potential here will be zero; but the difference of potentials between A and B must be

FIG. 119.

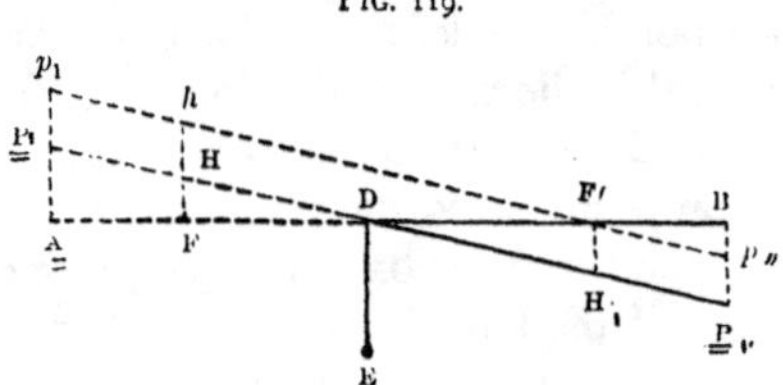

equal to nearly the whole E. M. F. of the battery, assuming the resistance between A and B to be large relatively to that of the battery. Hence A P_1 will be equal to half that E. M. F., and B P_{11}, a negative ordinate, will be equal to the same quantity. The sum of the lengths B P_{11} + A P_1 = A P, calling A P the E. M. F. of the whole battery as shown in Fig. 118. The ordinates F H and F_1 H_1 show the potentials at points F and F_1, one positive or measured upwards, the other negative or measured downwards. The difference of potentials between F and F_1 is the sum of H F and H_1 F_1. This difference will be exactly the same whatever point of A B be put to earth, if the battery is insulated. If F_1 were put to earth instead of D, then $p_1 p_{11}$ would be the line showing the potentials, and F h, the difference of potentials between F and F_1, is equal to F H $\div$ F_1 H_1.

§ **10.** Let us assume that the same difference of potentials is maintained by a battery between the ends of two conductors of different resistance represented by the lines A E and A_1 E_1, Fig. 120, and for simplicity's sake we will further assume that the potential at E is zero. If we now choose any two points B and B_1 so placed that A B : B E = A_1 B_1 : B_1 E_1, we shall have the line B D equal to B_1 D_1, showing that the potentials of B and B_1 are equal. Hence, if we join B and B_1 by a conductor, no current will flow from B to B_1; and if a galvanometer G were inserted in the wire joining B and B_1, it would remain undeflected, although the E. M. F. represented by A P and producing the currents through A E and A_1 E_1 might be very great and the galvanometer very sensitive. If, however, the wire or bridge, as it is called, joins B with a point in A_1 E_1 between B_1 and E_1, we shall have a current from A E which runs through the bridge; and on the contrary, if the bridge joins B with a point between B_1 and A_1, the

FIG. 120.

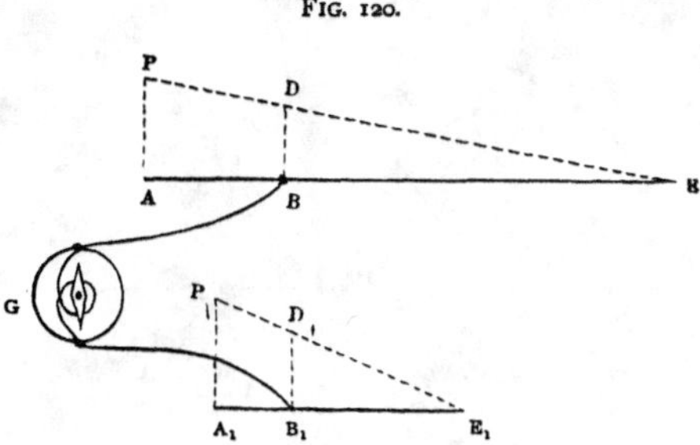

current will flow in the opposite direction through the galvanometer, i.e. from A_1 E_1 through the bridge.

If then we know the ratio A B to B E, as we shall do if

these two resistances are made up of graduated resistance coils, we shall be able to divide a resistance $A_1 E_1$ in the same ratio by simply seeking the point B_1 at which no current flows across the bridge.

And if A_1 B_1 is a known resistance, we can experimentally find a resistance $B_1 E_1$ which shall bear the same ratio to $A_1 B_1$ as B E does to A B.

§ **11.** The principles laid down in the two preceding sections give the most convenient method of measuring resistance. The *Bridge*, as it is technically called, is arranged as in Fig. 121.

Four conductors, A B, B E, A B_1, and B_1 E, are joined at A and E to the poles of a battery, the current from which flows round A B E and A B_1 E, corresponding to A B E and A_1 B_1 E_1 in Fig. 120. The difference of potentials between A and E depends on the battery used, but is obviously the same for the ends of the two circuits. The resistance between A and B we will call R; that between A and B_1, R_1; that between B and E, r; and that between B_1 and E, x the unknown resistance to be measured; R, R_1, and r are usually resistance coils.

A convenient constant ratio is chosen for R and r, such

FIG. 121.

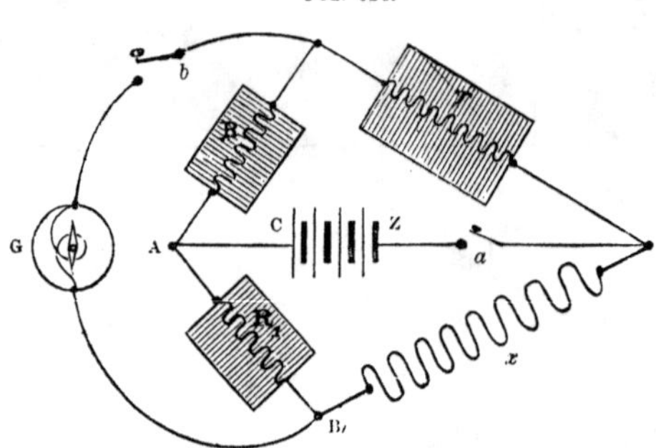

as equality, 1 to 10, 1 to 100, or 1 to 1,000; and then R_1 is adjusted until no current flows through the galvanometer G;

when this is the case we have $R : r = R_1 : x$ or $x = \frac{r}{R} R_1$; so that if $r = \frac{R}{100}$, x will be equal to $\frac{R_1}{100}$.

The convenience of this method is very great. Any galvanometer can be employed; but the more sensitive the instrument the more delicate the measurement of x. The constancy of the resistance of the galvanometer is of no consequence. The coils R, R_1, and r, are made of German silver or some other alloy varying little in resistance with a change of temperature. Two keys are inserted, one at *a* and one at *b*; the current is wholly cut off the four conductors until contact is made at *a*; and then, after the currents in the four conductors have come to their permanent condition, contact is made at *b* to test whether any current flows through the galvanometer. If none flows, making contact at *b* does not disturb the currents in the four conductors at all. R and r are usually so arranged as to give any decimal ratio between 1,000 to 1 and 1 to 1,000: the two keys at *a* and *b* are often arranged so that the same finger-piece moves both, making contact at *b* a little after contact has been made at *a*.

The three resistances R, R_1, and r, and the resistance of the galvanometer, should be small if x is small, and great if x is great. When x is very small, A B E is frequently made of a single wire of constant diameter; R_1 is kept constant, and the point B slipped along the wire A B E, until no current flows through G. Then the ratio of the resistances $\frac{r}{R}$ is the ratio of the actual lengths $\frac{BE}{AE}$ measured on a scale over which the wire A B E is stretched. An alloy of silver with 33·4 per cent. of platinum makes a good wire for this purpose. It must be a stout wire, or else the wear and tear of shifting the contact piece B will soon destroy the uniformity of its section and therefore of its resistance.

When x is small, great care is necessary to prevent the

resistance of mere connections between R, r, R_1, and x from being sensible. These connections may be made of stout copper rods $\frac{1}{2}$ centimètre diameter, and junctions made by dipping the ends of these rods in mercury cups, the ends of the rods being amalgamated.

The bridge is applied to measure the resistance of the gutta-percha sheath used to insulate the conducting wire of submarine cables: for this purpose E is connected with earth, the battery carefully insulated, and the wire to be tested is connected with B_1, but insulated at the other end instead of being connected with E; the insulated wire is submerged in an uninsulated tank or in the sea, and thus the only connection between B_1 and E is through the insulating cover or sheath. The resistance of this insulating cover is therefore the resistance x.

After the wires have been arranged thus we can, by joining the end of the conducting wire with E, measure the resistance of the copper conductor immediately before or after measuring the resistance of the insulator.

When no current flows across the bridge, the position of the battery and of the galvanometer may be interchanged, and no current will flow from A to E through the galvanometer.

§ **12.** *Kirchhoff's laws.*—If a number of currents $c_1\ c_2\ c_3\ \ldots\ c_n$ are flowing some to a point A (Fig. 122) and some from that point; then, since the whole quantity arriving at the point must be equal to that taken away, the sum of all the currents coming to the point must be equal to the sum of those going away from it: hence, calling the first series positive and the second series negative currents. the algebraic sum of all the currents must be equal to zero, a result written as follows,

$$\Sigma C = 0,$$

the letter Σ signifying that the sum of all the values of C are to be taken.

Let there be several sources $I_1\ I_2\ I_3$ of electromotive force in a circuit (Fig. 123), some acting in one direction and some in another, and joined by resistances $R_a\ R_b\ R_c$. Let the currents flowing through each be $C_a\ C_b\ C_c$. Let the difference of potential or E.M.F. between the two ends of R_a be $P_a - p_a$; that between the two ends of R_b, $P_b - p_b$; and that between the two ends of R_c, $P_c - p_c$.

Then by Ohm's law, $C_a R_a = P_a - p_a$, $C_b R_b = P_b - p_b$, $C_c R_c = P_c - p_c$ or $C_a R_a + C_b R_b + C_c R = (P_a - p_a + P_b - p_b + P_c - P_c) = (P_a - p_c) + (P_b - p_a) + (P_c - p_b)$.

Now $P_a - p_c$ is the difference of potentials produced by the electromotive force I_2; for however high or low the absolute value of the po-

FIG. 122. FIG. 123.

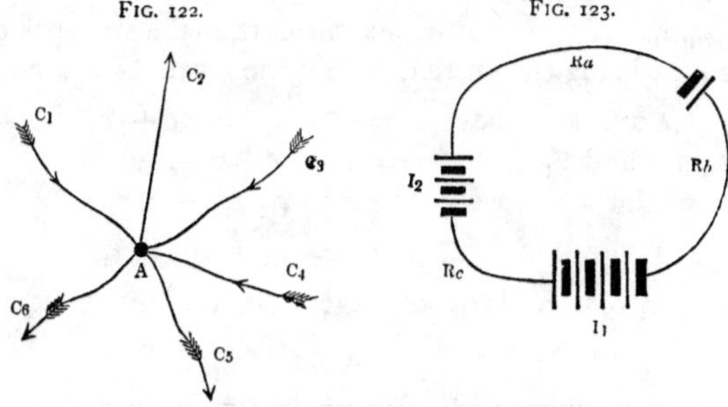

tentials P_a or p_c may be, by definition the difference of potentials must be equal to the electromotive force between them. Similarly $P_b - p_a = I_3$, and $P_c - p_b = I_1$:

hence $$I_1 + I_2 + I_3 = C_a R_a + C_b R_b + C R_c,$$

or $$\Sigma I = \Sigma C R \quad . \quad . \quad . \quad 9.$$

The sum of all the electromotive forces is equal to the sum of the products of each current into the resistance which it traverses.

One obvious application of this law of Kirchhoff's is to those cases in which the electromotive force in a circuit, instead of being due to a certain difference of potentials produced at one point of the circuit, as by a battery, is due to an E.M.F. distributed throughout the length of the whole or part of the conductor, as when the E.M.F. is due for instance to electromagnetic induction, where we only know for each part of the circuit that the E.M.F. is so much per centimètre of length. We now see that we need only add up all the electromotive forces in each unit of length, and then, knowing the whole E.M.F., we find that the current multiplied into the whole resistance of the circuit will be equal to the electromotive force thus calculated—in other words, Ohm's law is perfectly applicable to this case.

The results arrived at in sections 1 and 2 of this chapter are easily proved from Kirchhoff's equations.

§ **13.** The theory of the bridge may be proved as follows from Kirchhoff's laws :

Let five conductors r, r_i, r_{ii}, r_{iii}, r_{iv}, be arranged as in Fig. 124 with a battery I connected with A and E by conductors, as shown in the figure.

Let c, c_i, c_{ii}, c_{iii}, c_{iv}, C be the six currents, in the six parts of the circuit, c being the current in r, c_i the current in r_i, etc.

Then at A and E we have $C = c_i + c_{iii} = c_{ii} + c_{iv}$

„ at B and B_1 „ „ $c = c_i - c_{ii} = c_{iv} - c_{iii}$.

In the circuit A B B_i we have $c\,r = c_{iii}\,r_{iii} - c_i\,r_i$.

„ B E B_1 „ $c\,r = c_{ii}\,r_{ii} - c_{iv}\,r_{iv}$;

eliminating c_i c_{ii} c_{iii} and c_{iv} we have from the above equations :

$$c = \frac{r_{iii}\,r_{ii} - r_i\,r_{iv}}{(r_i + r_{iii})\,(r_{ii} + r_{iv}) + r\,(r_i + r_{ii} + r_{iii} + r_{iv})}\;C.$$

This gives the value of the current produced in the bridge r in terms of the whole current C produced by the battery. If there is no current in r, we must have

$$r_{iii}\,r_{ii} - r_i\,r_{iv} = 0 \text{ or } r_i : r_{ii} = r_{iii} : r_{iv}.$$

§ **14.** The *specific resistance* of a material referred to unit of volume is the resistance of the unit cube to a current

FIG. 124.

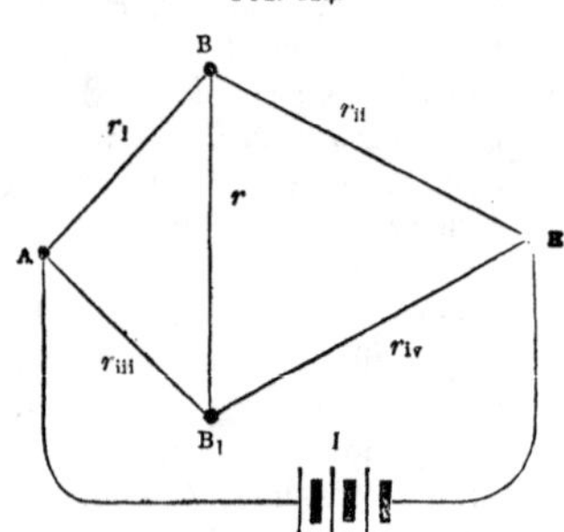

between two opposed faces. The following table contains the specific resistances of several metals and alloys at 0° C.

The specific resistances given are those of a cubic centimètre of chemically pure metals calculated from experiments by Dr. Matthiessen. The resistances of commercial metals are always higher, and frequently very much higher. It is not at all uncommon to meet with copper having 50 per cent. more resistance than that in the table. This is due to

Table. Specific Resistance of Metals and Alloys at 0° Centigrade, from Dr. Matthiessen's experiments.

NAMES OF METALS.	Resistance of one cubic centimètre to conduction between opposed faces. *Microhms.*	Resistance of a wire one mètre long and one millimètre in diameter. *Ohms.*	Resistance of a wire one mètre long, weighing one gramme. Ohms.	Resistance of a wire 1 foot long, 1/1000th of an inch in diameter. Ohms.	Resistance of a wire one foot long, weighing one grain. Ohms.
Silver annealed .	1·521	0·01937	0·1544	9·151	·2214
,, hard drawn .	1·652	0·02103	0·1680	9·936	·2415
Copper annealed .	1·616	0·02057	0·1440	9·718	·2064
,, hard drawn .	1·652	0·02104	0·1469	9·940	·2106
Gold annealed . .	2·081	0·02650	0·4080	12·52	·5849
,, hard drawn .	2·118	0·02697	0·4150	12·74	·5950
Aluminium annealed	2·945	0·03751	0·0757	17·72	·1085
Zinc pressed . .	5·689	0·07244	0·4067	34·22	·5831
Platinum annealed.	9·158	0·1166	1·96	55·09	2·810
Iron annealed . .	9·825	0·1251	0·7654	59·10	1·097
Nickel annealed .	12·60	0·1604	1·071	75·78	1·535
Tin pressed . . .	13·36	0·1701	0·9738	80·36	1·396
Lead pressed . .	19·85	0·2526	2·257	119·39	3·236
Antimony pressed .	35·90	0·4571	2·411	216·	3·456
Bismuth pressed .	132·7	1·689	13·03	798·	18·64
Mercury liquid . .	99·74	1·2247	13·06	578·6	18·72
Platinum silver . . Alloy hard or annealed, 2 parts silver, 1 platinum	24·66	0·3140	2·959	148·35	4·243
German silver hard or annealed .	21·17	0·2695	1·85	127·32	2·652
Gold-silver alloy hard or annealed, 2 parts gold, 1 silver . . .	10·99	0·1399	1·668	66·10	2·391

the presence of other metals in small quantities. Lead, tin, zinc, and cadmium, when alloyed with one another, conduct electricity as if the component parts had remained separate and were arranged as a bundle of conductors, each having a uniform section throughout. Alloys of bismuth, antimony, platinum, palladium, iron, aluminium, gold, copper, silver, mercury, and probably most other metals, have a much greater resistance than the mean resistance of their component parts. The resistance of a wire one mètre long, and one millimètre in diameter, is given in the table : this is equal to the specific resistance multiplied into $\frac{10000}{\cdot 7854}$ or 12732. The resistances of the wires are given in ohms, the specific resistances in microhms.

Notes.—In the above table the numbers underlined are direct observations by Dr. Matthiessen, B. A. Report, 1864.

The numbers given in Col. II. (mètre, millimètre) are obtained by calculating the value in Column III. for lead from the specific gravity 11·376 (Table II., Electric Conducting Power of Alloys) and making the other numbers in the column inversely proportional to the conducting powers given by Dr. Matthiessen, when hard drawn silver is 100, and gold silver alloy 15·03. Column III. is next filled in by calculating the values for zinc, platinum, iron, nickel, tin, antimony, bismuth, mercury, from Column II. by their specific gravities ; the three alloys from specific gravities given by Dr. Matthiessen ; the silver, copper, and gold, by proportion, from the hard drawn metals. Except in the case of lead, the underlined values do not agree with Column II., and the true specific gravity. Column IV. is calculated from Column II. by simply multiplying the numbers by 472·45 ; and Column V. from Column III. by multiplying the numbers by 1·4337. Column I. is calculated from Column II. by dividing the numbers in Column II. by 12,732.

§ 15. The *specific conductivity* of a material is the reciprocal of its specific resistance. Thus the specific conductivity of hard silver in ohms is $\frac{1}{\cdot 0000001652} = 605300$. There is a common but most reprehensible practice of referring conductivities to some material such as silver. The result has been that numerous most careful experiments by skilled electricians are found to be valueless, for no two of them take as their standard metal a metal with the same conductivity. Nor are the relative conductivities of the standards known. Even Dr. Matthiessen's experiments do not allow the construction of a perfectly satisfactory table.

It should be observed that while copper has the greatest conductivity or smallest resistance of any known metal relatively to its volume, aluminium has the smallest resistance for any length of a given *weight*, a matter frequently of considerable importance.

§ **16**. The specific resistance of all metals increases as the temperature increases, and for all pure metals except iron and thallium, Dr. Matthiessen found that the rate of increase was the same. The resistance R of a metal or alloy at the temperature t expressed in degrees Centigrade may be calculated from the resistance r at 0° Centigrade by the following formula:

$$R = r(1 + a\,t \pm b\,t^2) \quad . \quad . \quad . \quad 11^\circ$$

The following are the values of a and b:

	a	b
Most pure metals . . .	·003824	+ ·00000126
,, Mercury . . .	·0007485	– ·000000398
,, German silver . .	·0004433	+ ·000000152
,, Platinum silver . .	·00031	
,, Gold silver . . .	·0006999	– ·000000062

According to experiments by Dr. C. W. Siemens, the resistance r for any temperature up to one thousand degrees Centigrade is expressed by the general formula $r = \alpha\,T^{\frac{1}{2}} + \beta\,T + \gamma$ (Bakerian Lecture, 1871).

Very slight impurities increase the specific resistance of metals considerably, and they diminish the change of specific resistance with a change of temperature.

The copper wire obtained commercially for submarine cables has usually a specific resistance from 5 to 8 per cent. higher than that of pure soft copper. It is usually tested at 24° Centigrade, at which temperature the resistance of a foot grain of pure soft copper is 0·2262. The specified resistance of the French Atlantic cable at that temperature was 0·2456; the actual mean resistance per foot grain at 24° was 0·2388; calling R the resistance per knot, w the weight in lbs. per knot, and s the resistance per foot grain,

$$R = \frac{5293 \cdot 1\, s}{w} \quad . \quad . \quad . \quad 12$$

The resistance of iron used in telegraphy is given by Latimer Clark as 7 times that of pure copper, or at 24° Centigrade 1·58 per foot grain: different specimens vary considerably.

§ **17.** The specific resistance of insulating materials does not admit of being tabulated in the same manner as that of metals, because slight differences in the preparation of the materials cause great differences of specific resistance, and because of the effects of electrification* and of age. Gutta-percha and India-rubber as applied to insulate submarine cables have been the subject of an immense series of careful experiments. The resistance of a cubic centimètre of gutta-percha, a fortnight old, and tested at 24° Centigrade after one minute's electrification, varies from about 25×10^{12} ohms to 500×10^{12} or more. The mean value of the specific resistance of the gutta-percha employed for the 1865 Atlantic cable was 342×10^{12} (ohms) after one minute's electrification. India-rubber when in good condition has a still higher resistance. The Persian Gulf cable made by Hooper had a specific resistance of about 7500×10^{12} ohms.

Let R be the resistance of a length L of gutta-percha covering to conduction, from the wire inside to the water outside, that resistance being what is commonly called the *insulation resistance* of the covered wire or core of a submarine cable; let M be the specific resistance of the material referred to the unit of volume; and let $\frac{D}{d}$ be the ratio between the diameter of the covering and that of the covered wire: then,

* The effect of electrification or polarisation in causing an apparent increase of resistance is described in Chap. IV. § 10.

$$R = \cdot 3665 \frac{M \log \frac{D}{d}}{L} \quad . \quad . \quad . \quad . \quad . \quad 13^{\circ}$$

L and M must be expressed in the same system of units. The resistance R_k of a knot of cable is

$$R_k = \frac{M \log \frac{D}{d}}{506300} \quad . \quad . \quad . \quad 14^{\circ}$$

where M is the specific resistance referred as above to centimètres. The value of $\frac{M}{506300}$ adopted by Mr. Latimer Clark for gutta-percha at 75° F is 769, corresponding to a value for M equal to 389×10^6 megohms. This is a high value. The resistance of G.P. increases under pressure. Let R_p be the resistance at the pressure p expressed in pounds per square inch, and R the resistance at the atmospheric pressure: then, approximately,

$$R_p = R\,(1 + 0{\cdot}00023\,p.) \quad . \quad . \quad . \quad 15^{\circ}$$

The constant 0·00023 probably varies for different specimens and at different temperatures.

The resistance of G. P. also increases very considerably with age, if kept under water. This has not been observed with India-rubber. The resistance of some specimens of India-rubber tested by Dr. Siemens decreased under pressure.

§ **18.** We may calculate the resistance of an insulating material separating two conductors in the following way. Let a body of known capacity S measured in microfarads be charged to the potential P measured in any unit, and let it be gradually discharged through a great resistance R such as the gutta-percha covering of a submarine cable offers to conduction through the insulating envelope, from the wire inside to water outside—the potential of the water being zero. Let the potential of the charged conductor fall to p in the time t measured in seconds; then in megohms

$$R = \frac{t}{S \log_\epsilon \frac{P}{p}} = 0{\cdot}4343 \frac{t}{S \log \frac{P}{p}} \quad . \quad . \quad . \quad . \quad 16^{\circ}$$

The capacity in electrostatic measure of covered wire, neglecting the ends, is given by the equation 6, Chap. V.; to convert this into electro-magnetic measure, we must divide the value by v^2 (§ 2, Chap. VIII.); and to express the result in microfarads the quotient must be multiplied by 10^{15} (Chap. X. § 5): hence the value of S for one knot or 6087 ft. expressed in microfarads is

$$\text{S} = \frac{4{\cdot}2 \times 6087 \times 30{\cdot}48 \times 10^{15}}{4{\cdot}6052 \times (28{\cdot}8)^2 \times 10^{18} \times \log \frac{\text{D}}{d}} = \frac{0{\cdot}2038}{\log \frac{\text{D}}{d}} \quad . \quad . \quad 17^{\circ}$$

Substituting this value for S in equation (16), we have for the resistance per knot,

$$\text{R}_k = 2{\cdot}13t \frac{\log \frac{\text{D}}{d}}{\log \frac{\text{P}}{p}} \quad . \quad . \quad . \quad 18^{\circ}$$

This formula is the more convenient as D, d, P, p may be measured in any units as the ratios only are required. Moreover, $\log \frac{\text{D}}{d}$ is a constant for any one cable. The values of P and p may be observed on any electrometer, or by means of galvanometers, using the method described in the chapter on the Measurement of Capacity.

The specific resistance of very short specimens of wire insulated by different materials may be calculated by the above method, when the current traversing the material would be insensible even on the most sensitive galvanometer.

The method described in this section is only correct if R be constant throughout the experiment; we know that under electrification it actually increases from minute to minute, so that the result given by the formula is intermediate between the resistance when the experiment began and when it ended.

§ **19.** A rise of temperature invariably causes a decrease in the resistance of insulators. Within the limits of 0° and

24° Centigrade the law of the decrease for gutta-percha is approximately expressed by the following empirical formula:

Let r be the resistance of the material at the higher temperature, and R the resistance at the lower temperature, and let t be the difference of temperature in degrees Centigrade: then

$$R = r a^t \text{ or } \log \frac{R}{r} = t \log a \quad . \quad . \quad . \quad 19°;$$

where a is a constant varying with different specimens of gutta-percha and also with variations in the time of electrification. The value of log a increases as the time of electrification increases, and is also higher at the lower temperatures. The following table gives values of log a for different times of electrification and also for two ranges of temperature, from 0° to 12° and from 12° to 24°, derived from a series of experiments made on a knot of French Atlantic cable.

Time of electrification in minutes.	Between 0° and 12°.	Between 12° and 24°.
1	·0562	·0532
2	·061	·0544
5	·0657	·0554
10	·0686	·0560
15	·0706	·057
20	·0725	·0574
25	·0729	·0578
30	·0736	·058
60	·0765	·0600
90 or more	·0747	·0618

Thus the resistance R, after one minute's electrification at 0°, was 7,540 megohms. Then, to find the resistance r at 10° after the same time of electrification, we have $\log \frac{R}{r} = 10 \times 0{\cdot}0562$; whence $r = \frac{7540}{3{\cdot}648} = 2070$

The following is a table of the relative resistances at 0° and 24° after various times of electrification.

Minutes' electrification.	Resistance at 0°.	Resistance at 24°.
1	7540	369
2	9650	401
5	12300	457
10	14400	477
20	17400	493
30	18900	499
60	21900	509
90	24000	512

It should be observed that the difference in resistance produced by electrification is much greater at the low temperatures: or, putting the same statement in another form, there is a much greater change of resistance produced by a change of temperature after long electrification than with short electrification. Experiments have been most frequently made after one minute's electrification.

The following are a series of values of $\frac{R}{r}$ for the temperature of 0° and 24° from different observations.

Name of cable.	$\frac{R}{r}$.	Log a.
Persian Gulf	36·5	·0651
Cores in which thickness of G.P. does not exceed ·11 in.	23·62	·0572
French Atlantic . .	20·43	·0545
Willoughby Smith's improved G.P.	28·14	·0604
Silvertown India-rubber .	17·84	
Hooper's India-rubber .	3·01	·0199

The experiments on the Silvertown India-rubber seem to show that the increase of resistance does not follow the law expressed by equation (19). The resistance of Hooper's material on the contrary, according to Mr. Warren's experiments, does admit of being calculated by that formula up to the temperature of 38·33 Centigrade: the resistance is halved by a further increase of 18·33°.

The electrification of Hooper's material is still more remarkable than that of gutta-percha ; with one specimen the apparent resistance had increased fourfold at the end of 10 minutes, and after 24 hours' electrification the resistance was 23 times greater than at the end of one minute. According to Mr. Warren, if R_1 is the resistance after one minute, and R_t the resistance after the time t, the ratio $\frac{R_1}{R_t}$ is constant for all temperatures with this material.

§ **20.** The specific resistance of other insulating materials than India-rubber and gutta-percha has been very little tested; that of glass varies immensely in different specimens. Leyden jars may be found which do not lose more than $\frac{1}{400}$th of their charge per diem, and the greater part of this loss appears to be due to conduction over the surface, or creeping as it is called, rather than conduction through the mass of glass. The specific resistance of some kinds of glass is therefore nearly infinite ; but many specimens of glass, especially those which contain lead, hardly insulate as well as gutta-percha. Vulcanite, porcelain, and paraffin are good insulators, but I am aware of no experiments determining their specific resistance. Liquid paraffin and some oils are also good insulators.

§ **21.** Graphite and gas coke are used as conductors in batteries, and according to experiments by Matthiessen their specific resistance referred to the unit of volume is from about 1,450 to 40,000 times that of pure copper. Tellurium and red phosphorus have still higher specific resistances. The following table gives Dr. Matthiessen's results expressed in the units now adopted.

Specific Resistance of bad Conductors, computed from experiments by Dr. Matthiessen.

Materials.	Resistance in Microhms.	Temperature Centigrade.
Graphite, specimen 1 . .	2390	22°
,, 2 . .	3780	22°
,, 3 . .	41800	22°
Gas coke	4280	25°
Bunsen's Battery, coke . .	67200	26·2°
Tellurium	212500	19·6°
	ohms.	
Red Phosphorus . . .	132	20°

§ **22.** The specific resistance of liquid electrolytes is not very accurately known owing to the difficulty in measurement due to the phenomenon of polarization. A rise of temperature diminishes their resistance in all cases. Its effect has been studied by Becker ('Ann. d. Chem. u. Pharm.' 1850 and 1851) and by Beetz ('Pogg. Ann.' cxvii. 1862). Paalzow has endeavoured to avoid the difficulty caused by polarization by using composite electrodes consisting of amalgamated zinc plates in porous cells containing solution of sulphate of zinc ('Pogg. Ann.' cxxxvi. 1869). Kohlrausch ('Pogg. Ann.' cxxxviii. 1869) has used the rapidly alternating currents of a magneto-electric machine with electrodes of very large surface. J. A. Ewing and J. G. MacGregor ('Trans. R.S.E.' xxvii. 1873) have applied the 'bridge' method (§ 11), using a Thomson's 'dead beat' mirror galvanometer, which enabled them to observe the resistance before polarization had time to become sensible.

The saturated solution is frequently not the best conductor. This is the case with sulphate of zinc and chloride of sodium. Sulphuric acid when diluted with water has a minimum resistance when of specific gravity 1·25, or according to other experiments when 45·84 grammes of SO_3 are mixed with 100 cubic centimetres of water.

The following tables show the resistance of some of the solutions most employed in batteries. By the term 'specific resistance' is meant the resistance, expressed in ohms, of one cubic centimetre to conduction between opposed faces.

Sulphate of Zinc (at 10° Cent.)[1]

Density.	Specific Resistance.	Density.	Specific Resistance.	Density.	Specific Resistance.	Density.	Specific Resistance.
1·0140	182·9	1·1019	42·1	1·2709	28·5	1·3530	31·0
1·0187	140·5	1·1582	33·7	1·2891	28·3 min.	1·4053	32·1
1·0278	111·1	1·1845	32·1	1·2895	28·5	1·4174	33·4
1·0540	63·8	1·2186	30·3	1·2987	28·7	1·4220 }	33·7
1·0760	50·8	1·2562	29·2	1·3288	29·2	Saturated }	

The solution of maximum conductivity may be prepared by dissolving 73·5 parts of salt in 100 of water.

Sulphate of Copper (at 10° Cent.).[1]

Density.	Specific Resistance.	Density.	Specific Resistance.	Density.	Specific Resistance.
1·0167	164·4	1·0858	47·3	1·1679	31·7
1·0216	134·8	1·1174	38·1	1·1823	30·6
1·0318	98·7	1·1386	35·0	1·2051 }	29·3
1·0622	59·0	1·1432	34·1	Saturated }	

The resistance of mixtures of these salts 'is invariably less than the mean resistance of the components, being in many cases less than that of either.'[1]

Sulphuric Acid—diluted.[2]

Specific gravity.	0°	4°	8°	12°	16°	20°	24°	28°	Centigrade.
1·10	1·37	1·17	1·04	·925	·845	·786	·737	·709	Resistance
1·20	1·33	1·11	·926	·792	·666	·567	·486	·411	of one cubic
1·25	1·31	1·09	·896	·743	·624	·509	·434	·358	centimètre to
1·30	1·36	1·13	·94	·79	·662	·561	·472	·394	conduction
1·40	1·69	1·47	1·30	1·16	1·05	·964	·896	·839	between op-
1·50	2·74	2·41	2·13	1·89	1·72	1·61	1·52	1·43	posed faces,
1·60	4·82	4·16	3·62	3·11	2·75	2·46	2·21	2·02	expressed in
1·70	9·41	7·67	6·25	5·12	4·23	3·57	3·07	2·71	ohms.

[1] Ewing and MacGregor. [2] Becker.

Nitric Acid.

	2°	4°	8°	12°	16°	20°	24°	28°	Centigrade.
Specific gravity 1·36	1·94	1·83	1·65	1·50	1·39	1·3	1·22	1·18	Resistance of one cubic centimètre in ohms.

The specific resistance of water (res. of cubic centimètre) when pure is 9320 ohms, computed from experiments by Pouillet. The presence of $\frac{1}{20000}$th of sulphuric acid reduced this resistance to 1550. The temperatures were not given by Pouillet.

§ **23.** When the resistance of insulators is being measured, care must be taken to prevent conduction over the surface of the insulating material between the two conductors separated by that insulator. If, for instance, a conductor C, Fig. 125, supported by a long vulcanite stem, be charged, and the gradual fall of potential tested by observing the potential on an electrometer, the insulation resistance of *a b* will not really be tested, for conduction will take place almost wholly by creeping over the slightly damp or dirty surface from *a* to *b*. Similarly the insulation resistance of a short length of covered wire, Fig. 115, will be very incorrectly indicated by a galvanometer G, unless the surface of the gutta-percha near A separating the wire from the water is such as to allow no creeping. Surfaces have no special conducting power, but the slight film of damp or dirt conducts in proportion to its sectional area and the conducting power of the particular kind of dirt. Thus brass filings or salt with a little moisture form a highly conducting film. The surface of glass being hygrometric will always be covered with a conducting film, unless the atmosphere be artificially dried in the neighbourhood. The outer layers of gutta-percha, soon after being exposed to the air, become so far changed as to insulate badly, so that the surface should always be fresh cut when experiments are being performed. Old vulcanite is often found covered with a conducting film resulting from the decay of the material. The surface of old glass which has been exposed to the weather conducts better than new glass. Mr. Varley gives the following recipe for preserving and renewing the insulating power of ebonite or vulcanite supports :—

First, wash the ebonite with water, rubbing it well till dry; secondly, moisten the surface of the ebonite with anhydrous paraffin oil. To prepare this, put a quart of common paraffin and an ounce of sodium into a bottle.

FIG. 125.

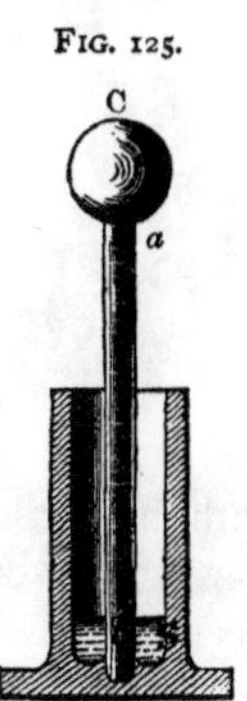

A glass support or the inside of a Leyden jar is best cleansed by being washed with distilled water and dried at a fire without being wiped. A stem such as *a b* may then be made to insulate admirably by setting it in a deep narrow tube with a little concentrated sulphuric acid at the bottom. To increase the resistance of the conducting film, its sectional area must be diminished as much as possible, and its length increased: hence a long rod *a b*, Fig. 125, will insulate better than a short one, and a rod of small surface better than one with a large surface.

The resistance of a film of dirt does not appear to follow Ohm's law. When the potential of the charged and insulated conductor is increased, the loss by creeping increases in a much higher ratio: probably the conduction is partly due to numberless small discharges from one speck of dirt to its neighbour.

CHAPTER XVII.

COMPARISON OF CAPACITIES, POTENTIALS, AND QUANTITIES.

§ 1. THE relative throw or swing of a galvanometer needle caused by the charging or discharging of two conductors gives a very convenient method of comparing their capacities when these are sufficiently large. Thus let *x y*, Fig. 126, represent the plates of a condenser separated by a dielectric from the opposed series of plates *a b*; let *a b* be connected

FIG. 126.

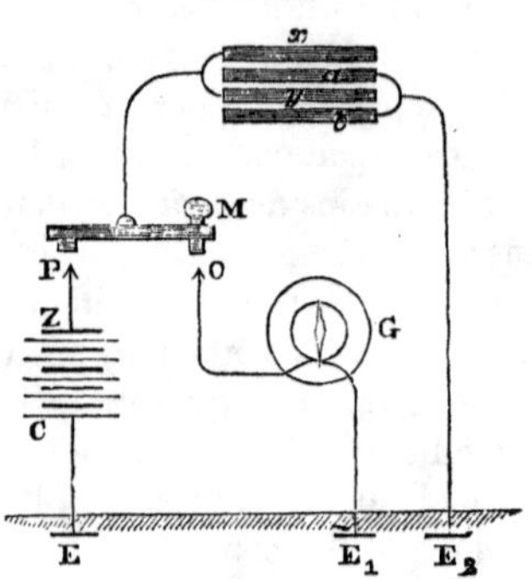

with the earth, and let *x y* be connected with the body of the key M; the contacts P and O of this key serve at will to connect *x y* with the zinc pole of the battery Z C, the copper pole of which is to earth, and with the one terminal of the galvanometer G, the other terminal of which is also to earth. If the handle at M be lifted, the condenser *x y* will be charged with negative electricity. On depressing M this charge will flow to earth through the galvanometer G; this flow will throw the needle of the galvanometer to one side by an impulse of very short duration. If the needle is impeded by no friction, calling S and S_1 the capacity of two condensers, which, when charged by the same battery, throw the needle to the angles i and i_1, we have

$$S : S_1 :: \sin \frac{i}{2} : \sin \frac{i_1}{2}$$

The current is proportional to the capacities, the impulse is proportional to the current, and the sines of half the angles are proportional to the impulses: hence we have the above proportion. Instead of observing the discharge we might have placed the galvanometer G between M and the plates *x y* of the condenser; in that case, on raising M we should observe the throw of the needle produced by the charge when flowing in instead of when flowing out; the throw in the two cases is the same if there is no leakage from *x y* to *a b*. We might substitute for the earth any other conductor, joining E E_1 and E_2 without in any way affecting the observation.

§ **2.** The galvanometer G may be shunted when one condenser is observed, and less shunted or not at all shunted when a second condenser is tested; but in that case it is necessary to take care that the resistance of the shunts bears the same relation to that of the galvanometer for transient currents as for permanent currents. The self-induction of the shunt and the galvanometer may be very different, and may seriously affect the proportion in which the current is subdivided between the shunt and the galvanometer.

§ **3.** A differential galvanometer may be made use of to compare two condensers, the capacities of which are nearly equal. The charges given to the two condensers by the same battery must, for this purpose, be passed simultaneously through the two coils of the galvanometer; the sine of half the throw will then be proportional to the difference between them. In making this experiment it is not necessary that the coincidence between the times occupied by the passage of the charges should be absolute; it is sufficient that both charges pass while the magnet is still sensibly at rest. A similar comparison may be made, using a simple galvanometer, by the following device:—

Pass a current from a battery C Z, Fig. 127, through a considerable resistance R R_1. Connect one point of the resistance R R_1 with earth at E, the rest of the system being insulated. Then two points A and B separated from E by equal resistances will be at equal and opposite potentials. Now let the two condensers to be compared be charged respectively by simultaneous contact with A and B, then if they are equal they will receive opposite and equal charges. Next connect the two condensers one with another (after removing both from

A and B); then the two equal charges will exactly neutralize one another, and no charge will be detected in either condenser. The absence or presence of a charge may be observed by galvanometer or electrometer. The proportion

FIG. 127

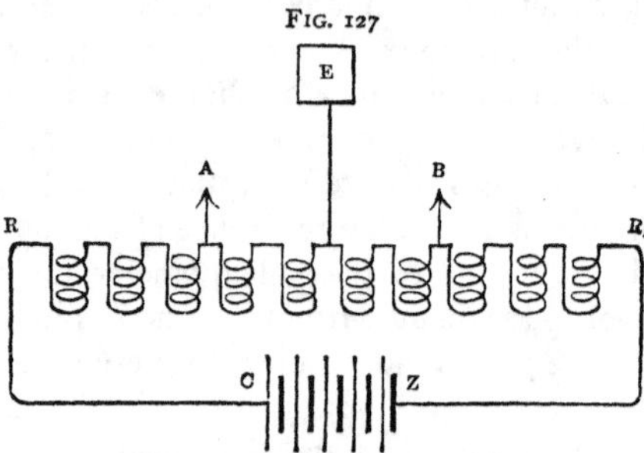

between two condensers may similarly be measured by observing the proportion between the resistances A E and E B required to produce charges which exactly neutralize one another. The capacities will be inversely proportional to the resistance A E and E B. These resistances must be considerable, or the potentials at A and B will be insufficient to charge condensers in such a way as to be measured by the electrometer or galvanometer.

The points A and B may be connected by sliding pieces to successive terminals subdividing R R_1.

§ 4. For small capacities Sir William Thomson's platymeter and sliding condenser may be used (vide Gibson and Barclay, spec. Ind., cap. Paraffin—Phil. Trans. 1871).

Let there be two equal condensers p and p_1, Fig. 128, the outer armatures of which are insulated and the inner armatures connected with an electrometer. Let A and B be the two condensers which are to be compared; connect the outer armatures of A and B with p and p_1 respectively, and their inner armatures with the earth.

Let A be so constructed that its capacity can be varied at will. Charge the outer armature of A positively, and at the same time connect the point q with the earth; the outer armature of p will take a positive charge, its inner armature a negative charge; p_1 will remain uncharged. Now break

FIG. 128

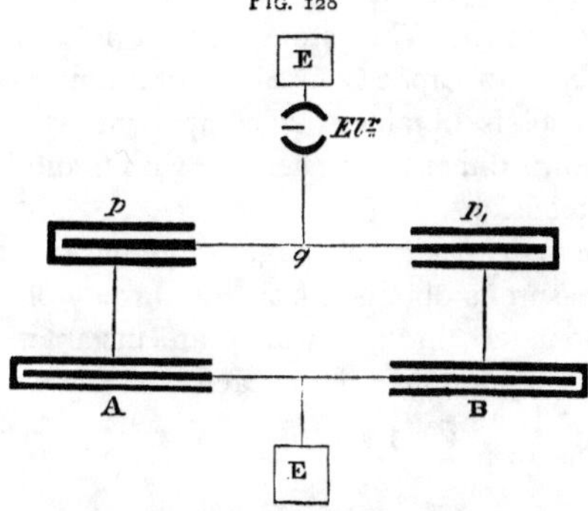

contact between q and the earth; the electrometer will not deflect, for the charge in p will be unaltered.

Connect the outer armatures of A and B; if the ratio of p to A is the same as that of p_1 to B, the potential of q will remain unchanged, and the electrometer will not be deflected; if $\frac{p}{A}$ is greater than $\frac{p_1}{B}$, the potential of q will be lowered; if $\frac{p}{A}$ is less than $\frac{p_1}{B}$, the potential of q will be raised by the connection of the outer armatures of A and B. The deflections of the electrometer due to the raising or lowering of the potential of q allow us to adjust the capacity of A until the ratio $\frac{p}{A} = \frac{p_1}{B}$, and if $p = p_1$, we shall then have A = B. A can therefore be adjusted until it is exactly equal to B.

This appears to be the best method for copying standard condensers, because it does not depend on the accuracy of any other instrument. Any error in the adjustment of p and p_1 can be detected and allowed for by reversing the position of A and B. The relation of equality is not required. In order that no deflection be produced by free electricity at q, it is sufficient if

$$p : p_1 = A : B.$$

The analogy with the Wheatstone's bridge is obvious.

§ **5.** The *absolute* capacity in electrostatic measure of any small condenser is obtained by comparison with that of a sphere of known dimensions enclosed within another sphere of known dimensions.

The absolute capacity of larger condensers in electromagnetic measure is obtained from the throw i of the needle of a galvanometer through which an instantaneous discharge is passed; we have the capacity,

$$S = 2\frac{t \sin \frac{1}{2} i}{\pi R_1} \quad \ldots\ldots\ldots \quad 1°.$$

Where t is half the period or time of a complete oscillation of the needle of the galvanometer when no current is passing, and R_1 the resistance of a circuit in which the E. M. F. used to charge the condenser would produce the unit deflection; i has the same meaning as in § 1. In a reflecting galvanometer half the deflection may be taken as equal to $\sin \frac{1}{2} i$. This formula follows from the formula for the impulse produced by the current on the magnet, and the formula for the throw produced by a given impulse. In order that it should be applicable, the impulse must be very short when compared with the time t, and the resistance of the air must be insensible. This latter condition is only fulfilled when successive oscillations of the needle are sensibly equal. A galvanometer with a heavy needle should therefore be used in making this observation. The absolute value of the difference between two condensers detected by the method described in § 3 can be determined in this way.

§ **6.** The comparison of *potentials* of two batteries may be made indirectly by observing the currents which the two batteries are capable of maintaining through known resistances; but this method has the defect that the electromotive force of most batteries varies when the resistance in circuit is changed, being higher with a large resistance and lower with a small resistance in circuit. The potentials can be directly compared by comparing the deflections which the two batteries produce on the same electrometer. If the difference is great, a graded electrometer must be employed, or the following method may be used: charge a condenser with the higher potential; insulate the condenser, and then diminish the potential in a known and convenient ratio by connecting a second condenser with the first, the ratio between the condensers being previously determined. In this way the reduced potential mav be brought within the range of the electrometer employed to measure the lower potential. If the condenser is large, the electrometer may be dispensed with and a galvanometer used to indicate the relative potentials, to which the condenser is successively charged by two batteries. The two discharges are proportional to sin $\frac{1}{2}$ *i*; and as the capacity of the condenser is constant, the potentials charging the condensers are proportional to sin $\frac{1}{2}$ *i*, or in the case of mirror galvanometers to the throw of the spot of light; by the use of shunts on the galvanometer this method is extended to the comparison of potentials differing 100 or 1000 fold.

§ 7. A *quantity* of electricity is seldom measured directly. A known current flowing for a given time conveys a definite quantity of electricity, and a body of known capacity charged to given known potential also contains a known quantity of electricity. The relative quantities per unit of surface on a conductor can be measured by the proof plane and an electrometer as already described. The quantity of electricity producing a given amount of heat or chemical action is best measured by the measurement of heat or of the weight of material electrolyzed. The quantity Q of electricity in a very short current flowing through a galvanometer is given in electromagnetic measure by the following formula:—

$$Q = 2 \frac{C_1 t}{\pi} \sin \tfrac{1}{2} t \; . \; . \; . \; . \; . \; . \; . \quad 2°$$

Where C_1 is the permanent current which produces the unit deflection on the galvanometer. This equation follows from equation 1.

CHAPTER XVIII.

FRICTIONAL ELECTRICAL MACHINES.

§ 1. THE simplest of these is the electrophorus, which consists of two parts: 1. a disc of ebonite, or similar material, A, cemented into a brass disc B, uninsulated; 2. a brass plate C which can be held in the hand by an insulating stem D. When the surface of the ebonite A is rubbed with flannel, silk, or a catskin, it becomes negatively electrified; if the disc C be now superposed on the electrified disc A, and connected with the earth by being touched with the finger, some of the negative electricity on A is conducted to earth. Some of the negative electricity remains on A, partly because there is not perfect contact all over the surface between A and C, and partly because the electricity on A is not wholly on the surface, but being attracted by the disc B, has penetrated the mass of the vulcanite in the manner indicated by the electrification described Chap. V. § 6. The negative electricity remaining on and in A attracts a positive charge to the lower surface of C. If the finger be now removed and the disc C lifted, it retains its charge of positive electricity, which may be

FIG. 129.

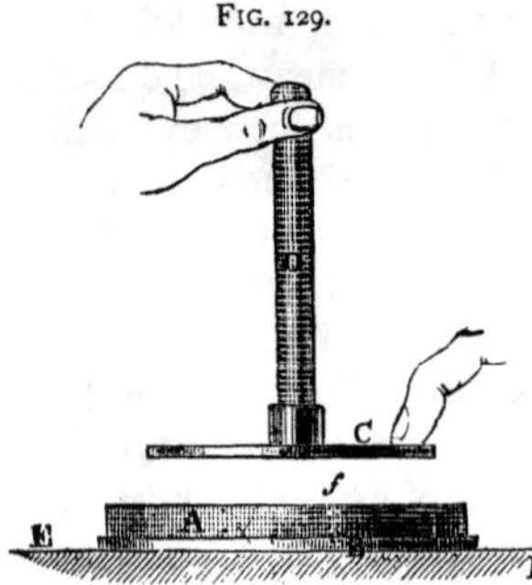

seen passing to earth in a spark if the knuckle or any other blunt conductor is brought near the edge of C. The discharged disc C may be again charged by being placed as before on the disc A and touched by the finger, and this process may be repeated until by gradual conduction to B and C the original charge on A is dissipated. It is certain that the electricity which is effective in inducing a charge on C does not lie on the surface of A, for the addition of one or two little brass pegs *f*, passing from the surface of A to B, improves the action of the electrophorus: this little brass peg serves to conduct any negative charge which may accumulate on the surface of A to the earth. The electrophorus therefore acts as if the parts were arranged as in Fig. 130, where the simple vulcanite disc A is replaced by a metal conducting disc *a a*, electrified with negative electricity, and separated from C by a thin layer of dielectric, and from B by a thicker layer of the same dielectric.

FIG. 130.

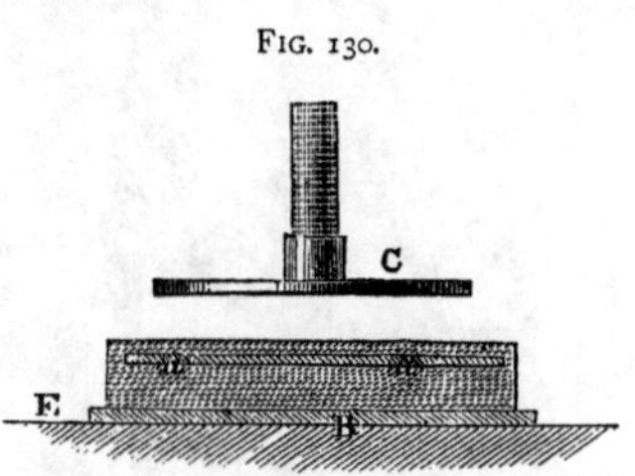

An electrophorus will continue to give sparks in rapid succession for a considerable period, and may be used to charge Leyden jars. A cheap electrophorus may be made by using a cake of resin instead of vulcanite, and wooden discs covered with tin foil instead of the brass pieces B and C.

§ 2. The frictional electrical machine, Fig. 131, consists of a vulcanite or glass disc or cylinder A, made to revolve between cushions or rubbers of leather or silk B B_1. By the friction the (silk) rubbers become negatively, and the glass positively electrified. The difference of potential depends on the substances used as rubbers and disc; if one of these be put to earth, the other will be raised or lowered in potential to twice the extent by which it would have been raised or lowered if both were insulated, having been at the potential of the earth before commencing the experiment. This action is precisely analogous to that which occurs

FIG. 131.

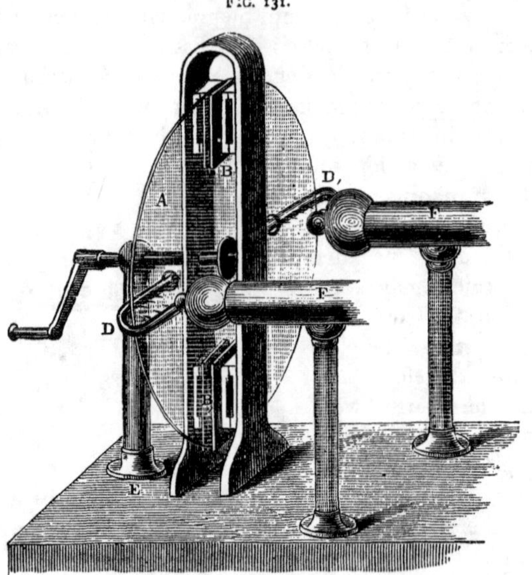

with a galvanic cell; when both poles are insulated, one is raised above the potential of the earth, and the other lowered beneath it. Let one pole be put to earth, the potential of the other is immediately doubled, the difference of potentials remaining what it was before. Let us assume that the rubbers in an electrical machine are put to earth, then the positive electricity of the glass is collected by a series of points D D_1, placed close to the glass, and connected with a conductor F or a Leyden jar. The glass points are sometimes described as acting by induction thus: the + electricity on A induces − electricity on the points, which springs across to the glass, neutralizing the + electricity on the glass, and leaving the conductor or Leyden jar positively electrified. There is neither theoretical nor practical difference between a negative spark passing from D to A, and a positive spark passing from A to D, and we may therefore correctly use the more simple statement given above. The positive electricity which the glass loses is supplied through the rubber; a stream of negative electricity flows from the rubber to the earth while the conductor or jar is being charged; and this is only saying in other words that positive electricity flows from the earth to the rubber, whence it crosses to the glass and so to the conductor F or to a Leyden jar. It is just as essential to the effective working of the electrical machine in charging a jar that the outside of the jar be to earth, as that the rubber be to earth; and if the outside of the jar and the rubber be connected, it is unnecessary that either should be to earth. It is necessary in order to charge a jar or conductor as highly as the machine is capable of doing, that the electric circuit should be complete, except across the dielectric used to insulate the conductor to be charged. It is of no importance whether the earth form part of that circuit. The parts must be arranged as in Fig. 132, where B represents the rubber, A the rubbed glass, G G_1 conducting wires or chains, F and C the two opposed coatings of the Leyden jar and D the dielectric; C may be a mere brass ball, F the walls of a room, and D the air of the room. The case will not differ from that of an ordinary Leyden jar except as to the capacity of the conductor C. The machine B A will produce the full difference of potential it is capable of producing between F and C. The charge given to C will simply then be proportional to its capacity. The circuit may all be insulated; it may be put

FIG. 132.

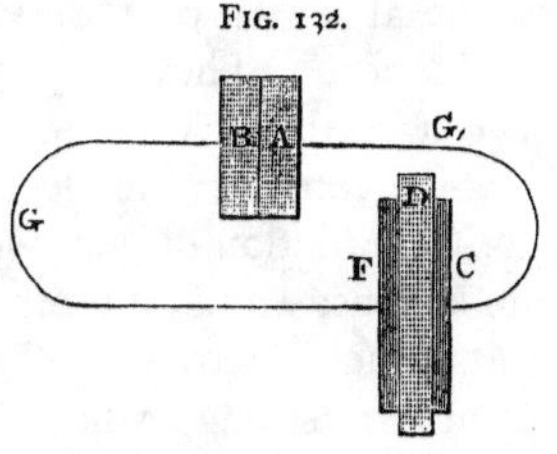

to earth between B and F, or it may be put to earth between A and C. The only effect of these changes will be to alter the absolute potential of F and C, but not to alter the difference. If, however, G and G_1 are both put to earth, the circuit is destroyed and no effect will be observed at F or C. Similarly, if G or G_1 are broken, the circuit will be destroyed; but in this case some less perfect circuit is generally completed, which will lead to the observation of some electrical difference between F and C if either G or G_1 are entire.

§ **3.** In electrical machines sold by opticians, large brass conductors F F, insulated on long stems, are usually connected with the collecting points D D_2 Fig. 131. These large conductors have a sensible capacity, and allow the machine to produce long sparks and other phenomena requiring the accumulation of a considerable *quantity* of electricity. The addition of a large pasteboard cylinder with rounded ends covered with tin foil insulated from the earth by a single long stem and connected to D D_1 by a wire through the air, allows the volume of the spark obtained from the machine to be greatly increased. The insulating stems are best made of vulcanite, and should be kept clean, as described in § 23, Chapter XVI. No points or sharp angles must form part of the system of conductors attached to D D_1, if phenomena requiring great differences of potential are to be observed. Glass stems and discs are old-fashioned. They are weak, hygroscopic, and when rubbed with hot cloths to dry them become covered with fluff which conducts the electricity to earth.

§ **4.** The friction of globules of pure water suspended in steam against wood and other insulators may be made use of to produce electricity. This fact was discovered by Sir William Armstrong, whose apparatus was made as follows:—

The steam issuing from a high-pressure boiler by the pipe A passes in a series of tubes (not shown) through the box B, which is supplied with cold water; from these tubes the steam charged with condensed globules issues through the jets *c c c*. These jets are lined with wood. The friction charges the steam with positive electricity, which is gathered by a series of points at D attached to the insulated conductor F. The globules of water must be pure, or only charged with insulating materials. The resistance of pure water is so great that it may be looked upon as an imperfect insulator of the same class as flannel; the material against which the water rubs exercises, as might have been anticipated, a great influence on the amount and sign of the electricity produced. When turpentine is mixed with the water, the vapour becomes negatively electrified.

FIG. 133.

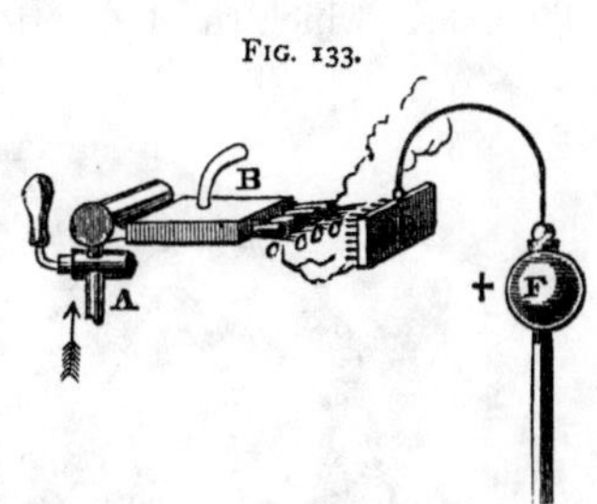

CHAPTER XIX.

ELECTROSTATIC INDUCTIVE MACHINES.

§ 1. THE action of the electrophorus, described in § 1, Chapter XVIII. may be imitated by arrangements no part of which requires to be electrified directly by friction; and, moreover, the apparatus can be arranged so that the inducing charge shall be continually strengthened by the action of the machine. Inductive machines of this kind have been invented by Bennett, Nicholson, Mr. Varley, Sir William Thomson, and others. Mechanical energy in these instruments is converted directly into an accumulation of electricity at different potentials, the work done being expended in overcoming electrostatic forces. The following is Mr. Varley's design:—

A series of metal conductors, *c*, *c*, *c* (Fig. 134), which will be called carriers, are attached by means of a vulcanite disc *b* to the axle *a*, which can be made to rotate at pleasure. The

FIG. 134.

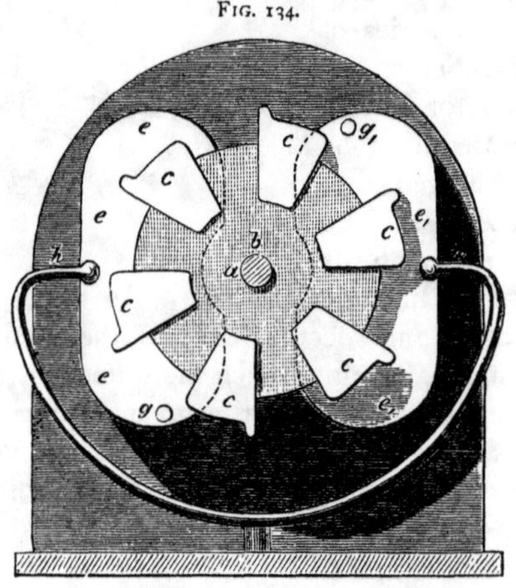

disc and carriers rotate between two pairs of metal insulated cheeks, *e* and e_1, which will be called inductors. The knobs *h* and h_1 are in connection with the earth, and are grazed by the carriers *c c* as they revolve. There are also contact pins at *g* and g_1, which put each carrier successively in contact with *e* and with e_1 for an instant in passing.

Let a small charge of positive electricity be communicated to *e*, the rest of the apparatus being at the potential of the earth. The plate *e* will induce a negative charge in *c* as it rises past *h*, the positive electricity flowing to the earth through *h*. The carrier *c* conveys this negative charge to g_1, giving up almost the whole of it to the surrounding inductor plates e_1. This redistribution of the charge leaves *c* almost neutral, and the inductor e_1 next induces a positive charge in *c* as it descends past h_1; the carrier conveys this to *e* through the pin *g*, and so augments the original positive charge.

When it again passes h, it receives by induction a greater negative charge than before, which again augments the negative charge in e_1, and this induces a new positive charge on c, which is transferred to e. Each turn thus augments the charge on both inductors in a continually increasing ratio; and the only limit to the charge which can thus be accumulated on the inductors is that determined by the escape of electricity from them in the form of sparks or brushes. A continuous supply of sparks may be drawn from e or e_1. The knobs h and h_1 need not be in connection with the earth, provided they are in connection with one another. In that case, when c passes h, and c_1, immediately opposite, passes h_1, c and c_1 are connected for an instant. A positive charge is induced in c_1, and a negative charge in c. When this arrangement is adopted, one of the inductors may be in connection with the earth.

The arrangement adopted by Sir William Thomson to replenish Leyden jars, Chap. XIV. § 2, in which he wishes to maintain a constant potential, is very compact. The inductors are metal plates ee_1 bent so as to form cylindrical surfaces, as in Fig. 135. The axis A supports two carriers C C_1, which are also parts of cylinders not exactly concentric with the inductors. In the fig. the axis and carriers are shown removed from their positions inside the inductors. The connectors are shown at h and h_1. The springs g and g_1 correspond to the pins with the same letter in Mr. Varley's arrangement. In the *mouse mill*, another arrangement used by Sir William Thomson to give a rapid succession of sparks, the inductors are parts of cylinders and the carriers are long strips like the staves of a barrel. The smallest conceivable charge on one inductor of these machines is sufficient to start them; indeed, it is difficult, if not impossible, so completely to reduce e and e_1 to the same potential that after a few turns of the carriers they shall not be highly charged.

§ 2. Holtz's electrical machine is an inductive machine in

which the carriers are replaced by the imperfectly conducting film which usually covers a disc of glass, or by the external film of the glass itself considered as a body capable of

FIG. 135.

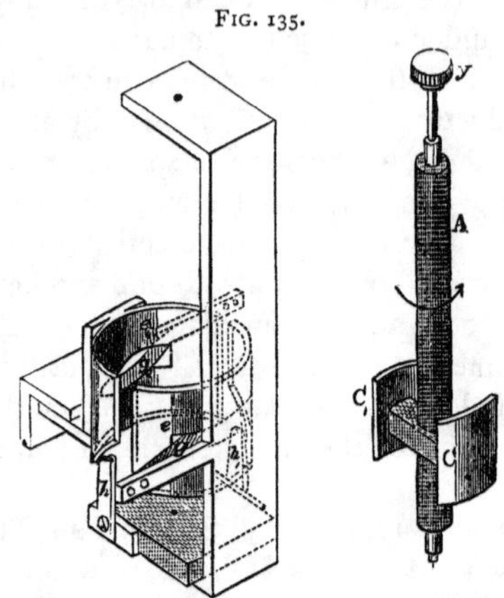

receiving a charge, though not of conducting electricity. This film must be a sufficiently good insulator not to allow the escape of the charge it has taken. The theory of the machine will be more readily understood if we replace the imaginary film by a series of insulated carriers similar to those described for Mr. Varley's apparatus.

Let there be a fixed disc, Fig. 136, of insulating material B and a rotating disc of insulating material A; on each side of the disc A let there be a series of metal carriers *c* and *d* all insulated from one another. On the disc B let there be two inductors *e* and e_1, the first positively and the second negatively charged. *e* and e_1 cover both sides of disc B for a short distance, and there are two openings F and F_1, as shown. The fixed rods *h* and h_1 serve to join successive pairs of carriers *d* and d_1 as they come opposite *e* and e_1. The rods *h* and h_1 are shown with a couple of little balls, which can be separated to show sparks passing along the connecting rods h and h_1. There are also shown two springs g and g_1, which serve to connect each carrier c in succession with e and e_1. Now let the disc A revolve so that the side nearest us moves in the direction of the arrow; when c is opposite e, and c_1 opposite e_1, d and d_1 being connected by h and h_1, there will be a positive charge induced on the external surfaces of d and c_1, a negative charge on the external surfaces of d_1 and c. As the rotation continues, each of these carriers will become disconnected from h and h_1, and will carry with it its charge of electricity without any considerable change in the distribution. d_1 and c_1 will, after a fraction of a revolution, come opposite F, where they are shown as c_a and d_a. The positively charged carrier c_a will come in contact with the spring g; at the same time c and d will have come to the position c_b and d_b, and the negatively charged carrier c_b will come in contact with the spring g_1. There will then be a redistribution of electricity. The capacity of c_a and c_b is diminished by the absence of the plate B at F and F_1, and the result of the redistribution is to remove the greater part of the positive electricity from c_a to e, of the negative electricity from c_b to e_1, to set free negative electricity on d_a and positive electricity on d_b. When, therefore, d_a comes under h into the position of d, the negative electricity flies to d_1, or, in other words, positive electricity flies from d_1 to d, and the cycle of operations recommences. The rods h, h_1, the carriers c, c_1, &c., the inductors e, e_1, and the contact springs g, g_1, all play exactly the same part in Holtz's machine as in Varley's, with the exception that in the new

FIG. 136.

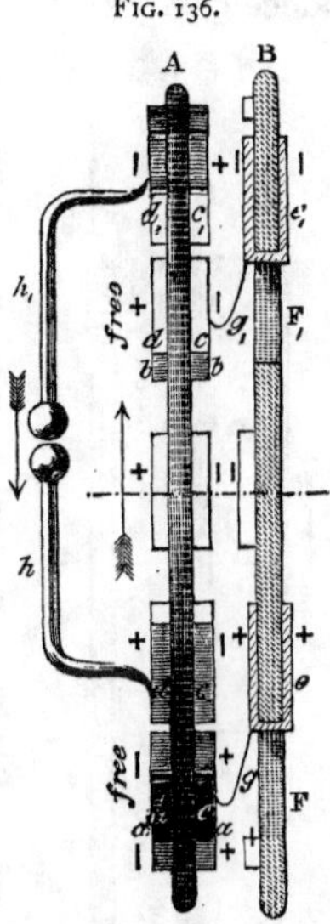

arrangement the connectors h, h_1, instead of joining c, c_1 directly, join a new set of carriers d, d_1, &c., on which c, c_1 induce charges.

FIG. 137.

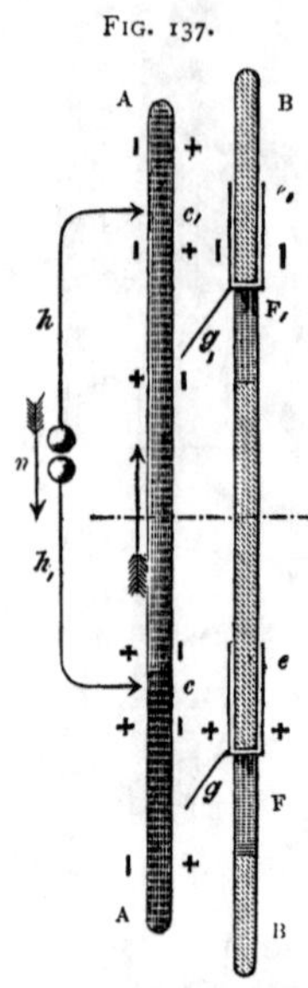

The actual Holtz's machine has no carrier. There is a fixed disc of glass B and a rotating disc of glass A. At the openings F and F_1 there are the inductors e and e_1, made of paper; the connecting-pieces g and g_1 are also of paper, and merely point at the place where the carriers should be; the connectors h, h_1 are brass rods ending in points opposite c and c_1; the part of the carriers is played by the surface of the glass; the action is identical with that described for carriers. The openings at F_1 and F serve to insulate the positive from the negative parts of B as well as to alter the capacity of each portion of the surface of A as it passes them; the rods h and h_1 are arranged so that they can be withdrawn, leaving a space at n across which sparks pass; if the space be gradually increased between h and h_1 at n, after the machine has been set in action by charging e or e_1, a splendid violet brush of some inches in length may be observed passing at n. If Leyden jars are hung on h and h_1 to increase their capacity, this brush is replaced by a torrent of brilliant sparks. With large Leyden jars on h and h_1 one spark of extraordinary length and volume passes at sensible intervals of perhaps one or two seconds.

In the figures the openings F and F_1 are shown as if they were near together, because the whole series of inductions can thus be better brought into one view. In the machine itself, as shown in Fig. 138, the openings are diametrically opposite one another, and the electricity is collected from the glass by a comb or series of points H and H_1 attached to the rods h and h_1. The openings F and F_1 are behind the transparent plate A, though shown in the full lines.

FIG. 138.

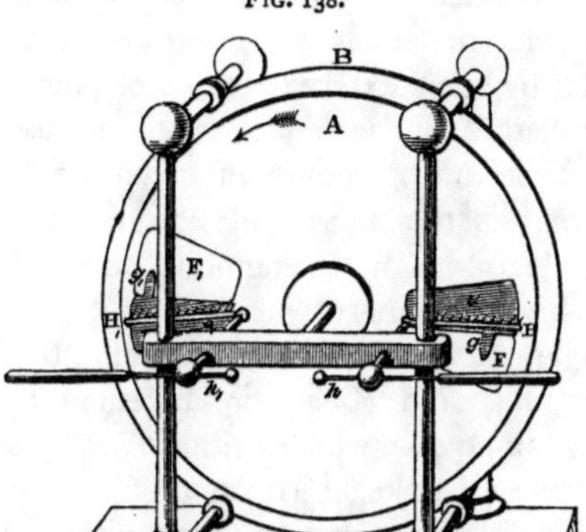

The dark portions of the figure e and e_1 are the paper armatures which are on both sides of B. The gear is omitted by which A is driven. The plate B is carried by four supports touching its edge.

CHAPTER XX.

MAGNETO-ELECTRICAL APPARATUS.

§ 1. THE phenomenon described in Chapter III. §§ 18 and 19, and more fully explained in Chapter IX., is often described as *magneto-electric* induction when the current is induced by the motion of a wire in a field produced by a *magnet*, the term electro-magnetic induction being reserved for the case in which an electric current induces magnetism. The [1] distinction in this sense is rather popular than scientific,

[1] An essential and scientific distinction can be drawn between the two cases by applying the name, magneto-electric induction, to all those

but it is convenient to retain the name magneto-electrical apparatus for those arrangements in which powerful electric currents are induced in wires moved across a magnetic field produced by permanent magnets or electro-magnets.

In magneto-electric apparatus the moving coils of wire must be driven by some external source of power.

The term electro-magnetic apparatus is used, on the contrary, for those arrangements in which the battery producing a current is the source of power which produces motion. An electro-magnetic engine is one which may be employed to drive machinery.

§ **2.** Arrangements giving electric currents by the relative motion of magnets and coils were invented by Pixii and Ritchie. The apparatus which will be now described is generally known as Clarke's: In front of a powerful horse-shoe magnet A, Fig. 139, there are two bobbins B and B_1 of insulated wire; these two bobbins are carried by one frame V, which rotates round a horizontal axis, being driven by a pulley. The two coils of wire are continuous, so that a single current may flow round both; but they are so joined that the current flows in a right-handed direction round one and flows in a left-handed direction round the other. Each bobbin has a core of soft iron, and these cores are joined by iron at the back: that is to say, at the ends farthest from the horse-shoe magnet. Two ends of the wire on B and B_1 are directly joined, but the two other ends are connected through a set of springs rubbing on suitable contact pieces on the axis, with two fixed terminals T and T_1, and the circuit is not complete till these are joined. We will suppose this to be done. As the coils rotate, each soft iron core is successively magnetised in opposite directions; thus coil B, when opposite a north pole, has its south pole near the magnet and its north pole at the back, and this

cases which require relative motion, and using electro-magnetic induction to denote only those phenomena of induction which result from the change of currents or magnetism without relative motion.

arrangement of the magnetism is reversed when B is opposite the south pole; thus in every revolution a magnet is, as it were, introduced into B, withdrawn, and replaced with its poles in the opposite direction, and again withdrawn.

FIG. 139.

The withdrawal of a magnet having its north pole at one end of B, and the introduction of a magnet having its south pole at the same end, both tend to induce a current in one direction; but the withdrawal of this second magnet, and the introduction of the reversed magnet, induce a current in the opposite direction. Thus from the instant the coil B begins to leave the pole S, to that instant at which it arrives opposite N, a current in one and the same direction is being induced; but as soon as B begins to leave N and return to S the direction of the current is reversed, and continues reversed until opposite S. Thus two equal and opposite currents are induced in B during each revolution. The same statements hold good of B_1, but when the current induced in B is right-handed that in B_1 will be left-handed. When the coils are joined as described, the two currents are added to one another; the currents can be observed and utilised on that portion of the circuit which is interposed between T and T_1. With the connections as described the currents will be reversed between T and T_1 at every half-revolution; but it is easy to arrange a set of contact pieces in the axis so that although the currents must necessarily be reversed in the coils, they flow always in one direction between T and T_1.

§ **3.** Even when flowing in one direction the currents between T and T_1, must rise to a maximum and decrease to a minimum once during each half-revolution.

The maximum current occurs at those points where the armature (as the soft iron continuous core may be termed) resists the motion mo t strongly. At these points the greatest change of magnetism is taking place in the armature. The motion of the coils alone without a core would give rise to similar but much weaker currents. The best length and thickness of wire depends on the resistance through which the current is required to flow between T and T_1. If this resistance is small, the coils B and B_1 should be made of thick wire; if the external resistance is great, then the coils should be composed of many turns of thin wire.

§ **4.** Instead of a simple pair of bobbins and a single horse-shoe magnet, we may arrange any convenient number of bobbins on a ring moving in front of the poles of a series of magnets also arranged in a circle. Still better, we may let the ring of coils rotate between two rings of magnets, each coil having its own core, which is alternately magnetised in opposite directions; each coil being then connected with its neighbour, so that the current flows alternately in a right-handed and left-handed direction, we add the electro-motive forces due to all the coils.

The coils may be joined in series, or the pairs may be joined in multiple arc, the former plan being adopted if the object is to get a great E. M. F. between T and T_1; the latter plan if our object is to obtain a moderate E. M. F., with a very small resistance in that part of the circuit which forms part of the magneto-electric machine. Great heat would soon be developed with the latter plan. With the former (coils in series) very perfect insulation is required between the separate layers of the coils, or sparks will perforate the insulating substance and destroy the action of the coils. The following is a description of a machine of this class constructed by Mr. T. Holmes, and successfully used by him to produce the current for a large electric lamp :—

The coils, eighty-eight in all, are fixed in the rim of a wheel about five feet in diameter, with their axes all parallel to the axis of the wheel. They are arranged in two rings, each containing forty-four equally spaced bobbins. The centre of each bobbin in one ring corresponds with the centre of the space between two bobbins in the other ring. This wheel is driven at about 110 revolutions per minute. Horse-shoe magnets are fixed in a frame round the circumference of the wheel in three planes, or rings, containing twenty-two each. The two poles of each magnet are in the same plane, or ring. The distance between their poles is equal to the distance between the bobbins, or coils. The magnets in the two outside rings have similar poles opposite one another. The magnets in the inner ring are placed with opposite poles facing the two similar poles of the outer rings. The two outside rings have compound magnets of four steel plates; the magnets of the inner ring between the two sets of bobbins have six plates. The weight of each plate is six pounds. Alternate coils have their iron cores magnetised in opposite directions, but the wires are so connected in series that the induced currents flow all in the same direction relatively to the wire. The length of the hollow iron core inside each bobbin is $3\frac{1}{2}$ inches. Its external diameter, $1\frac{1}{2}$ inch; its internal diameter, 1 inch. Two copper wires, ·148 inch in diameter, forty-five feet long, are wound round each core and connected in double arc. These wires are equivalent to one wire ·2 inch in diameter of the same length. The iron core and brass bobbin surrounding it are split; that is to say, an open slit is left down one side of each cylinder. This prevents the induction of currents in the bobbin and wire where they are not wanted.

Each ring induces forty-four distinct currents during one revolution of the wheel, and the maximum current from one ring coincides with the minimum current from the other; and as each current lasts a very sensible time, and by a commutator is transmitted always in one direction, their combination does not produce a series of sparks, but a nearly constant and uniform current. One and a quarter horse-power is required to drive the machine when in action, and much less when the circuit is broken so as to stop the induced current. This machine offers a striking example of the transformation of work into a current of electricity.

§ **5.** If the change of magnetisation could take place instantaneously, there would be no limit to the electromotive force which these machines could produce, except the limit imposed by the difficulty of insulating the wire and of driving the coils against a great mechanical resistance; the electromotive force induced in the coils would increase in direct proportion to the speed at which they were driven. Practically owing to the coercive force of even the softest iron and the self-induction of the wire on the bobbins, the change of magnetisation and of direction of the current occupies a very sensible time, and if the speed be increased beyond that at which the greatest change of magnetisation occurs, the electromotive force will fall off instead of increasing. The effect of the coercive force is diminished as stated above by making the core hollow, and the effect of useless induction is diminished by splitting it from end to end.

§ **6.** Obviously the magnets used to induce the currents might be electro-magnets; but if these were excited by an independent battery, the induced current would be obtained at a much greater cost than would give the same current directly from a battery.

Mr. Wilde conceived the happy idea of using a current induced by permanent magnets to excite a large electro-magnet which is used to induce a second current, which can be so much greater than the first as the electro-magnet is more powerful than the permanent magnet. The second current may be used to excite a second electro-magnet still more powerful than the first, and this second electro-magnet used to induce a third current greater than either of the two others. Dr. Siemens and Professor Wheatstone simultaneously invented a further extension of the same idea. They use the current induced by the permanent magnet to convert this magnet itself into an electro-magnet. The effect is very remarkable. However weak the permanent magnetism in the inducing magnet may be in the first instance, a few rapid turns of the coils with their armatures induces a current which increases in geometrical proportion, increasing the magnetism of the inducing magnet at the same time, until the resistance of the armatures as they pass the poles is such as to balance the driving power. The current in the main circuit may be directly utilised, or one portion of it may be shunted for use while the other branch maintains the magnetism of the electro-magnet. Mr. Ladd modifies this arrangement by having two distinct coils on his armature, one of which is used to excite the electro magnet, while the other conveys the induced current which is to be utilised outside the machine. Ladd's, Wilde's, and Siemens' machines will produce currents capable of fusing an iron rod an inch in diameter and a foot long. The armatures and coils become themselves so hot that they must be artificially cooled, or the machine can only be worked for short periods without being permanently injured.

§ 7. The armature used in these new machines is generally of the form introduced by Messrs. Siemens, which is much superior to that in Clarke's apparatus.

The compound horse-shoe magnets are arranged in a pile of considerable depth, each separated from its neighbour by a sensible space, as shown in Fig. 140. The armature A A_1 rotates round the axis X Y between the poles in a position where the magnetic field is much more intense than that

FIG. 140.

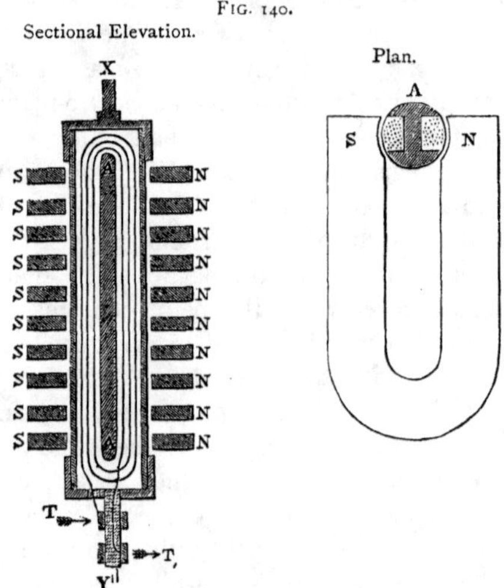

occupied by Clarke's armature. This armature is a long bar of soft iron of an ⊢⊣ section, as shown in plan at A (Fig. 140), and is magnetised transversely. The wire is wound round it longitudinally, passing up one side and down the other.

As this armature rotates round the axis X Y its magnetism is reversed, and at each reversal a current is induced in the enveloping wire. The intensity and uniformity of the magnetic field in which the wire is placed cause this arrangement to give much better results than those obtained by Clarke's arrangement.

§ **8.** It is unnecessary that the armature either of Siemens' or Clarke's or any magneto-electric machine should complete one or more revolutions in order to induce a current:

the smallest motion about the axis is sufficient to produce some electromotive force, because it will change the intensity of the field in which the armature is placed. With Siemens' armature especially a very small deviation in one direction from the position shown in the plan, Fig. 140, will give a powerful current. The wires of the coil move almost directly across the lines of magnetic force, and the armature will be so magnetised as to help the induction so produced. A slight motion in one direction will induce a positive current, a slight motion in the opposite direction a negative current. Keys for sending electric signals without batteries are constructed on this principle.

§ **9.** The *Inductorium*, or *Ruhmkorff's* coil, is strictly speaking an electro-magnetic apparatus, inasmuch as the inducing magnet is not moved, but is magnetised and de-magnetised by the passage and interruption of a current from a battery. It is used to obtain by induction a great electromotive force from a battery of small electromotive force. The inductorium consists of an electro-magnet excited by a comparatively short coil of thick wire called the primary coil : a long coil of fine wire, called the secondary coil, is wound round the same electro-magnet; the primary circuit, which is completed by a battery of small resistance such as Grove's, is alternately made and broken with great rapidity; the secondary circuit is always complete, or interrupted only by such a space that the electromotive force induced in the secondary is sufficient to cause the passage of a spark. When the primary circuit is closed, the electro-magnetism of the core induces a current in the secondary wire in a direction opposed to that of the primary circuit. When the primary circuit is interrupted, the diminution of the magnetism in the core induces a current in the same direction round the wire as the primary current, and therefore in a direction through the secondary coil opposed to the current previously induced.

The electromotive force per foot of the wire in the

secondary coil depends on the intensity of the magnetic field produced and on the rapidity with which it is produced. The sum of the electromotive forces thus induced in a long coil is enormously greater than the E. M. F. of the inducing battery; the longer the secondary coil the greater the electromotive force.

§ **10.** Sparks many inches in length can be obtained from the secondary circuit of a large inductorium, but in such apparatus the greatest care is requisite in the insulation of the secondary coil. Each wire must be insulated from its neighbour by layers of some hard insulator which a spark will not easily pierce, and care must be taken so to wind the coil that no two portions of the secondary coil at very different potentials are near together: this is effected by winding the coil in successive compartments A, B, C, as in Fig. 141, where each compartment is insulated from its neighbour by discs

FIG. 141.

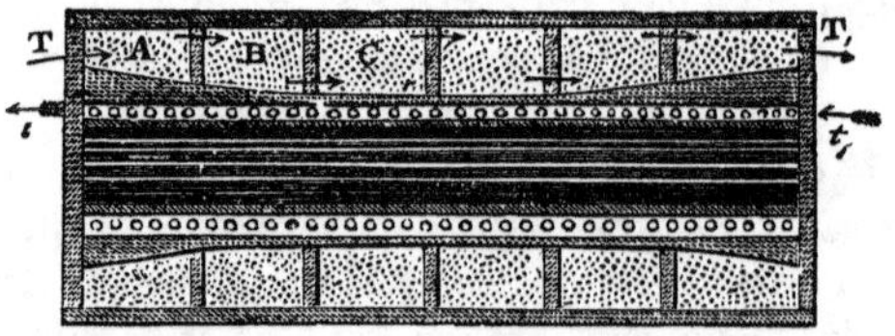

of vulcanite. In order to facilitate the rapid change of magnetism, the core should be either a hollow split cylinder or a bundle of iron rods insulated from one another.

The making and breaking of the primary current is generally effected by a little oscillating hammer having a small armature of soft iron at its head: this hammer is placed so as to be attracted when the iron core is magnetised; by its motion towards the core it breaks the primary circuit; the core being no longer magnetised allows the little hammer to fall back and so once more to complete the primary circuit; this re-magnetises the core, and the hammer again breaks the circuit, and this action repeats itself indefinitely. There are adjustments by which the rapidity of the oscillations of the hammer can be regulated until the best result is obtained. The limit to the speed at which the successive currents can be induced depends on the coercive force of the iron core and the self-induction of the secondary coil. The work done in the secondary coil by the induced current is necessarily less than that done in the primary coil by the battery, however much greater the electromotive force may be.

The following is a description of an inductorium made by Messrs. Siemens :—The core is made of iron wires 1·3 m.m. diameter and 95 centimètres long. These are cemented together and form a cylinder 60 m.m. diameter. Two layers of copper wire 2·5 m.m. diameter form the primary coil. This coil and the iron core weigh 35 lbs. They are placed in a tube of hard vulcanite 26 m.m. thick at the ends, and 12 m.m. thick at the middle : along this tube 150 thin discs of vulcanite are fixed at equal intervals, and the ends are covered with thick discs of the same material. Each subdivision between the little discs is filled with a coil of fine silk-covered and varnished copper wire 0·14 m.m. diameter : these coils are connected in series, so that the current flows from the outside to the inside of one compartment and from the inside to the outside of the next, in order that no two portions of wire at greatly differing potentials may ever be in close proximity. The length of the secondary coil is 10,755 mètres, and it makes 299,198 turns round the cylinder. The weight of the copper wire is 58 lbs. and its resistance about 155,000 ohms.

There is some difficulty in arranging a good make and break piece acted upon by the hammer on account of the large sparks which pass between the contacts tending to fuse them together and oxidise them. Messrs. Siemens make contact between a platinum point and a platinum or silver amalgam covered with alcohol.

When long sparks are wanted, the make and break apparatus is driven slowly, by clockwork or by a separate electro-magnetic engine, so as to give a long contact, which is then suddenly broken. The above apparatus will give sparks of from one to two feet in length, with six large Grove's elements in the primary circuit: 50 miles of fine wire have been used in some induction coils.

§ **11.** A Leyden jar or some other form of condenser is frequently attached to the secondary circuit when this is used to give sparks. The one armature of the condenser is connected with one end of the secondary wire and the other armature with the other end of the same wire, near the opposed points across which the spark is to pass; the effect of this arrangement is that a considerable accumulation of electricity takes place near the points before the difference of potential is sufficient to cause the spark to pass, and consequently the number of sparks observed in a given time is less with the condenser than without, but each spark conveys more electricity and is much more brilliant. An electromotive force in the coil insufficient to cause any spark to pass may nevertheless help to charge the armatures of the condenser, and thus some portions of the inductive action may be utilised with the condenser which without it would be wasted. The dielectric must be thick and strong, or it will be pierced by the spark.

A condenser is also frequently employed, connected with the primary circuit.

§ **12.** The Inductorium may be used to give the sparks required for examination by the spectroscope or to give an electric light, which is, however, comparatively feeble. It may be used to charge Leyden jars and produce physiological effects; it may be used to produce the beautiful luminous effects which occur when electricity is passed through rarefied gases. The gases are enclosed in glass tubes having platinum electrodes soldered into the glass and terminating in balls at a considerable distance apart: instead of the spark observed in air, a diffused light is seen differently coloured in various gases and beautifully stratified. These appearances have been carefully studied by Gassiot, Plücker, and others. The tubes enclosing the gases may be bent into very complicated shapes, and filled in different parts with different gases, so as to produce a striking and pretty appearance when the current from the inductorium passes: they are generally called Geissler tubes. The induction of a magnet, or of a current of electricity, or of a simple conductor outside the tubes, can be observed on the luminous current within, causing it to be distorted or move in those directions in which the inductive force would act on a solid wire conducting a similar current: for this experiment the tube must be wide or nearly spherical, so that the luminous current occupies only a portion of the enclosed space.

CHAPTER XXI.

ELECTRO-MAGNETIC ENGINES.

§ 1. THE most elementary arrangements by which electricity can be made to produce regular motion by electro-magnetic force are those in which a short wire or rod conveying a current is made to rotate by the direct and continuous electro-magnetic attraction to or repulsion from some fixed conductor conveying the same or another current.

Let O P, Fig. 142, be a wire capable of rotation round O,

FIG. 142. FIG. 143.

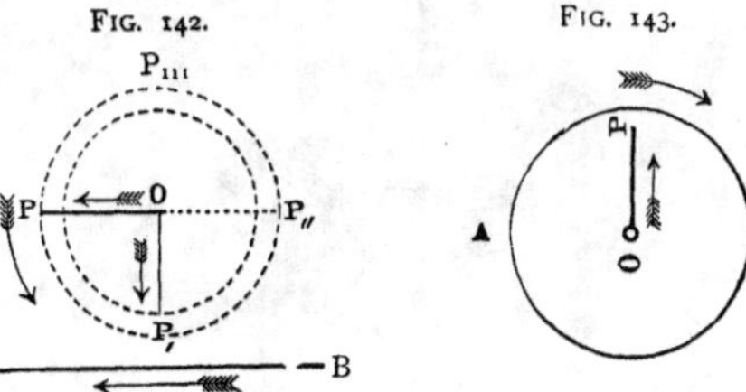

and conveying a current from the centre to the circumference of a ring-shaped trough of mercury into which the end of the wire P dips. Let the same current or another be conveyed in a straight wire A B near the edge of the mercury ring. Then the wire O P will be attracted by A B until P reaches the position P_1, Chap III. § 6; the wire will then be repelled till it reaches the position P_{111}, when it will be again attracted, and thus continuous rotation may be produced in the direction shown by the arrow, if the other portions of the circuit are arranged so as not to neutralise the series of actions described. The force available even with very powerful currents is small.

Again, let the fixed current flow in the circle A B as shown by the arrow, Fig. 143; the moveable wire O P in which a current flows from the centre to the circumference will be continuously impelled to rotate in a direction opposed to that of the fixed current. The force will be very small, but we may multiply it by using a coil of many turns for the conductor A B. No convenient way has yet been found of multiplying the conductor O P, and the power given out by this arrangement is therefore still very small.

A horizontal circular current also tends to produce continuous rotation in a vertical current approaching it or receding from it. Thus let a moveable system P M P_1, Fig. 144, be

FIG. 144.

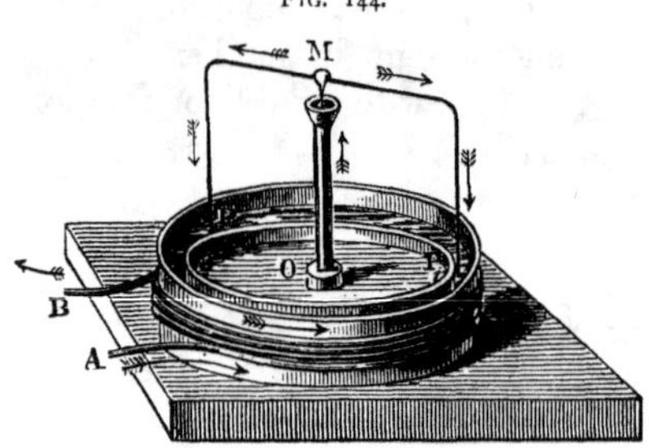

placed in the centre of a fixed ring A B, through which a current flows as shown by the arrow. Let the ends P and P_1 dip in a mercury trough, by which the circuit through

O P and O P_1 may be maintained : both vertical currents descending to P and P_1 are acted upon in one direction by the fixed current, and tend to turn P M P_1 in a direction opposed to that of the current in A B.

§ 2. Currents can be made to rotate by magnets, and magnets by currents, under the influence of continuous electro-magnetic attraction and repulsion. Let a magnet N S, Fig. 145, be weighted so as to float upright in a vessel filled

FIG. 145.

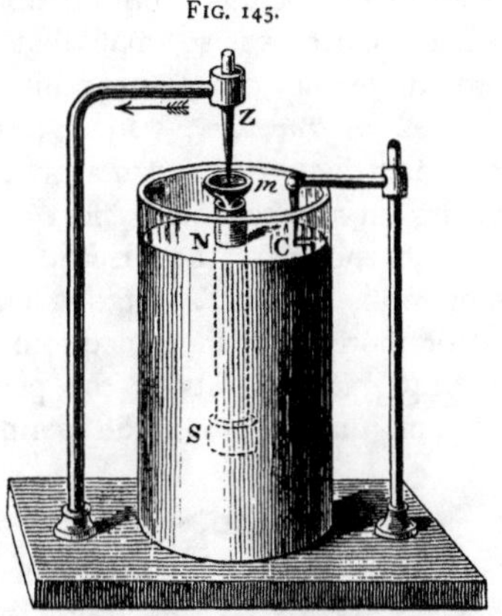

with mercury, and let the upper end of the magnet carry a little capsule *m* of mercury, serving to connect the magnet with one pole of a galvanic battery by the point Z, and yet leave it free to rotate ; the magnet should be well varnished, except at its lower end. Let the other pole of the battery be brought to the mercury near the magnet by a wire C : the magnet will rotate so long as the circuit is complete. The cause will be obvious if we consider the magnet to be a kind of solenoid, for then a force will act between each ring of the solenoid and the current going from the centre to the circumference, as in the second experiment of the last §. The force in this case will cause the ring (the solenoid or magnet) to rotate, the current flowing from centre to circumference being fixed.

If the magnet be fixed and a little wire frame similar to that in Fig. 144 be pivoted upon it with the two vertical ends P dipping into the mercury near the magnet, the frame will be caused to rotate by the magnet. This is explained by the third experiment of § **1**, if we look upon the magnet as a solenoid.

§ **3**. The power to be obtained from the above arrangements of magnets and currents is so small that they cannot be employed to drive any other apparatus, and cannot therefore be termed *electromotors*. By alternately magnetizing and demagnetizing electromagnets we can construct electromotors giving out as mechanical effect a considerable fraction of the whole energy of the electric current. The simplest electromotor is Froment's rotating engine. This consists of one or more horse-shoe electromagnets, A A_1, fixed as in Fig. 146, radially outside the periphery of a drum, D, capable of rotation. On the periphery of this

FIG. 146.

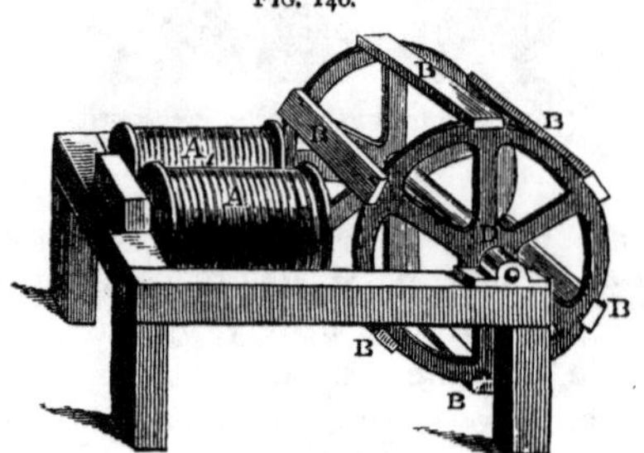

movable drum there are a series of soft iron bars or armatures, B B B, etc. As the drum revolves it completes a circuit, by suitable make and break pieces, sending a powerful current through each electromagnet as each armature approaches its poles within 15° or 20°: the electromagnet then attracts the armature and so drives the drum forward. The circuit is interrupted, and the magnet therefore unmade, just as the armature passes the poles; the drum continues its rotation by inertia or by the action of another electromagnet, until a second armature approaches the poles of the first electromagnet, when the circuit is made as before. The make and break pieces and successive electromagnets are so arranged that the current is not cut off from one circuit till it can flow through the next. This has the double advantage of tending to produce uniformity in the driving action and of preventing the passage of sparks when the contacts are made and broken. These sparks tend to burn the contacts, and gradually to prevent them from closing the circuit.

Another form of electromotor is constructed, resembling the ordinary beam steam engine; the piston is represented by a magnet which is alternately sucked into a hollow coil, and repelled as the current in the coil is reversed; sometimes a soft iron piston is used, which is alternately attracted and set free.

§ **4.** Much more attention would be directed to electromotors than they have hitherto received were it not for the fact that they are necessarily at least fifty times more expensive to maintain in action than the ordinary steam engine. Zinc is the cheapest material by the consumption of which electricity is produced. The energy evolved by the consumption of one grain of zinc is only about $\frac{1}{10}$th of that developed by the consumption of a grain of coal. A large fraction of the energy in the case of the zinc can be converted into an electric current, whereas we have not yet discovered any means of obtaining the energy of coal except as heat, and we necessarily waste a great part of this heat in the process of transforming it into mechanical energy. In the transformation of energy into mechanical effect the advantage lies with electricity. The whole of the energy either of heat or of an electric current can never be transmuted into mechanical effect. In the best steam engines not one quarter of the heat is so transformed; more frequently about a tenth is so used. It is probable that larger fractions than these of the total energy could be transformed by an electromotor into mechanical effect; but this advantage, even if realised, cannot nearly counterbalance the disadvantage entailed by the cost of zinc, which is 20-fold that of coal weight for weight, and 200-fold that of coal for equal quantities of potential energy. In estimating as above that the zinc motor may be only 50 times as dear as the coal motor, I assume that the electromagnetic engine may be four times as efficient as the heat engine in transforming potential into actual energy.

CHAPTER XXII.

TELEGRAPHIC APPARATUS.

§ 1. THE instruments used in telegraphy may be divided into two great classes:—I. Those which transmit signals representing the alphabet by signs of a purely conventional character. II. Those which transmit signals shown or recorded in some ordinary printed alphabet.

In the first class the apparatus is simpler, because the symbols representing the alphabet are chosen with reference to the indications most easily produced by electricity in a telegraphic circuit. The advantages of the second class of instruments are, that the chances of error which result from the translation of telegraphic symbols into ordinary writing are avoided, and that no special training is required to read the messages as they are received. Each class is best suited to a special kind of work. For the general business of the country, carried on by a special staff, the first class is almost wholly employed, and will probably retain this pre-eminence. For private telegraphs read by untrained persons, and for large stations where highly-trained mechanics and electricians can be employed, the second class of instruments, which show messages in letters or print them in type, will probably also continue to be employed.

Both classes may be subdivided into those instruments in which a *galvanic battery* generates the current, and those in which the current is induced by a *magneto-electric* arrangement.

§ **2.** A telegraphic circuit, when a battery is used, consists of (1) an insulated *wire* connecting the transmitting and receiving stations, (2) the wire of the *receiving apparatus* at the station where the message is to arrive, (3) *the earth*, which conveys the received current back to the sending station, (4) the sending *battery*, or other *rheomotor*,* which is alternately allowed to transmit its current into the line, and insulated from that line by the manipulator who works the *sending apparatus*.

The sending apparatus is commonly some contrivance for making or breaking the connection between the battery and the line; so that when the circuit is completed, its resistance is the sum of the resistances of the battery, the line, the wire in the receiving apparatus, and the tract of earth connecting the two stations. When a magneto-electric sender is used instead of a galvanic battery, the resistance of its coils takes the place of the resistance of the battery. In land lines the distinctness of the signals depends, other things being equal, on the strength and uniformity of the currents transmitted; and in order to save the expense of employing batteries or magneto-electric arrangements of great electromotive force, it is desirable to keep the resistance of all the parts low. Thus, the thicker the wire the better will be the signalling with all classes of instruments; but the size of the wire is of much greater importance on long lines than on short ones. The larger the plates of the battery the better, but on long lines the resistance of this part of the circuit sinks

* Rheomotor is the name given by Professor Wheatstone to any apparatus which can generate an electric current.

into insignificance in comparison with that of the line. The less the resistance of the receiving apparatus the better ; but this also forms a small percentage of the whole resistance on long lines. The resistance of the earth between most stations is insensible if care be taken to make the two connexions with earth at the two stations by large plates buried in damp earth. Occasionally, however, it may be necessary to take a wire a long way from the signalling station before a suitable spot for a good earth connexion can be found. Signals are sometimes stopped altogether by a failure in the earth connexion.

CLASS I.

§ **3.** All signals are made by the alternate transmission and interruption of currents, and these currents may be either positive or negative ; that is to say, they may be sent from the copper or zinc pole of the battery into the line, the other pole of the battery being necessarily put to earth at the same time. The following are the elements out of which every telegraphic alphabet must be compounded in Class I.

1°. The relative length or duration of the currents sent.

2°. The relative strength of the currents.

These strengths may range from zero upwards through all strengths of positive current, and from zero downwards through all strengths of negative current.

The simplest symbols are those which record merely two lengths, one long and one short ; and those which record merely two strengths, one positive and one negative. The *Morse alphabet* is the standard example of the former class, and the *single needle* alphabet the standard example of the second class.

§ **4.** Morse signals are sent by a simple key, which the operator depresses when he wishes to send a current, and raises when he wishes to interrupt it. Fig. 147 shows a common form. The insulating parts are generally made of dry wood, the resistance of which is amply sufficient.

A short depression or mere tap sends the short elementary signal technically called a *dot*; a longer depression sends the second elementary signal technically called a *dash*. The Morse alphabet is formed by a combination of dots and dashes, separated by equal intervals. The letters are

FIG. 147.

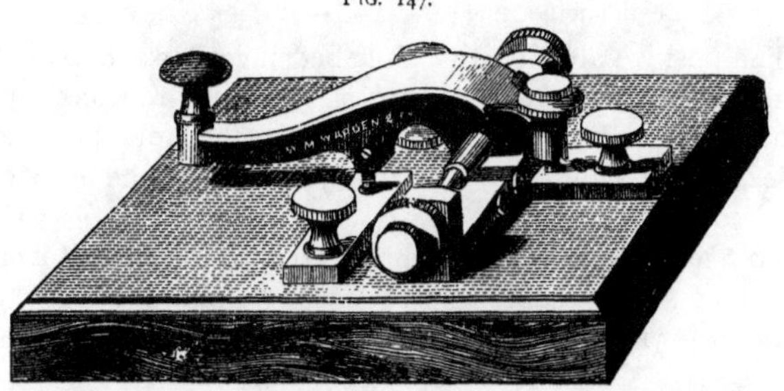

separated by longer pauses, and words by still longer intervals.

The following table gives the Morse alphabet.* The short lines are dots, the long lines dashes.

A · —	J · — — —	S · · ·
Ä (æ) · — · —	K — · —	T —
B — · · ·	L · — · ·	U · · —
C — · — ·	M — —	ü, ue · · — —
D — · ·	N — ·	V · · · —
E ·	ñ — — · — —	W · — —
é · · — · ·	O — — —	X — · · —
F · · — ·	ö, œ — — — ·	Y — · — —
G — — ·	P · — — ·	Z — — · ·
H · · · ·	Q — — · —	Ch — — — —
I · ·	R · — ·	

Full stop (.) · · · · · ·	Note of admiration (!) — — · · — —
Colon (:) — — — · · ·	Hyphen (-) — · · · · —
Semi-colon (;) — · — · — ·	Apostrophe (') · — — — — ·
Comma (,) · — · — · —	Parenthesis (— · — — · —)
Note of interrogation (?) · · — — · ·	Inverted Commas (" ") · — · · — ·

1 · — — — —	6 — · · · ·
2 · · — — —	7 — — · · ·
3 · · · — —	8 — — — · ·
4 · · · · —	9 — — — — ·
5 · · · · ·	0 — — — — —

Bar of division — — — — — —
Call signal — - — - — - —
Understand message - - - — -
Repeat message - - — — - -
Correction or rub out - - - - - - - - -
End of message - — - — - - — -
Wait - — - - -
Cleared out and all right - — - - — - - — -
Begin another line - — - — - -

The positive and negative alphabet may be exactly similar to the above; the dash, or long signal, being replaced by a mark on the right side of the paper, or by the motion of some index to the right, and the dot by a mark on the left side, or a motion to the left.

§ 5. Ink marks similar to those printed above are made on a long strip of paper at the receiving end of a line, by the device shown in Fig. 148.

FIG. 148.

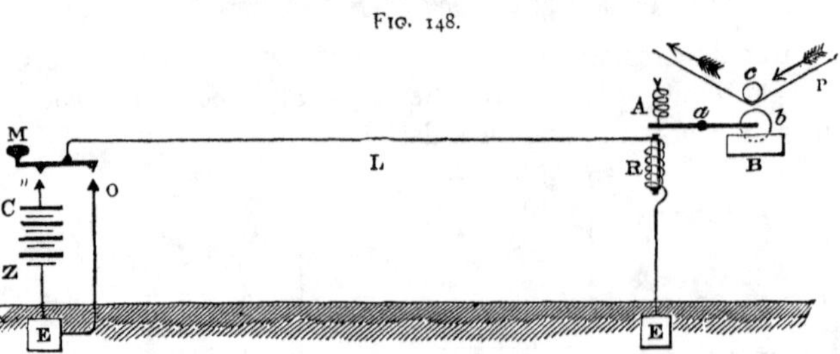

Let M represent the Morse sending key; L the insulated line, reaching from the sending station to the receiving station, where the conductor is connected to one end of the wire of an electro-magnet R, the other end of that wire being directly connected with E, the earth. Let A be a soft iron armature hinged at *a*, and having a narrow roller *b* continually revolving in an ink trough B. Let the strip of paper P be continually moving in the direction of the arrows. Then when M is depressed, making contact at *m* with one pole of a battery C Z, the other pole of which is to earth, a current will flow through the whole circuit and make the core of R magnetic. The end A of the armature will be depressed, the little roller pressed against the paper, and a black mark made, the length of which will depend on the rate at which the paper is moved, and the time during which M remains depressed. On raising the handle M so that the contact is now made at O, the current will cease to flow ; the core of R will lóse its magnetism : A will rise, pulled up by a little spring, and the ink mark will cease on the paper. Thus a short depression of M will make a short mark or dot ; a long depression of M will make a long mark or dash. The handle M is in the diagram shown in a neutral position, making contact neither at O nor at *m* ; in practice it is never in this position, but makes contact at O when not depressed by hand.

Fig. 149 shows a complete Morse ink writer as made by Messrs. Siemens Brothers. The following is a description of the instrument almost in their own words :—E is the electro-magnet, through which the received current passes. N is a handle by which the clockwork is wound up.

The clockwork placed inside the instrument turns a small milled roller W, and the printing disc D. The friction roller W_1 is pressed, by means of a spring V, upon W, and turns with it.

The disc of telegraph paper S is placed upon the horizontal wheel P, which turns on a hardened pivot *a.* Horizontal wheels for paper were first introduced by Mr. Stroh, and are much superior to vertical wheels. The end of the strip of paper is led round the roller S^1, turning on a vertical axis, thence under the roller S^{11}, over the roller *s*, and under the small steel roller *i*, where it is struck by the printing disc D, on the armature *e* being attracted by the electro-magnet E. From the small roller *i* the strip of paper passes between the friction rollers W and W_1, which, when they revolve, draw the paper forward in the direction of the arrows.

The roller W_1 can be lifted by the small handle X ; and it will be found convenient to lift it in this manner when introducing the paper between the friction rollers W and W_1.

Fig. 149.

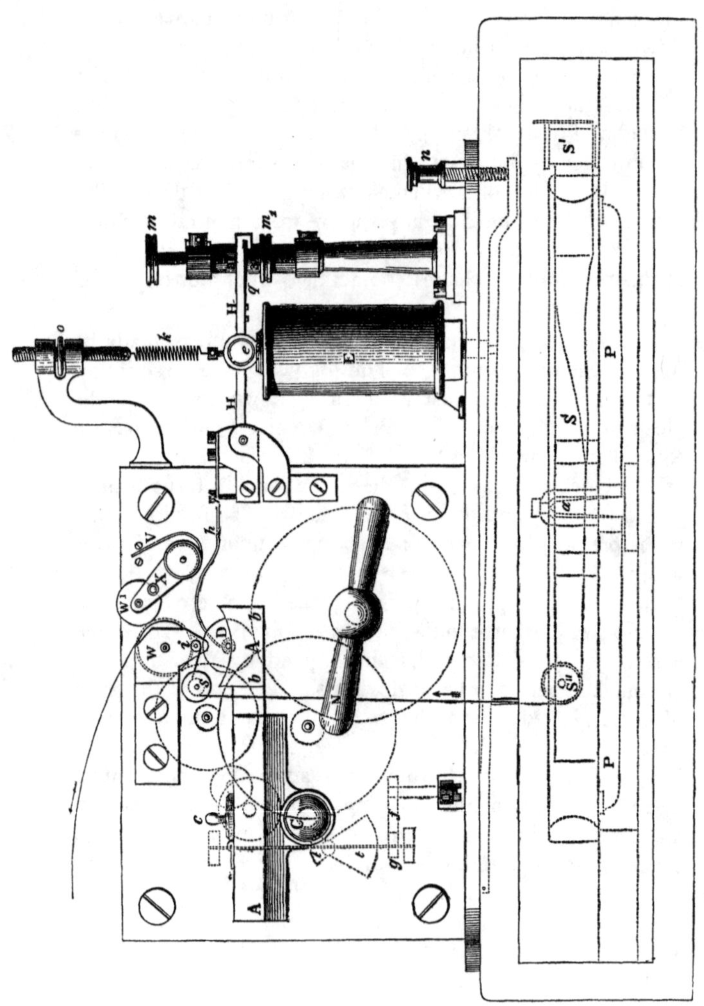

A A is a brass vessel for holding a supply of printing-ink, the opening to which for putting in the ink is supplied with a cover *c* to prevent dust from getting into it; the vessel terminates in an open cup or trough *b b*, in which the printing disc D revolves. The vessel A A is fastened to the side of the apparatus by means of a screw with a milled head C, so that it can be easily removed for refilling or cleaning. The spindle on which the printing disc D is fastened revolves in an eye at the end of the continuation *h* of the printing lever H H. The spindle is made to revolve by being joined, at the end furthest from the printing disc, by a species of universal joint, to the end of a short spindle carrying a cog-wheel in gear with the clockwork. The printing disc is thus kept revolving, although free to follow the motions of the printing lever.

Should it be wished to stop the clockwork of the instrument, the handle Q must be pushed to the right, by which the spring *f* is pressed against the small metal collar *g* of the regulator *t*. The release of the clockwork is effected by moving the handle Q in the opposite direction.

The cores of the electro-magnet are of soft iron, united by a cross-bar and surrounded by the wire coils. The lever H H moving between the points 2 and 3 of the screws *m* and m_1, carries on one arm an armature of iron *e*, and at the other end the continuation *h*, in an eye at the end of which revolves the end of the spindle which carries the printing disc D.

The contact screws *m* and m_1 limit the play of the printing lever H H. In order to draw the lever back to its normal position as soon as a current has ceased, a spring *k* is provided, the degree of tension of which can be regulated by means of the nut *o*. Another adjustment has been adopted, in addition to the above, by which the electro-magnet E has been made moveable, and can be raised or lowered by means of the milled headed screw *n*, thereby increasing or decreasing the distance between the cores of the magnets and the armature *e* of the printing lever H H.

When the circuit, Fig. 148, is closed at *m* a current from the copper of the distant battery, after traversing the line, enters the printing instrument R, passes through the coils E of the electro-magnet, Fig. 149, and leaving the instrument returns through the earth to the zinc of the distant battery. As long as the current lasts, the iron cores are converted into magnets, the free ends of which will attract the armature *e* and thus set the printing lever H H in motion. The continuation *h* of the printing lever H H consequently presses the disc D against the paper band, upon which it produces a dot or a dash, according to the length of time during which the armature is attracted by the cores.

There are many modes of receiving and recording the Morse signals besides that just described. In many old instruments the roller *b*, Fig. 148, is replaced by a mere steel pointer or style, which makes a little indented line when pressed on the paper by the depression of A. In Bain's chemical telegraph, Fig. 150, the electro-magnet R is wholly dispensed with. The depression of M sends a positive current

FIG. 150.

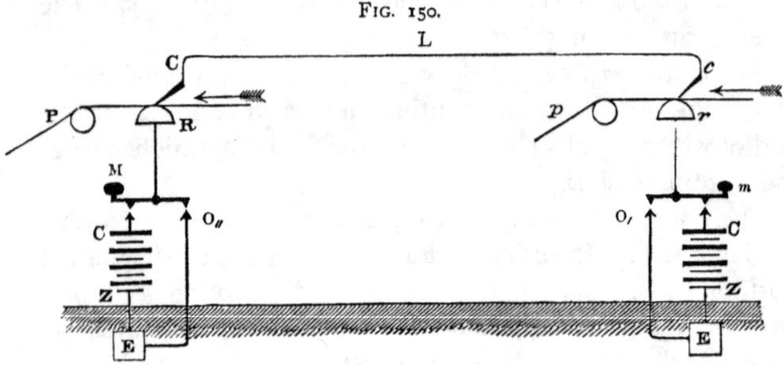

through R, C, and L, and then at the receiving station through a steel style *c*, pressing on a band of paper *p*, which has been soaked in a mixture of equal parts of saturated solutions of ferrocyanide of potassium and nitrate of ammonia. The current next flows to *r* and through *m* to earth, the handle of *m* being raised. The diagram shows the connections so arranged that all signals can be sent from either end. At the receiving station the keys M or *m* make contact at O or *o*. Prussian blue is deposited so long as the current passes through the paper, and thus the long and short signals are recorded by short or long blue marks. There should be a slight excess of carbonate of ammonia in the solution of nitrate.

Sometimes the Morse signals are indicated to the ear or eye without being recorded. Thus, even if the paper at P, Fig. 148, be removed, the mere sound of the armature as it rises and falls is intelligible to the ear of a skilled operator. The *sounder*, as it is called, is coming into extensive use and consists of a Morse receiver without clockwork or paper or inking roller. The sound is produced by the tapping of the lever H, Fig. 149, against the stops *m* and m_1. The mere deflection of a galvanometer needle, included in the circuit at R, will be equally intelligible to the eye. It is only necessary to make the needle light and confine its motion within narrow limits, so that each current in passing produces a single well-marked depression lasting for a longer or shorter time, and not a series of unchecked oscillations.

§ **6.** The simplest form of receiving instrument for positive and negative signals is a little galvanoscope, the index of which can deflect only a short distance to right or left of its zero, being checked by stops. The inside of one of these instruments is shown in Fig. 151. I I are the coils fastened to the back of a little door which opens to allow the works to be got at ; A is a support in which one pivot of the needle works ; N P are the keys used in sending ; the needle S N and pointer *a b* are shown in Fig. 152. The key by which the positive and negative signals are sent from one and the same battery is better shown in Fig. 153. L and E are two springs connected respectively with the line and with earth. They, when untouched by the hand, press against the upper bar C, which is connected with the copper pole of a battery. Either spring can be depressed by the finger so as to come in contact with the bar Z, which is connected with the zinc pole of the battery. If L is depressed, a negative current flows into the line; if E is depressed, a positive current flows into the line. The galvanoscope at the other end is so connected that the depression of the left-hand key causes a deflection to the left; a depression of the right-hand key a deflection to the right. The form of galvanoscope used is

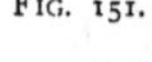

FIG. 151.

called the *single needle* instrument, and the alphabet the single needle code. The Morse code given above is often used, a dot being a deflection to the right and a dash a deflection to the left.

Sir Charles Bright introduced the *bell instrument* as a substitute for the single needle. His instrument contains two bells struck by the depression of the armatures of two electro-magnets, one working each bell. Each electro-magnet was worked by its own relay ; one of the relays worked when a positive current was received, and the other when the received current was negative. This instrument is falling into disuse.

§ **7.** The connections shown above are most suitable for comparatively short lines. On longer lines more complex arrangements are generally adopted, involving the use of relays. The *Relay* is an instrument which retransmits the original signal from a fresh battery : it may be used either to send this signal to a distance along a second section of line, or simply to send a strong current from a local

FIG. 152. FIG. 153.

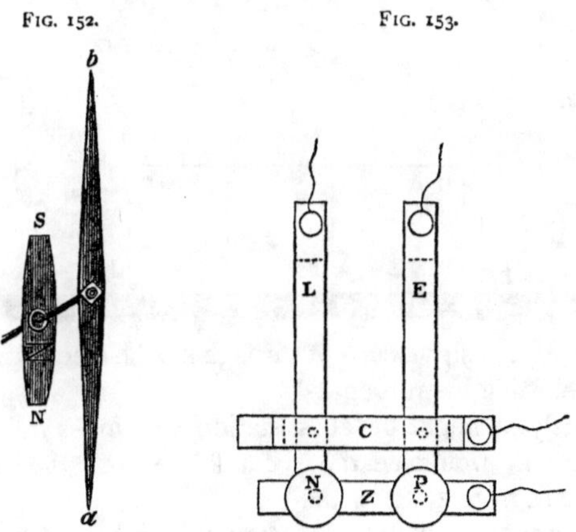

battery through the receiving instrument. The current received from a distance is often so diminished by leakage that it is insufficient to work the electro-magnet which marks the paper, or to give legible or audible signals, and yet it may be sufficiently strong to move an armature with sufficient force alternately to make and break an electric contact, and thus indirectly to work the receiving or recording instrument. Fig. 154 shows the connection for a Morse system with relays at each end, worked by single currents.

Corresponding parts at the two stations are indicated by the same letters, capitals being used for one station and italics for the other. R is the relay, and C Z the sending battery; R_1 is the Morse instrument, and C_1 Z_1 the local battery used to work it. The depression of the key *m* making contact at *o* sends a positive current through the line L to M, and through the contact P to the electro-magnet R of the relay and thence to earth. The electro-magnet R attracts the armature of the relay, making contact at N and

FIG. 154.

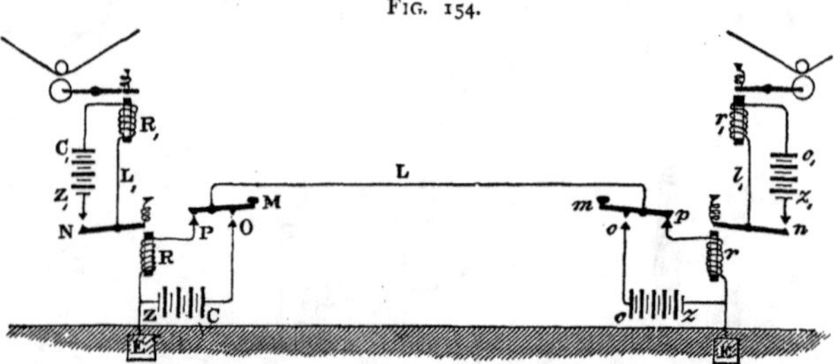

thus sending a positive current through R_1, the electro-magnet of the recording instrument.

Obviously R_1 might be at a station 100 miles from R, in which case L_1 would be the second line, and the portion of the circuit from Z_1 to R_1 the earth.

Relays are constructed so that a very slight difference in the strength of a current determines whether the moveable tongue or armature makes contact at N, or rests against an insulated stop. Care is also taken to provide such adjustments that the tongue may be made to move with any desired strength of current: thus the relay may be set so that with zero strength the tongue rests on the stop and makes contact when the current reaches the strength unity, or it may be set so that it rests against the stop when the current has a strength 100, and makes contact when the current has a strength 101.

Relays are also often made so that the tongue moves only with a current of one sign, remaining unaffected by a current of the opposite sign; the core of the electro-magnet may in this case be a hard steel magnet, the polarity of which is never reversed by the currents received. Other relays are made so that when the tongue has once been deflected to make contact, it will not return until a reverse current has been sent through it. The best known form of this species is the polarized relay made by Messrs. Siemens, and shown in Fig. 155. S is the south pole of a hard steel magnet, the north pole of which is bifurcated and ends in the two pieces n n_1, between which the tongue a of the relay oscillates, pivoted at D. The coils are wound round the two north branches of the magnet in opposite directions, so that a current in one direction tends to make n_1 north and n south, while the reverse current would make n_1 south and n north. The tongue a, made of soft iron, becomes a south pole by contact with S S.

FIG. 155.

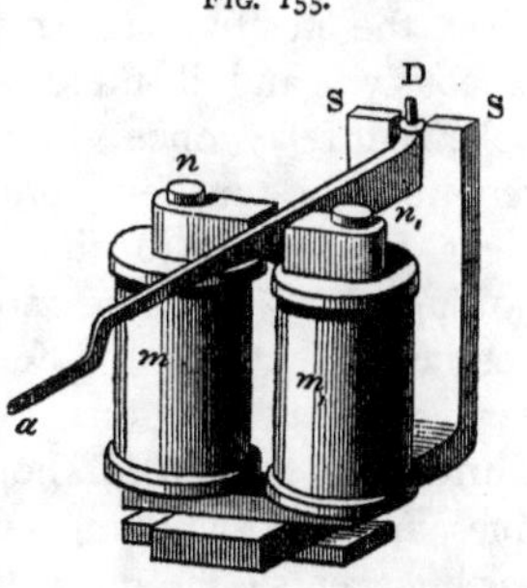

Relays can be arranged so as to send positive and negative currents corresponding to positive and negative currents received.

The Morse ink-writer can easily be arranged so as to act like a relay, the armature being employed to make the necessary contacts instead of to mark paper. With instruments of this class Messrs. Siemens, on the Indo-European line, work from London to Teheran, a distance of 3,800 miles, without any retransmission by hand. There are five relay stations in this circuit.

§ **8.** In ordinary Morse signals and in all others where only one current is absolutely required, there is nevertheless some advantage in using the negative current to draw back the armature and so terminate each signal. This system was introduced by Mr. Varley. It considerably simplifies the adjustment of the relays and has other advantages. Where these reverse currents are not used, the relay tongue must be pulled back by a spring or by magnetic attraction, and their adjustments require to be continually altered. This spring requires continual adjustment to suit the strength of the received current, which varies much during each day as the insulation of the line varies. With a polarized relay and reverse currents, no such adjustment is required, because the positive and negative currents decrease simultaneously; and if there were no *earth currents*, a good polarized relay once set for reverse currents would never require to be touched; practically, all relays require adjustment from time to time. Earth currents are currents flowing along the line, not sent by the batteries, but depending either on a difference of potential between the earth at the two stations or on induction from passing clouds. Currents often flow for hours in one direction through the lines, and the signalling currents are superposed on these earth currents; the relays then have to be set, so that when no signal currents are passing the armature is attracted more strongly by one armature than by the other, and the amount of this bias must be regulated as the earth currents vary.

§ 9. With the connections as shown in Fig. 154, although no current is sent direct from the battery through the home relay circuit, every signal sent causes the relay at the sending station to work, if the line is long and well insulated, or if it includes many miles of underground or submarine wires. This action is due to the statical charge which accumulates on the line L. When contact is made by the key M at O, the line L becomes statically charged. When contact is broken at O, and made at P, part of this statical charge flows to earth through the relay R, the other portion flowing on through the distant relay *r*; thus the key M as it makes and breaks contact causes intermittent currents to flow through the home relay which will work the local Morse instrument R_1. This action is not only unnecessary, but is detrimental, because the currents returned in this way are often so strong as to alter the permanent or residual magnetism of the relay, which then requires readjustment when signals begin to arrive from the distant station, and moreover the local battery C_1 Z_1 is put in action by these return currents when not required. The return current is especially great when any portion of the line L is formed of wire coated with india-rubber or gutta-percha, because lines so formed have a much larger electrostatical capacity than the ordinary aërial land line. Where this inconvenience exists, each station may be provided with an apparatus called a switch, by which the connections are altered at will, so that when the station M, fig. 154, for instance, is sending the relay, R is not in the circuit between P and E, which points are then directly connected.

The sending key M is sometimes so made as to put the line to earth for a short time between the two positions where it makes contact respectively with O and P.

A still better arrangement for discharging may be employed, in which the action of the current sent from the home station puts P to earth by means of a separate relay, and keeps P to earth by residual magnetism for a very short time after the key M has broken contact at O and made contact at P. With this arrangement the distant station can at will interrupt the sender.

§ **10.** The following points must be attended to in the construction of telegraphic apparatus :—

The core of the electro-magnet should be arranged so that its magnetism changes rapidly at the commencement or cessation of a current; otherwise rapidly alternating changes produced by rapid signals will not be registered by the armature. With this object, if soft iron is used, the mass should not be great; the core should be hollow, and split longitudinally; and the iron should be carefully selected with as little coercive force as possible. The highly magnetized cores of polarized relays gain and lose the small increments of magnetism due to feeble currents with less delay due to coercive force than is experienced with soft iron. The coercive force in the armatures is another source of delay in rapidly alternating signals. These armatures should, therefore, be made light, and must not pass through very different states of magnetization. If allowed, for instance, actually to touch the core of the electro-magnet, they become so highly magnetized that when the electro-magnet is weakened by the cessation of the current, they often adhere to the core under the influence of residual magnetism, requiring a very strong spring to pull them back, and consequently a very powerful current to pull them against the spring to the electro-magnet. The most delicate relay is that in which, other things being equal, the armature moves in a nearly constant magnetic field, which is alternately weakened and strengthened by the received current. The alteration produced in the magnetic field of the electro-magnet by the passage of a current should, however, be the greatest which that current can produce, and this condition requires that the iron or steel core should not be very small; moreover, some little pressure must be exerted at the contacts, or the tongue of the relay will be made to tremble by the mere passage of the local current, which exercises a repulsion on itself; to obtain the necessary force, the armature must have considerable bulk: these two last conditions are antagonistic to those first mentioned, and experiment alone can determine the best proportions. The form of the electro-magnet should be such as to give the strongest and most uniform field possible with a given intensity of magnetization. This condition is entirely violated in the common relay or ink-writer, where the armature stretches across the poles of an ordinary horseshoe magnet. It is much more nearly complied with in the Siemens polarized relay described above. The form of the iron or steel core and the distribution of the core on the magnet should be such as to give the maximum intensity of magnetization per cubic centimètre of core consistent with a given current passing through a given length of wire. This condition is probably very imperfectly fulfilled by any relay yet constructed.

The mass of the armature should be so distributed that its moment of inertia may be the smallest that is consistent with the necessary weight of the armature and position of the pivots; any increase in the moment of inertia produces a proportional diminution in the angular velocity with which the tongue will move under a given force, and the rate at which a relay will work depends on this angular velocity. If the moment of inertia be doubled, the force remaining the same, the angular velocity acquired in a given time will be halved, and the angle traversed in that time will be halved; but to traverse the same angle, i.e. to pass from one contact to the other, will not require double the time, but only 1·414 times the time required by the lighter armature, because $1{\cdot}414 = \sqrt{2}$. The moment of inertia is the sum of the products of the weight of each particle into the square of its distance from the pivot round which the mass rotates: it is therefore not only desirable, when rapid motion is to be produced by a weak force, that the weight should be small, but also that it should be near the pivots.

No harm is done, however, by putting the pivots far from the points of contact, because we thereby diminish the angle through which the armature has to move between the contacts; so that if we halve the angle and double the moment of inertia, the one change exactly compensates the other.

The wire on the electro-magnet (or in the coil of the single needle instrument) should have a moderate resistance relatively to that of the whole circuit:[1] thus on short lines a

[1] One authority says $\frac{5}{16}$ of the resistance of the whole circuit; this seems very large.

thick, short wire should be used for the electro-magnet; but on long lines, relays with long, thin wires are required. The reason for this is the same as that for using galvanometers with long coils to test insulation, and galvanometers with short coils to observe currents in circuits otherwise of small resistance. The common single needle instruments have a resistance of about 200 ohms, the coil being made of No. 35 wire.

The direct ink-writer used for short lines may be coiled with No. 35 wire (0·005 inch diameter), and have a resistance of about 500 ohms.

The electromagnets in local instruments (no line wire on circuit) are made with wires of from ·022 inch to 0·012 inch diameter (Nos. 24 to 30).

A Siemens polarised relay may be made with No. 40 copper wire, and have a resistance of 500 to 700 ohms. These relays sometimes have a resistance of 3,500 ohms.

All contacts must be made by platinum points, platinum being the only metal which is not oxidized or dirtied by the passage of the little spark which accompanies the making and breaking of the circuit. This spark wears out even the platinum contact pieces in time: it may be avoided by connecting permanently the two contact pieces through a resistance so large that the current passing when contact is broken is small enough not to be injurious. The same object is gained by placing a small condenser between the contact pieces, each contact piece being connected with one of the two armatures.

§ **11.** In place of a voltaic battery, a magneto-electric arrangement may be employed to send currents. Thus a Siemens armature worked by hand may be employed to send Morse signals, the motions of the hand being similar to those required for the Morse key. The depression of a handle moves the armature in one direction, and sends, say, a positive current, which by a polarized relay causes an ink-writer to begin marking the paper. So long as the armature and handle remain depressed the ink-writer continues to mark, though no current is flowing through the relay, the tongue of which is held over by the permanent magnetism of its magnet; when the handle is raised and the armature moved back to its original position, another short current is sent in the opposite direction to the first. This second current throws back the tongue of the relay, and the ink-writer ceases to mark. The current produced is the equivalent of the power employed to work the armature; considerable force must therefore be exerted to send a current suitable for a long circuit. Other magneto-electric arrangements are used to send + and − signals for the single needle receiver. The induced currents are of very short duration; and hence, although the E. M. F. which produces them may easily be made much greater than that of the batteries usually employed to signal, yet the actual quantity of electricity transmitted for each signal is generally much less than is sent by a battery.

On a long line the received current is longer in duration than the sent current, and proportionately feebler. On a short line the received current and that sent are both so short, that even when strong they may fail to move an armature which would work freely with a feebler current prolonged for a longer time. The E. M. F. produced by the magneto-electric arrangement is so great near the sending station, that the leakage is much greater in proportion to the whole quantity of electricity sent than when a battery is used. This would not be the case if the resistance of the faults where electricity escapes followed Ohm's law, but the resistance of faults seldom follows Ohm's law. More especially surface conduction, which is the chief cause of leakage on land lines, allows much more than double the current to pass when the E. M. F. is doubled. On underground or submarine lines the high potential produced for a short time by the magneto-electric sender tends to send minute sparks through the insulating material, and so to cause faults.

Magneto-electric senders, owing to the above causes, are not much used on long or important lines.

§ 12. The simple Morse or + and – key can be worked at the rate of from twenty-five to thirty-five words per minute by a skilled operator. Receiving instruments can, however, record even more than 100 words per minute (of five letters each). Automatic transmitters have therefore been adopted in which the messages are prepared by several operators, being represented by punched paper or metal types, and these types or paper strips passing through the transmitter determine the required succession of currents. Sir Charles Wheatstone's automatic transmitter is the most successful yet used. In this instrument the messages are represented by three rows of holes in a strip of paper. For + and – signals a hole on the right-hand side represents a + signal or dot, a hole on the left-hand side a – signal or dash. Uniformly spaced central holes serve to move the paper on at a constant speed. The right and left-hand holes determine the contacts made and signals sent very much as the cards in a Jacquard loom determine the pattern in woven stuff. The contacts are determined by the position of two little plungers, which are either kept down by the unpunched paper or come up through the holes. Whenever a plunger rises through a hole a current is sent into the line; a + current when the hole is on the right side; a — current when the hole is on the left side. The contacts are pressure contacts, with a slight slip at the moment of making contact, which are superior to any contact in which the surfaces merely slide one on the other. By a somewhat more complex arrangement of similar character, the long and short Morse signals are sent. A full description of this instrument is given in the Fifth Edition of Mr. R. S. Culley's Hand-book of Practical Telegraphy.

CLASS II.

§ **13.** The elementary signals used in those telegraphic systems which show or print letters are produced, as in Class I., by the alternate transmission or interruption of currents, sometimes all of one sign, and sometimes both positive and negative; but these transmissions and interruptions are not themselves the subject of direct observation or record: they are used to work the escapement of clockwork in what may be termed 'step by step' instruments, or to connect synchronous actions in the sending and receiving instruments, which are driven with similar motions at the two ends of the line.

The 'step by step' instruments sometimes print the messages, but more frequently show the required letters in succession on a dial. The synchronous instruments all print the letters, but they effect this by various distinct inventions, the more striking of which are Hughes's, Caselli's, and Bonelli's.

All 'step by step' instruments are very much alike. A ratchet wheel on an axis bearing the pointer is worked by a propelment which, as each current passes, turns the ratchet through a segment of a circle corresponding to one tooth or half a tooth of the ratchet. Fig. 156 shows a form now made by Messrs. Siemens Brothers, and very similar to that first introduced by Sir Charles Wheatstone: *n s* are two poles of a polarized electro-magnet, similar to that used in their relay (§ 10 above). The soft iron tongue T works between these, pivoted at *t*, being attracted to *s* by one current, and to *n* by the reverse current. The tongue T carries at its other extremity one end of the axis of the ratchet wheel D, having thirteen teeth; the other end of the axis is on a fixed bearing, and carries the pointer. The play of T is limited by two stops, q, q_1, and the rotation of the ratchet is determined by two stops p, p_1, and four springs, h, h_1, h_2, h_3, two of which, h and h_1, have a catch at their end, adapted to hold the ratchet. The tongue T is shown drawn towards *n*; the ratchet is locked by the spring *h*, so that it cannot turn to the right neither can it turn to the left, because it is locked by the stop *p*. The position of the pointer is therefore perfectly definite.

FIG. 156.

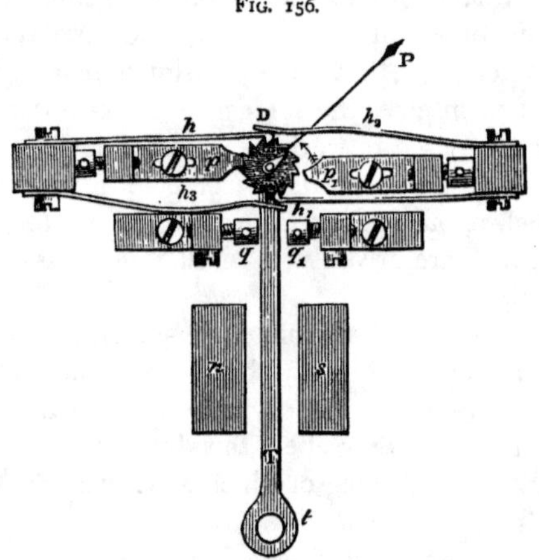

The next current received will attract T to *s*; the spring *h* will turn the ratchet $\frac{1}{26}$ of a revolution, and it will then be locked by the spring h_1 and the stop p_1; the following current will turn the ratchet an equal distance by moving it towards *n*, and thus each alternate current will carry the pointer forward by $\frac{1}{26}$th of a revolution over the dial, on which there are twenty-five letters and one blank.

These thirteen positive and thirteen negative currents will cause the index to make one complete revolution. Let us assume that the index is at the letter A, then one current will move the index to the letter B, three currents more will move it to E, and seven currents will send it to L; by sending the right number of currents and then pausing for an instant, the index will be made to travel from letter to letter, and to pause at each letter required to be read. The index may be driven by clockwork and the teeth of an escapement wheel liberated by the currents, or the escapement wheel may, as in the above example, be replaced by a propelment wheel, such that each motion of the armature causes it to move on one tooth. The latter is the plan now most in use.

The right number of currents is sent by means of a dial at the sending station, and an index with a handle which can be turned from letter to letter; the letters on the sending dial correspond in number and arrangement to those on the receiving dial. The handle always moves in one direction and sends one current (positive and negative alternately) as it passes each letter. When the index of the receiving instrument and the handle of the sending instrument have once been set opposite the same letter, the sending operator has merely to turn his handle at a moderate speed to each letter in succession which he wishes to send, and by so doing he will send just the number of currents required to bring the receiving index step by step to the same letter. Should any current or currents fail to move the receiving index, the sender and receiver, finding that the signals are not understood, put their instruments to one letter or mark (sending no currents) by a mechanical arrangement contrived for the purpose, and recommence the message from the point at which it became unintelligible. The currents sent by the handle as it is turned round may come from a battery, or, as is more commonly the case, from a magneto-electric arrangement. Fig. 157 shows the magneto transmitter used by Messrs. Siemens.

The handle H is fastened to the spindle A carrying the toothed wheel L, which latter gears into the pinion T of the cylindrical armature or keeper E. This armature E is mounted vertically upon pivots between the poles of a series of permanent magnets G G G. One revolution of the wheel L, or of the handle H fixed thereto, causes the pinion of the armature E to revolve thirteen times, as the teeth of the former are in the proportion of thirteen to one of the latter. As one full turn of the armature produces two currents of opposite directions in a coil of insulated wire forming part of the cylindrical armature E, twenty-six currents, alternately positive and negative, are generated during one revolution of the handle; the dial is divided, as above stated, into twenty-six parts, viz. twenty-five letters of the alphabet (I and J being taken as one) and one blank.

FIG. 157.

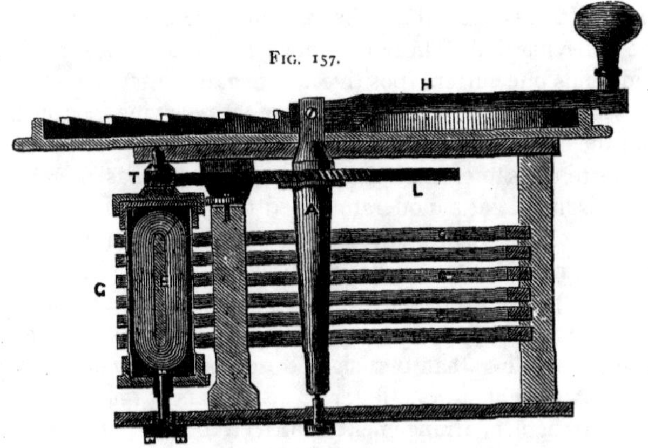

Sir Charles Wheatstone's magneto-electric letter-showing dial step-by-step instrument is perhaps the best yet introduced.

When a radial arm is employed to drive the armatures of magneto-electric induction coils, the induced currents are generally very unequal in strength, because the operator naturally begins and ends the motion comparatively slowly. Sir Charles Wheatstone, therefore, drives the magneto-electric armatures continuously, and regulates the number of currents admitted into the line by a series of stops, corresponding to thirty letters and symbols arranged round a dial. The propelment in the receiving instrument is admirably light and accurate, and its workmanship very perfect. These little instruments are chiefly used for short private lines, but have been employed on circuits of more than 100 miles in length.

§ **14**. The 'step by step' printing instrument is made on a plan differing little from that of the letter-showing instrument. The pointer is replaced by a ring on which the types of the required letters and symbols are placed; this ring is turned by the propelment or by an escapement and clockwork, so that each required letter is brought in turn opposite the paper on which the symbol is to be impressed; the paper is then struck against the letter on the ring by some special device differing in different instruments. In one the mere pause of the dial suffices to allow the striking or printing hammer to act. In another positive currents alone are used to work the escapement, and a negative current, sent when the desired letter is reached, determines the impression by the stroke of a hammer. In a third a second line wire is used to give the blow which prints the letter. The paper then moves on one step. These instruments have not come largely into use. It will be observed that the number of alternating currents required for each letter in the 'step by step' instruments greatly exceeds the number required by instruments of Class I.

§ **15**. The Hughes printing instrument is the typical synchronous printer. The principle on which it is based may be stated as follows:—Two type-wheels, having letters on their periphery, one at the sending and one at the receiving station, revolve with equal velocity, and are moreover so placed that the same letter in each wheel passes corresponding fiducial marks at the same time. The fiducial mark in the receiving instrument is opposite a little roller, carrying a strip of paper which is struck against the edge of the rotating wheel by the release of the armature of an electro-magnet whenever a current is received; a letter is printed by the blow without stopping or sensibly retarding the wheel; the paper is then pulled on a step by clockwork, the armature replaced on the electro-magnet, and all is in readiness for the next letter. The letter which is printed depends on the letter of the wheel which happens to be opposite the roller and paper at the moment when the current arrives. A series of keys like the keys of a pianoforte, and each lettered to correspond with the letters of the alphabet, are so arranged relatively to the sending wheel that the depression of the key A causes a single current to be sent when A is opposite the fiducial mark at the sending station; the current occupies no sensible time in reaching the other station, and strikes up the paper when the A on the receiving wheel is at the fiducial mark. The letter A is therefore printed; if the operator next touches the key N, the sending wheel causes a current to pass when N is opposite the fiducial mark; at the same instant N is opposite the paper and roller at the receiving station, and the letter N is accordingly printed. This action can be repeated indefinitely with any series of letters so long as the two wheels keep perfect time. Each wheel is driven by clockwork, and regulated so as to keep very nearly perfect time, by a spring pendulum, which vibrates with extreme rapidity, and regulates a frictional governor connected with each wheel; any trifling deviation from perfect synchronism is corrected by every current sent. The act of printing slightly accelerates the receiving wheel if it is behind time, and slightly retards it if it is too fast. This is done by a little wedge which, whenever a letter is printed, is forced between the teeth of a star wheel fixed to the type wheel. This wheel is not rigidly connected with the axis on which it is centred but is maintained in its position by friction. This position can therefore be corrected without sensibly affecting the speed of the clockwork. This instrument is the best of the printing instruments hitherto introduced: it has the great advantage that only one current is required for each letter.

§ **16.** Bakewell's and Caselli's copying telegraph apparatus requires synchronous motion at the two ends of the line. The principle on which their instruments are constructed may be explained as follows.

The message is plainly written in common ink on a sheet of paper, A, covered with thin tin foil, Fig. 158. A corresponding sheet of paper, B, is chemically prepared, so that

FIG. 158.

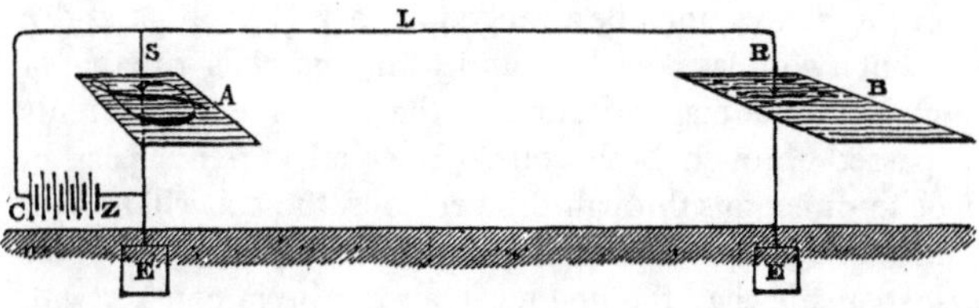

when a current passes through it from a pointer R to earth, a mark is made similar to that used in Bain's instrument. The pointers S and R are drawn across the papers A and B in a succession of parallel equidistant lines with a perfectly synchronous motion. A battery is connected with the tinned paper, the line L, and the earth, as shown in the sketch. When the pointer S touches the tin, the battery is short-circuited through the tin; no sensible current reaches B, and R leaves no mark; but when S crosses the ink on A the current from C Z flows through L, and so long as S remains insulated from A by the ink a line is drawn by the point R.

FIG. 159.

It is easy to perceive that the result must be as accurate a copy of the original writing as can be produced by a series of fine lines interrupted in the proper places, as in Fig. 159.

The synchronism required is in Caselli's instrument obtained by a pendulum at each receiving station; one beat of the pendulum corresponds to each line drawn across the paper; the one pendulum controls the other by a current which it transmits from the sending station through a special circuit temporarily connected with the line.

§ **17.** By various differential arrangements messages can be sent simultaneously in both directions through one line. The currents sent f om the two stations do not really travel simultaneously in opposite directions through the line, but the effect of the signals on each receiving instrument is precisely the same as th ugh the line were being worked in only one direction.

Let the connections be arranged as in Fig. 160. R and *r* represent two relays, each wound with two coils capable of producing equal magnetization in the core if equal currents are passed through both coils. If equal currents pass in opposite directions through the two coils, the coil will neither be magnetized nor demagnetized. M and *m* are two Morse keys, so made that the line must always be in contact with the earth or the battery, or (for a very short time, as the key moves) with both. When the handle at M is untouched, there is unbroken connection from the line round the inner coil of the relay to earth through the contact O and the wire V. There is a second connection between the line and the earth from the point N, through the outer coil of the relay, and through the resistance coils W. The condenser D is connected, as shown, with this branch.

When the handle M is depressed, contact is made at P,

FIG. 160.

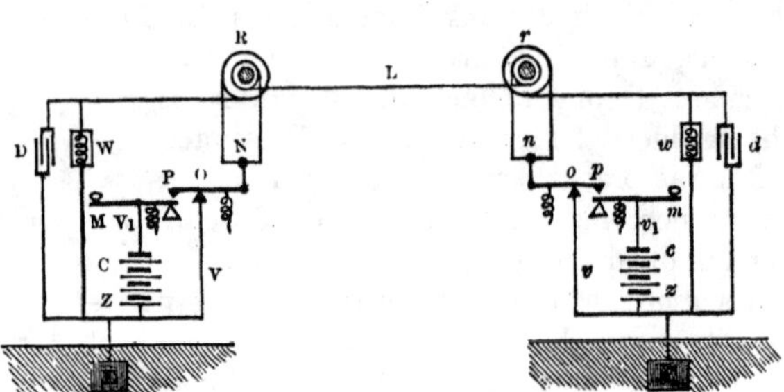

which for an instant short-circuits the battery C Z through the wires V_1 and V, and immediately afterwards contact is broken at O, so that the battery C Z is connected with W and thence with two circuits, one through the line to the distant station and one through the outer branch of the relay to earth at the Home station through W.

The resistance of W is made equal to that of the line L, added to that part of the circuit by which L is connected with earth at the distant station; the capacity of D is so chosen that W and D may represent an artificial line in all respects equivalent to the real line.

Thus there may be nine arrangements of the positions of the keys M and *m*.

1. Let M be depressed and *m* untouched. The battery C Z sends a current round both coils of R, which does not work, as the currents flow in opposite directions; it also sends a current through the line L, and thence round the inner coil of *r* and to earth through *o*; the relay *r* works and gives a signal.

2. Let M be depressed and *m* also depressed. The currents which each battery would send through the line neutralise one another, but each battery sends a current through the outer coil of its own relay; both relays work, and signals are received at both stations. The current sent through the outer coil of each relay is equal to that which the battery would send through the line and inner coil of the distant relay.

3. Let *m* be depressed and M untouched. This case is similar to the first case; a signal is indicated by the relay R.

4. Let neither key be depressed, both batteries are cut off the line and no signal is indicated by either relay.

5. Let both M and *m* be in the intermediate position, contact made at P and *p* but not broken at O or *o*. No signal will be given at either station.

6 and 7. Let the key at M or *m* be in the intermediate position and the other key not depressed; no signal will be indicated at either station.

8. Let the key at M be in the intermediate position when *m* is depressed, the current produced by *c z* will be un-

altered, and the signal will be received through the inner coil of R.

9. If the key at *m* is in the intermediate position, and M depressed, a signal will be received by the inner coil of *r*.

In every arrangement of the keys M and *m*, the effect produced on the relays is such that when *m* is depressed R receives a signal, when M is depressed *r* receives a signal.

This arrangement is a modification of that introduced by Messrs. Siemens and Frischen, and is due to an American, Mr. Stearns.

Mr. Stearns finds it advantageous to introduce two resistance coils, V and V_1; V is made equal to V_1, + the battery resistance; and V_1 is chosen sufficiently large to prevent the polarization of the battery when momentarily short-circuited through V and V_1.

By short-circuiting the battery, Mr. Stearns is able to avoid insulating the point N when the key M is in its intermediate position. If N were insulated, the received current would pass round both coils of the relay and would pass to earth through the resistance W. At first sight this latter arrangement (which was that used by Messrs. Siemens and Frischen) seems perfect, for we have the current diminished to one-half by a doubled resistance and at the same time acting with double force per unit of current on the relay. This reasoning does not take into account the inductive retardation (Chap. XXIII.) produced by artificially lengthening the line. Mr. Stearns, in all positions of the key, signals through a line of constant length and capacity.

BELLS.

§ 18. Bells may be classed as a distinct kind of telegraphic apparatus. Besides the bells which have already been described, in which each signal sent causes the hammer to strike one blow, there are two kinds of electric bells:—*First*, those in which the hammer is driven by a weight and clockwork; the clockwork remains at rest so long as a certain detent or trigger restrains it, but runs down, striking the bell, so long as the detent is held back by the armature of an electro-magnet actuated by the received current. While the current is maintained, the weight runs down and the bell continues to ring. *Secondly*, those in which the hammer is attached to the armature of the electro-magnet, and is furnished with contact pieces (as in Ruhmkoff's coil), such that when the armature is attracted to strike a blow, the contact is broken, and the current ceasing, the armature returns to its original place, makes contact again, and is again impelled to strike a blow. This action is repeated so long as a current is sent from the sending-station. The second form of bell, sometimes called a *trembler*, is the more convenient, and is used for household and hotel purposes.

Electric bells may with especial propriety be introduced into hospitals, and may be employed even in private houses by invalids. The effort required to ring the electric bell is that of making contact at one part of the circuit. This can be done by the smallest pressure on the little button of a handle or little box, which can be held in the hand in bed, and attached by flexible wires to the wall. This arrangement allows the patient to assume any posture without losing command of the bell. Electric bells are also used for railway signalling, and in all telegraph stations to call the attention of the clerks.

CHAPTER XXIII.

SPEED OF SIGNALLING.

§ 1. ELECTRICITY cannot properly be said to have a velocity. It is true that when a circuit is completed at any one point, electrical effects are not produced at other points of the circuit until a sensible time has elapsed; so that, for instance, when a signal is sent through the Atlantic cable, it does not produce any effect in Newfoundland simultaneously with the depression of the key in Ireland. The distance divided by the time occupied in the transmission of the signal may be called the velocity with which that particular signal was transmitted; it might even be termed the velocity with which a certain quantity of electricity traversed the cable, but it is not the velocity proper to or peculiar to electricity, for under different circumstances the same quantity of electricity may be made to traverse the same distance with almost infinitely different velocities.

For about two-tenths of a second after contact is made in England, no effect can be detected in Newfoundland even by the most delicate instrument: after ·4″ the received current is about 7 per cent. of the maximum permanent current which will ultimately flow equally through all parts of the circuit. The current will gradually increase until, 1″ after the first contact was made, the current will have reached about half its final strength, and after about 3″ it will have attained nearly its maximum strength; during the whole time the maximum current is flowing into the cable at the sending end. The velocity with which the current travels even in this one case has therefore no definite meaning; the current does not arrive all at once like a bullet, but grows gradually from a minimum to a maximum. The time required for any given similar electrical operation on various lines is directly proportional to the capacity of the unit of length of the conductor, to the resistance per unit of length, and to the square of the length intervening between the sending and receiving station. Fig. 161 shows the curve representing the law of increase of the received currents, which is the same on all lines. The vertical ordinates parallel to O Y represent strengths of current, the maximum or permanent current flowing through the circuit after equilibrium has been reached being called 100.

The horizontal ordinates parallel to O X represent intervals of time, measured from the time at which contact was first made, and expressed in terms of an arbitrary unit, a, different for different circuits, but constant for any one circuit. For a uniform line of the length l, the resistance per unit of length

FIG. 161.

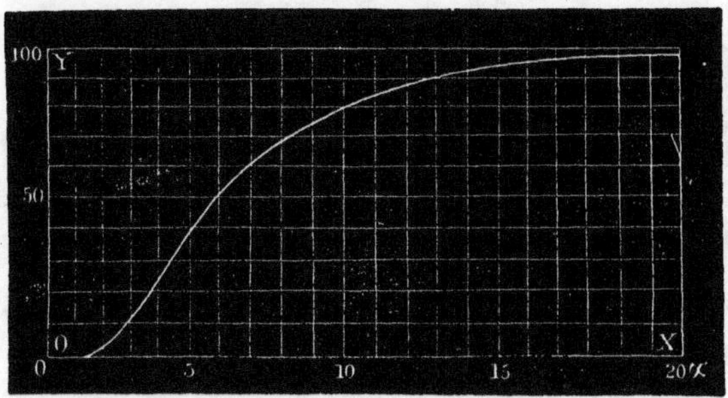

R and the capacity per unit of length S, the value of a is given in seconds by the expression

$$a = \frac{S\,R\,l^2}{\pi^2} \log_e (10^{\frac{1}{10}}) = \cdot02332\ S\ R\ l^2 \ \ldots\ldots\ 1^\circ.$$

In this expression absolute measure (gramme mètre second) is used. When S_1 is measured in microfarads per knot, R_1 in ohms per knot, and l_1 in knots, the above expression becomes

$$a = \cdot02332\ S_1\ R_1\ l_1^2 \div 10^6 \ \ldots\ldots\ 2^\circ.$$

For the French Atlantic Cable we have $S_1 = 0{\cdot}43$ $R_1 = 2{\cdot}93$ and $l_1 = 2584$; and hence for a the value ·196 second.

In terms of a the arrival curves for the received current of all lines are identical, and the same curve shows the law

FIG. 162.

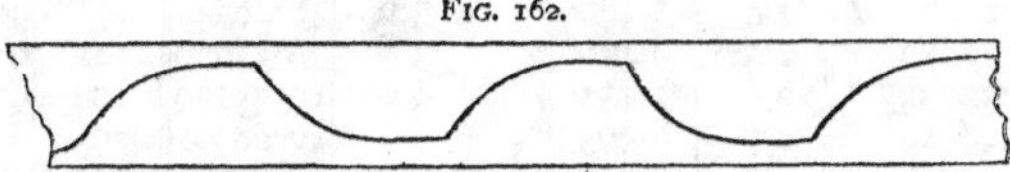

according to which the current at the receiving end dies away when at the sending end the line has been put to earth. A succession of contacts with a battery and with earth at the sending end prolonged each for times equal to about 25 a would produce the series of changes in the received current shown in Fig. 162, each curve being a complete arrival curve.

Fig. 162a.

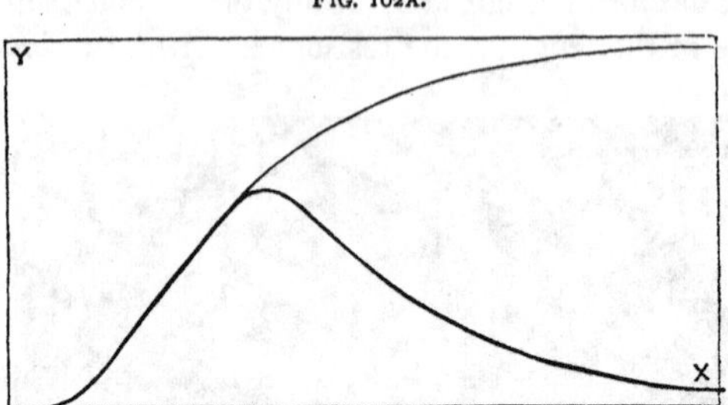

The annexed table shows the value of the vertical ordinates corresponding to successive multiples of a, the maximum current being 100.

t in terms of a	Strength of current in percentages.	t in terms of a	Strength of current in percentages.	t in terms of a	Strength of current in percentages.	t in terms of a	Strength of current in percentages.
·4	·00000000271	1·1	·04140636	3·5	18·48434	7·8	66·95995
·5	·00000051452	1·2	·08927585	3·6	19·84366	8·0	68·42832
·55	·0000033639	1·3	·1704802	3·7	21·21342	8·5	71·82887
·60	·000016714	1·4	·2959955	3·8	22·59017	9·0	74·87172
·62	·000029252	1·5	·476336	3·9	23·97071	9·5	77·59133
·64	·000049412	1·6	·720788	4·0	25·35217	10·0	80·02000
·66	·000080817	1·7	1·036905	4·2	28·10757	10·5	82·18760
·68	·00012835	1·8	1·430252	4·4	30·83807	11·0	84·12139
·70	·00019845	1·9	1·904356	4·6	33·52902	12	87·38402
·72	·00029937	2·0	2·460812	4·8	36·16892	13	89·97752
·74	·00044152	2·1	3·09969	5·0	38·74814	14	92·03836
·76	·00063776	2·2	3·81846	5·2	41·26032	15	93·67565
·78	·00090371	2·3	4·61560	5·4	43·70028	16	94·97631
·80	·00125804	2·4	5·48661	5·6	46·06449	17	96·00951
·82	·00172272	2·5	6·42695	5·8	48·35070	18	96·83023
·84	·00232333	2·6	7·43163	6·0	50·55770	19	97·48215
·86	·00308919	2·7	8·49536	6·2	52·68501	20	98·00000
·88	·00405358	2·8	9·61264	6·4	54·73314	21	98·41134
·90	·00525387	2·9	10·77797	6·6	56·70294	22	98·73809
·92	·00673158	3·0	11·98582	6·8	58·9502	23	98·99763
·94	·00853247	3·1	13·23087	7·0	60·41164	24	99·20379
·96	·01070646	3·2	14·50800	7·2	62·15439	25	99·36754
·98	·01330764	3·3	15·81233	7·4	63·82523		
1·00	·01639420	3·4	17·13921	7·6	65·42636		

When the line is put to earth at the sending end before the maximum current is reached, the falling curve is superimposed on the ascending one, and a derived curve is produced as shown in Fig. 162 A, which gives the effect of making contact for 5 a and then putting the line to earth. At the time 6 a from the beginning of the operations the strength of current will be 50·55770 − ·01639 = 50·54131; and at the end of 7 a it will be 60·41164 − 2·46081 = 57·95083; and in this manner the whole of the derived curve can be traced. If now the line be put in contact with the battery again at the end of 7 a, the third curve can be derived by again superimposing the original curve on the first derived curve; so that at the end of 8 a the strength would be 68·42832 − 11·98582 + ·01639420; and in this manner the effect of any number of operations can be computed.

§ 2. It follows from the above, that the result of a series of short equal contacts alternately with earth and a battery at the sending end will produce a small series of rises and falls in the strength of the current, which grow smaller and smaller as the length of the contacts diminishes: the mean strength of the current will be half the permanent maximum produced by a permanent current; and when the alternate contacts are made short compared with a, no sensible variation can be detected in the current which flows from the cable at the receiving end. As the contacts are lengthened, the amplitude of variation increases. The following table gives some amplitudes due to a succession of simple *dots* or equal contacts with the earth and with a battery.

Length of pair of contacts in terms of a. . . .	2·9	3·0	3·5	4·0	5·0	0·0	7·0	8·0	9·0	10
Amplitude of variation of current in percentages of maximum	2·69	2·97	4·52	6·31	10·42	14·85	19·67	24·42	29·11	33·68

The theory of the speed of signalling was first given by Sir William Thomson, read before the R. S. May 24, 1855, published in the Proceedings, and reprinted in the Phil. Mag., February 1856.

§ **3.** Signals sent through land-lines last so long relatively to the exceedingly short value of a for such lines, that in all ordinary cases the current rises almost to its maximum, and falls to zero at each dot. The capacity in electrostatic measure of wire of diameter d suspended at a height h above a flat plane, and remote from all other conductors, is

$$s = \frac{l}{2 \log. \frac{4h}{d}} \quad . \quad . \quad . \quad . \quad 3^{\circ}$$

Taking $h = 3$ mètres and $d = 0{\cdot}004$ mètre, we have $s = 0{\cdot}062$, or in absolute electro-magnetic measure $s = \frac{0{\cdot}062}{(28{\cdot}8 \times 10^9)^2}$ or about ·013 microfarad per statute mile. There is experimental reason to believe that the actual capacity is about double this amount, or even a little more, owing to the induction between the wire and the posts and insulating supports. Even taking s as ·03 microfarad, and the resistance of a mile of ·004 mm. wire as 15 ohms, we have for a line 350 miles long

$$a = {\cdot}00126 \text{ second.}$$

This value is so small that even with 20 a for each contact and 40 a for each dot, the dot would only occupy ·05″, or 20 dots could be made in a second; and for every dot the current would rise almost to its maximum and fall almost to its minimum. The above speed would give about 80 words per minute as a speed at which the effect of what is called retardation would be insensible in diminishing the rise and fall of the received current.

Instruments intended for use upon land-lines are therefore invariably constructed on the hypothesis that the received current will at each signal rise and fall through a considerable percentage of its maximum strength. The spring attached to the armature of the electro-magnet is adjusted so that at some one strength of received current the armature will rise, and at another strength differing little from the former it will fall: in order to work such an instrument safely, the received current must rise much above the first and fall far below the second strength, and this is the case even when 100 words per minute are sent by Professor Wheatstone's automatic sender from London to Edinburgh.

§ **4**. On submarine lines any such condition as a great and regular rise and fall in the received current limits the speed of transmission very seriously: 40 a for the French Atlantic cable corresponds to nearly 8 seconds, and two minutes would be required for the transmission of each word, if this interval of time were required for each dot; whereas from 15 to 17 words have actually been sent through this cable in a minute. The duration of a dot at the speed of 15 words per minute must have been about ·27 second, or about 1·38 a. Many of the dots can have produced no more variation in the received current than is equivalent to $\frac{1}{1000}$th of the permanent current; the theory of superimposed signals shows us that the exact effect of any one positive or negative dot depends on the 20 or 30 preceding signals, so that even very regular sending produces irregular results at the receiving end. Signals such as these cannot be received by any arrangement of armatures or other apparatus which moves at a fixed strength of current, but require some arrangement which shall be capable of following and indicating or recording every change in strength of the received current. Sir William Thomson, by his invention of the mirror galvanometer so constructed that it could fulfil this condition, rendered submarine telegraphy commercially practicable. The spot of light wanders over the scale, following every change of current, and the clerks by degrees acquire sufficient skill to interpret the seemingly irregular motions. One dot will cause the light almost to cross the scale, the second moves it a little farther, the third or fourth hardly cause a perceptible motion, but the clerk by experience knows that the four very different effects each indicate a simple dot, each sent by the clerk at the other end in a precisely similar manner.

§ 5. Sir William Thomson's syphon recorder actually draws on paper the curves which we have learnt to construct theoretically. Ink is spurted from a fine glass tube on to paper

FIG. 163.

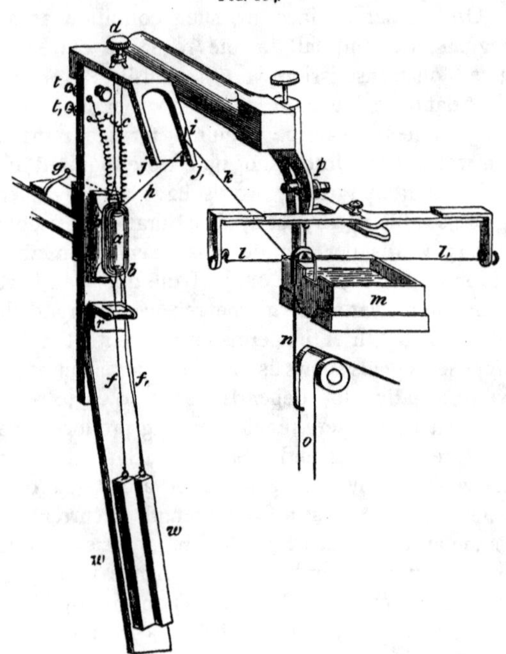

drawn past it with a uniform motion : the glass point of this tube moves to the right or left through distances proportional at each instant to the strength of the current, and thus the signals are drawn on the paper in the form of curves representing the strength of the current at each instant of time. The glass tube *n* (Fig. 163) is pulled backwards and forwards by being connected through the threads k h and lever i with a very light movable coil b b, placed between the two poles of a very powerful electro-magnet, not shown.

A soft iron fixed core a is placed in the centre of the coil. The coil oscillates about a vertical axis, being directed by a bifilar arrangement f f_1. The received current passes through this coil from the terminals t t_1 : the vertical arms of the coil are impelled across the magnetic field in one direction or the other according to the sign and strength of the received current. The magnetic field in this arrangement is very intense and very uniform, which gives great sensibility to the apparatus. The glass syphon n is strung on the wire l l_1, the shorter end dips in the ink-trough m, and the longer end is opposite the paper o; the syphon can be withdrawn from the ink by the slide p; the spring g keeps the threads k h taut; the directing force of the bifilar arrangement is adjusted by varying the position of the bracket r; the two weights w w_1 hang from the coil by the two directing threads.

If the coil is shunted so that there is a comparatively short circuit through which the current induced by its motion can flow, the electro-magnetic induction of the magnet on the coil tends to check rapid oscillations not due to the signals.

FIG. 164.

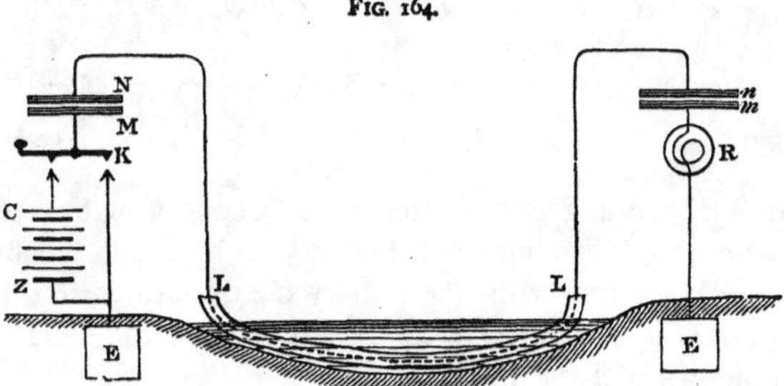

A certain portion of the received current is lost through the shunt, which is, however, rarely required. for the capacity of the cables connected with the coil is such that a very sensible induction takes place even without the shunt.

The ink is electrified by an induction machine similar in principle to that described in Chapter XIX. § 1, and is thus made to fly to the oppositely electrified strip of paper in a succession of fine drops.

§ **6.** If it were necessary to allow the recording point to travel over the whole possible range of the received current, it is clear that practically dots of only $\frac{1}{1000}$ of the maximum strength would correspond to $\frac{1}{1000}$ of the breadth of the paper, and could not be made legible with any practicable breadth of paper. They *are* legible on the mirror galvanometer because the light can range over a length of some feet, but $\frac{3}{4}$ inch is a broad paper strip for any recording instrument. Mr. Varley's mode of signalling by condensers supplies the means of keeping the light of the mirror galvanometer always at one part of the scale, and the glass tube end of the recorder within a very narrow strip of paper.

The line L, Fig. 164, is attached to the insulated armatures

FIG. 165.

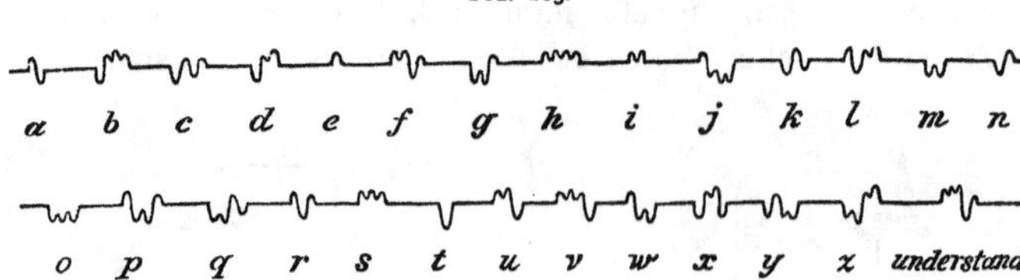

N and *n* of two large condensers; the second armature M at the sending end is connected to a key K, by which it can at will be connected with the battery C Z or with earth; the armature *m* is permanently connected through the receiving instrument R with earth.

When by the key K, M is connected with the positive pole, N is rendered negative by induction; a current flows from N to *n*; *n* becomes positive and *m* negative by induction, and to charge *m* negatively, a short current flows from *m* to E through R, making the desired signal in one direction; the current sent through R begins suddenly, is very small, and would gradually die out, even if M were not put to earth: the fall in the current is, however, accelerated by raising the key and putting M to earth. A negative signal is given by connecting M with the zinc instead of the copper pole of the battery.

With this arrangement no electricity flows into or out of the cable but by induction: the charge in the cable is re-arranged at each signal. The current received through the instrument R never increases beyond that due to the first signal.

Fig. 165 shows the alphabet, and Fig. 166 shows a message sent with condensers and received by the recorder.

FIG. 166.

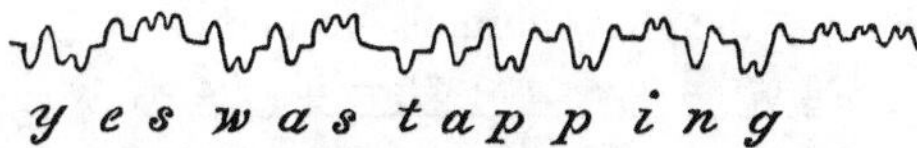

Mr. Varley's system has the additional advantage that no permanent earth currents can flow through the line, for the line is not connected anywhere with earth. A sudden change of potential in the earth at either end will induce a current, but sudden changes are much rarer than slow changes, and the latter, however great, are quite cut off by the condensers.

§ 7. The time of every electrical operation is proportional to a, or to S R l^2; and consequently, whatever instrument is employed to record or receive the messages, the speed of working must with that instrument be inversely proportional to S R l^2, and with any cables of uniform construction the speed must be inversely proportional to the square of the length.

The speed will, however, differ enormously, according to the nature of the electrical operation required for working the instrument. Thus the Morse instrument probably requires that the dots should occupy a time of from 15 to 20 a, and is therefore about 14 times slower than the mirror galvanometer, which will show dots of 1 or 1·2 a. The speed of the syphon recorder is nearly equal to that of the mirror.

The speed depends on the weight per knot w of the copper and on the weight per knot w of the gutta percha employed, and may be calculated from the following formula, where L is the length of the cable in knots.

Speed by mirror in words per minute—

$$=0{\cdot}2325\, w \frac{\log (70{\cdot}4\, w + 480\, \mathrm{w}) - \log 64\, w}{\mathrm{L}^2} \times 10^6$$

If Mr. Willoughby Smith's material is used instead of gutta percha, the multiplier ·275 may be used instead of 0·2325; and for Hooper's material, if the specific gravity is such that its weight per knot is $\frac{\mathrm{D}^2 - d^2}{400}$ lbs., and its specific inductive capacity 3·3, the above formula becomes

$$\cdot 295\, w \frac{\log (70{\cdot}4\, w + 400\, \mathrm{w}) - \log 64\, w}{\mathrm{L}^2} \times 10^6$$

The speeds given correspond to 13 words per minute through the French Atlantic Cable. As many as 17 have occasionally been sent. For Morse instruments the above speeds must be divided by 14.

It will be observed that when a constant ratio is maintained between the weights per knot of dielectric and conductor, the speeds of working are directly proportional to the quantities of material used.

CHAPTER XXIV.

TELEGRAPHIC LINES.

§ **1.** A TELEGRAPHIC line is an insulated wire reaching from station to station. On land an iron wire is generally used, supported on stoneware, porcelain, glass, or vulcanite insulators carried by wooden or iron posts. Sometimes underground wires are used, and these are generally made of copper insulated with gutta percha or india rubber, and protected by tape, leaden or iron tubes, wooden troughs filled with bitumen, or an iron wire serving. Submarine lines invariably have a copper conductor insulated with gutta percha or some preparation of india rubber, forming what is called a core. This core is served with hemp or jute, and covered helically with iron or steel wires, which are further covered in many cases with hemp and tar, or a bituminous compound. It is desirable that the conductor of a telegraphic line should have a small resistance, and that it should be well insulated. The smaller the resistance of the line, the smaller the battery required to work it, and with a given insulation the smaller the leakage. On submarine lines the speed attainable is increased by diminishing the resistance of the conductor. Bad insulation or great leakage involves the use of large batteries, frequent adjustment of the receiving instruments to suit variations in the received currents, resulting from variation in the resistance; bad insulation also involves greatly increased difficulty in ascertaining by electrical tests the position of any injury occurring to the line. The following paragraphs relate chiefly to the modes practically adopted for securing moderate resistance and high insulation:

§ **2.** The iron wire used in land lines is in this country generally No. 8, B.W.G. $\frac{1}{6}$ inch diameter.

The following table (p. 340) gives some of the other sizes adopted. The weights per statute mile are taken from Mr. Clark's tables. There are considerable differences in the weights given by different authors, and I am not aware that any one set of tables are authoritative.

Mr. Culley gives No. 8 wire as 0·17 inches diameter; its resistance 13·5 ohms, and that of No. 4 as 7·8 ohms. There is great difference in different specimens. The strength of good iron wire varies from 20 tons per square inch for large gauges such as No. 1 to 40 tons per square inch for No. 8

and smaller sizes. Mr. Culley gives 1,300 lbs. for No. 8, and this corresponds by the above table to 36·7 tons per square

Size of wire, B.W.G.	Diameter in inches.	Weight in cwts. per statute mile.	Resistance in ohms at ordinary temperatures per statute mile.	Strain corresponding to 10 tons per square inch. (cwt.)	Where used.
1	·3	1245	4·16	14·13	India. Some long lines in England.
2	·284	1117	4·57	12·66	
4	·238	783	6·51	8·89	
6	·203	570	8·96	6·47	Germany
8	·165	376	13·6	4·27	England and Germany
10	·134	249	20·5	2·82	England, short lines
4 millimètres	·157	340	15·0	3·86	France
3 millimètres	·118	192	26·7	2·18	,,

inch. The iron wire should be galvanized, and should be capable of being bent round itself and unbent without injury. It should also stand bending four times, first one way and then the other, to a right angle, being held in a vice. The wire is stretched 2 per cent. cold before being used. This process is called *killing*, and not only detects weak places, but makes the wire less springy and more manageable. It should be painted or varnished in smoky places.

From 25 to 20 poles per mile may be used on straight lines, but 16 poles per mile are sometimes used if no more than four wires are required. On sharp curves as many as 40 poles per mile may be required. The fewer the poles the better the insulation. For 10 wires or less the diameter of wooden poles may be 5 inches at the top; for a larger number of wires 6 inches. Creosoted larch is the best material; and the batts should be charred and baked to prevent decay, and tarred if well-seasoned. The pole above ground should be painted.

The distance between the wires should not be less than

12 inches vertically, and 16 inches horizontally, with 20 poles per mile.

§ **3.** No line can be perfectly insulated. On land lines no leakage occurs from the wire to the air, but at every pole there must with the best construction be some leakage, or, in other words, at every pole there is a connection with the earth. The resistance of this connection is very great when the wire is well insulated, and small when there is bad insulation.

The wire is always separated from the wooden pole by an *insulator*, and the insulation of the wire depends on the design, material, and condition of these insulators. Glass of certain kinds offers the greatest resistance to conduction through its substance of any known material, but it does not answer well for telegraphic insulation, because surface conduction plays by far the greatest part in the leakage from a line, and glass is highly hygroscopic, i.e. it will be found covered with a moist film in most states of the weather. Ebonite (hard vulcanized india-rubber) has a high insulation resistance and does not readily become damp, but rain wets it easily, and therefore when employed for insulators it is generally covered with a cap of some other material: it soon becomes dirty and spongy on the surface.

Porcelain of certain qualities insulates well; it is not nearly so hygroscopic as glass, and rain runs readily from its highly glazed surface. The glaze insulates still better than the substance of the porcelain, but in some specimens is liable to crack with old age, when its value is lost.

Brown stoneware is an excellent and cheap material for insulators: its glaze does not crack, but its substance has not so great a specific resistance as highly vitrified porcelain. The point of chief importance in all insulators being the condition of the surface, porcelain and stoneware are the favourite materials; they keep clean, do not change with age if well selected, and do not harbour insects.

The form most used approaches that of a bell, or of

several bells one inside another. In Fig. 167 No. 1 shows Latimer Clark's double-bell insulator; No. 2 Varley's insulator, made in two pieces; No. 3 the French cup insulator, a very rudimentary design; and No. 4 Siemens' insulator, protected and supported by an iron cap.

FIG. 167.

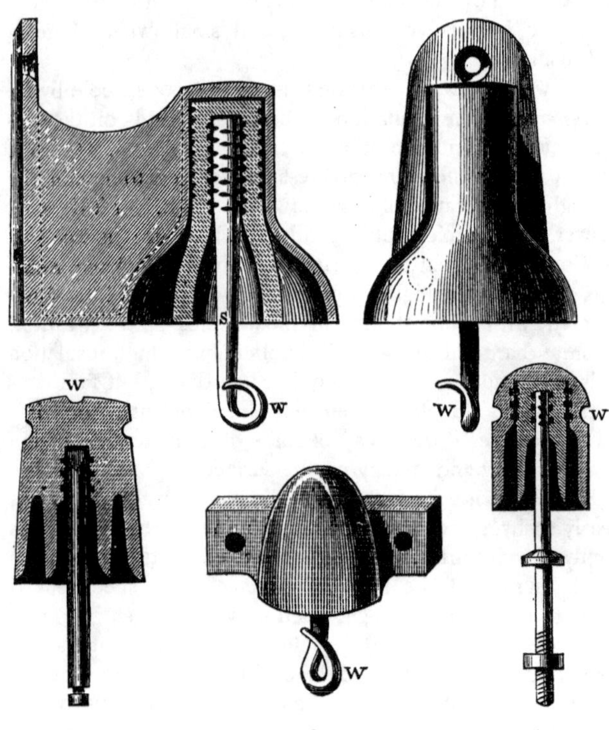

The objects aimed at in each design are the following:—

1. To make any conducting film which may be deposited on the surface of the insulator between the wire and the pole as long as possible, because, other things being equal,

its resistance increases directly as its length. This object is attained by the series of bells, for the electricity has to run down outside and up inside each, in succession, before getting from the wire to the pole.

2. To make the cross section of the conducting film as small as possible. With this object the insulator is kept as small in diameter as is consistent with other conditions of excellence.

The thickness of the deposited conducting film depends on external conditions, but the larger the diameter of our bells the larger will be the cross section of the film, i.e. the ring of moisture which we should find outside and inside each ring of insulating material if it were sawn across horizontally.

3. To expose one portion of the insulator to the rain, so that it may be cleansed by rain from dust, salt, smoke, spiders' webs, &c.

4. To protect another portion of the insulator from rain, so that when the outside is wet the inside may still insulate. These two conditions are fulfilled by the forms 1 and 2.

5. To prevent the failure of part of the insulator from destroying the insulation. With this object some good insulators are made in three parts, as shown in Fig. 2—two distinct cups and a vulcanite covering to the iron supporting pin.

6. To prevent insects from settling in recesses. This object is difficult of attainment, and limits the depths of the recesses under the bells.

7. To provide strength and protection against malicious injury. This leads to the adoption of metal caps as in Fig. 4.

§ **4.** Besides leakage from the wires to the earth, wires on poles are subject to the defect of more or less electrical connexion one with another, by the surface conduction from one insulator to another. To prevent this very serious inconvenience a wire from the earth is led up the pole and across every portion of it by which electricity could be conducted from one insulator to the other. A short circuit or line of no sensible resistance is thus provided, so that all leakage finds its way at once to the earth; simple loss weakening the transmitted currents causes much

FIG. 168.

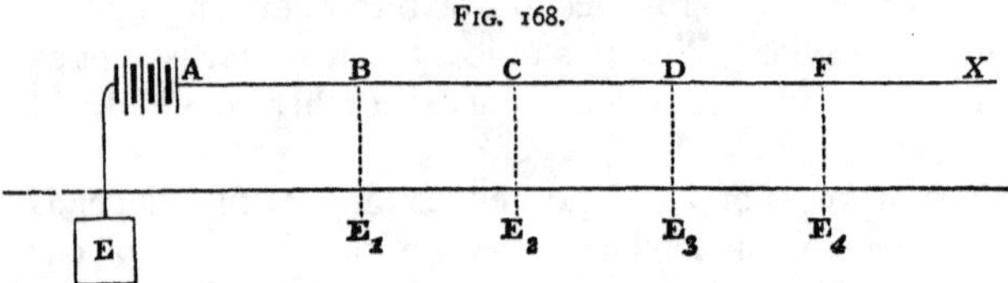

less inconvenience than cross connections by which the message on one wire finds its way partly into its neighbour. The earth wire is carried above the pole and forms a lightning conductor.

§ 5. The insulation resistance of a line is measured by measuring the resistance experienced at the end A when the end X is insulated, Fig. 168.

FIG. 169.

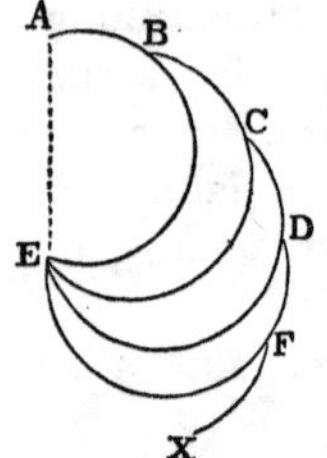

The resistance thus measured is not the sum of the several insulation resistances B E_1, C E_2, D E_3, &c., but is the resistance due to the circuits A B E_1, B C E_2, C D E_3, &c. arranged in multiple arc as in Fig. 169. We can calculate this total resistance if we know the resistance of each elementary part. First find the resistance between the points D and E due to a double arc; next add this resistance to that between D and C; next compound the resistance so found with that due to the arc C E; this will give the resistance due to all the conductors between C and E; add C B and proceed as before till the resistance due to all conductors between A and E is obtained.

When the resistance *m* of each part of the line between two poles is constant, and the insulation resistance *i* at each pole is also constant, we can calculate the difference between the current Q_0 sent into the line and that received at the further end Q_n by the following formula.

Let n be the number of poles, and let $z = e^{n\sqrt{\frac{m}{i}}}$ where $e = 2{\cdot}718$,

$$\text{then } Q_n = \frac{2\,Q_0}{z + \frac{1}{z}} \quad \ldots\ldots\ldots \quad 1^\circ$$

Mr. Varley considers no line well insulated for which the fraction $\frac{m}{i}$ is greater than $\frac{1}{80000}$. This fraction may also be defined as the ratio of the resistance of the conductor per mile to the insulation resistance of each mile. Q_n will be 46 per cent. of Q_0 in a line of 400 miles with the above value of $\frac{m}{i}$

§ **6**. On submarine and underground circuits, the insulation depends wholly on the resistance to conduction across the sheath of the gutta percha or india rubber covering. Surface conduction can only occur at the two extremities of the line, and unless by gross neglect, or on very short lines, cannot be a sensible cause of leakage.

Equation 1° is applicable to submarine lines, calling m the resistance of the conductor per mile, i the insulation resistance of each mile, and n the length of the line in miles.

The conductor is invariably a copper strand, and the resistance can be calculated for pure copper from the Table, § 14, Chap. XVI. In practice from five to eight per cent. extra resistance must be allowed for on account of impurities.

The smallest conductor in practical use for sea lines weighs 73 lbs. per nautical mile of 2,029 yards ; the largest yet employed (French Atlantic) weighs 400 lbs.

The large cores require nearly an equal weight of gutta percha as a covering, and the lighter conductors require a still larger proportion of insulator ; the 73 lbs. of copper is generally covered with 120 lbs. of gutta percha. Hooper's india rubber is sometimes used in smaller quantities than gutta percha.

The electrical tests applied to ascertain the quality and

condition of the materials employed in the case of submarine cables are—the measurement of the resistance of the core; the measurement of the resistance of the insulator to conduction from the copper inside to water outside; and the measurement of the capacity of the insulated conductor in microfarads. The methods of making these tests have been already described.

The insulation resistance R of a length L of the insulating core measured in centimètres is given in terms of the resistance R_s of one centimètre cube to conduction between its opposed faces by the following formula:

$$R = R_s \frac{\log_\epsilon \frac{D}{d}}{2\pi L} = \frac{\cdot 3665 \, R_s \log \frac{D}{d}}{L} \quad . \quad . \quad . \quad . \quad 2°$$

where $\frac{D}{d}$ is the ratio of the external diameter of the insulator to that of the enclosed conductor. From this equation we have the resistance R_k of one knot of insulating envelope:

$$R_k = \frac{1 \cdot 975 \, R_s \log \frac{D}{d}}{10^6} \quad . \quad . \quad . \quad . \quad 3°$$

R_s is what was called in Chap. XV. the specific resistance of the material.

The following table gives the value of R_k and R_s for some important cables at 24° C. after 1 minute's electrification.

	$\frac{D}{d}$	R_k megohms.	R_s megohms.
Malta Alexandria (first) . . .	2·95	115	4×10^6
Persian Gulf, mean	3·48	193	10×10^6
Second Atlantic, mean . . .	3·28	349	342×10^6
French Atlantic, mean . . .	2·92	234	256×10^6
Hooper's Persian Gulf (india rubber), mean	—	8000	7572×10^6

The specific gravity of gutta percha is between 0·9693 and 0·981. The weight W_k of gutta percha per knot in any case is

$$w_k = \frac{D^2 - d^2}{480} \text{ lbs.} \quad . \quad . \quad . \quad . \quad . \quad 4^\circ$$

Where D and d are measured in thousandths of an inch. The specific gravity of Hooper's rubber is about 1·176, and the constant divisor for the weight of Hooper's material in the above formula is 400 instead of 480. The weight per knot w_k of a copper strand of 7 wires such as is used for submarine lines is in lbs.

$$w_k = \frac{d^2}{70{\cdot}4}$$

§ 7. The capacity in electrostatic measurement s of any length of wire for a submarine cable may be calculated by equation 6, Chap. V. The electromagnetic capacity S is more commonly required, and we know (Chap. VIII. § 2) that $S = \frac{s}{v^2}$, where $v = 28{\cdot}8 \times 10^9$. Hence in absolute electromagnetic measure

$$S = \frac{K\,L}{4{\cdot}6052 \times 28{\cdot}8^2 \times 10^{18} \times \log \frac{D}{d}} = \frac{K\,L}{3820 \times 10^{18} \log \frac{D}{d}};$$

and calling S_M the capacity in microfarads, we have

$$S_M = \frac{K\,L}{382 \times 10^4 \log \frac{D}{d}} \quad . \quad . \quad . \quad . \quad 5^\circ$$

This value of S^M is given in terms of L measured in centimètres: practically it is convenient to measure the length in knots; and as one knot is equal to 185,526 centimètres, (6087 feet), we have, calling L_k the length in knots,

$$\text{Cap. of cable} = \frac{{\cdot}04857\, K\, L}{\log \frac{D}{d}} \quad . \quad . \quad . \quad 6^\circ$$

Taking the value of K for gutta percha as 4·2 (vide Chap. V. § 5), we find the capacity of the French Atlantic cable to be about 0·43 of a microfarad. This value agrees with the result of direct experiment by the ballistic method (vide § 5, Chap. XVII.).

§ 8. Fig. 170 shows a cross section and a projection of the component parts of the Anglo-American Atlantic cable drawn full size. In the centre is the copper strand of 7 wires: round this we have the gutta percha envelope covered by a serving of jute, outside which there are ten wires of

FIG. 170.

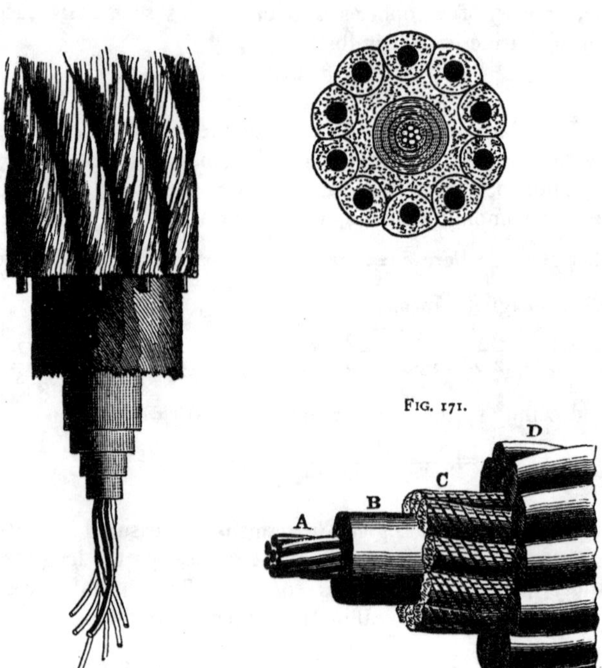

FIG. 171.

what is called homogeneous iron, each enveloped in fine strands of Manilla hemp.

Fig. 171 shows the more common type of cable, in which the hemp-covered steel wires are replaced by iron wires of considerable size. These iron wires, laid on as shown in Fig.

171, are often covered with one or two outer servings of jute and a compound of mineral pitch, silica, and tar, known as Clark's compound.

CHAPTER XXV.

FAULTS IN TELEGRAPHIC LINES.

§ **1.** ANY impediment to signalling due to the condition of the line is a fault. Faults are of three kinds :—1. A defect producing bad insulation. 2. A defect producing want of continuity in the line, or excessive resistance. 3. Contact between two neighbouring conductors used for separate messages.

Defective insulation in land lines may be due to cracked, dirty, or otherwise defective insulators, or to contact between the line and some conductor in connexion with the earth. In the first case the defect may be distributed over a great length of line. We can determine its importance by electrical measurements. In the second case the fault has a definite position, and we can determine its importance and its position by electrical tests. In submarine cables, defective insulation is always due to connexion between the sea and the internal conductor at one or more definite points. The second class of fault implies a rupture in the conducting wire of the line or in the connexions at the stations, or in the connexions with the earth at the stations. In many cases its position can be ascertained. Frequently the first and second faults co-exist : i.e. the line is broken and its end is in contact with the earth. The third class of fault seldom arises except on land lines. When the connexion arises from the actual contact of one wire with another, its position is easily found.

Tests for the position of faults can generally be made more accurately on submarine lines than on land lines, because the insulation of the undamaged portions of the line is generally better. The following descriptions refer especially to submarine faults, but the same principles are applicable to land lines.

§ 2. Let there be a fault in an otherwise well-insulated conductor, involving loss of insulation at one point, at the distance A B, Fig. 172, from station A.

If the connexion at B with the earth has no sensible resistance, we have only to measure the resistance A B, and divide by the resistance of the line per mile, to obtain the distance A B in miles. This measurement may be made by the Wheatstone balance, connected as shown. A D and D F are the two arms of the balance, F E is the box of resist-

FIG. 172.

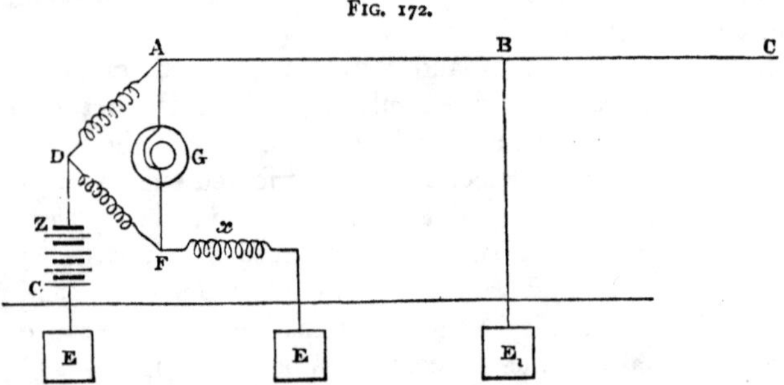

ance coils. If A D is $\frac{1}{10}$ of D F, and the plugs in the box between F and E arranged so as to give 1,500 units when the galvanometer G remains undeflected on the completion of the circuit, then A B E_1 has a resistance of 150 units; and if the line has a resistance of 5 units per mile, B is 30 miles from A. It is always desirable to insulate the end of the line at C during this test. We can easily ascertain whether the resistance of B E_1 is sensible or not, by repeating the test from C. If by the second test we find a distance B C, which, added to A B, makes up the whole length of the line, B E_1 can have no resistance. If, on the other hand, the sum of the measurements from C and from A gives a greater length than A C, this can only be due to the *resistance of the fault*; for we have not really measured the resistance of A B and B C, but of A B + B E_1 and B C + B E_1. If then the sum of the two measurements exceeds the resistance A C, the excess will be equal to twice the resistance of the fault. Let m be the resistance measured at A, n the resistance measured at C, and L the resistance of the whole line.

$$\text{Then A B} = \frac{L + m - n}{2} \text{ or B C} = \frac{L + n - m}{2} \quad . \quad . \quad 1$$

This method would be perfect if the resistance of the fault were really constant while the resistances m and n were being measured; but faults usually vary very much, owing to polarization; and hence, except with great faults of small resistance, this method is defective.

§ **3.** A second method of determining the resistance A B is given by the following test, on the assumption that the resistance of the fault is constant:—Measure at A the resistance m of the line when C is insulated, and measure the resistance e when the end C is put to earth.

$$\text{Then A B} + f = m; \text{ A B} + \frac{1}{\frac{1}{f} + \frac{1}{\text{B C}}} = e \text{ and A B} + \text{B C} = L$$

$$\text{therefore A B} = e - \sqrt{(L - e)(m - e)} \quad . \quad . \quad . \quad 2°$$

This test is even less trustworthy than the preceding one. By taking a large number of values of m n and e with different poles of the battery, and different strengths of battery, and choosing the smallest values obtained as those corresponding with one and the same minimum value of f, some approach to accuracy can be made. Great experience is required in testing to enable the observer to judge of the nature of a fault. By noting the polarization obtained with positive and negative currents of different strengths the character of a fault can generally be determined, and a guess made at its probable resistance.

§ 4. When there is a well-insulated return wire from the distant station C back to A, the position of a leak can be determined with great accuracy by what are called loop tests. The observer has then both ends of a complete metallic circuit before him, and the ratio between the two parts which intervene between the two ends and the fault can be determined by several methods, all independent of the varying resistance of the fault.

Mr. Varley uses a differential galvanometer to ascertain when an equal current runs into both ends of the metallic circuit and out at the fault. This will only be the case when the resistance between the galvanometer and the fault is the same by both roads. This condition is fulfilled by adding a resistance *r* between one coil of the galvanometer

FIG. 173.

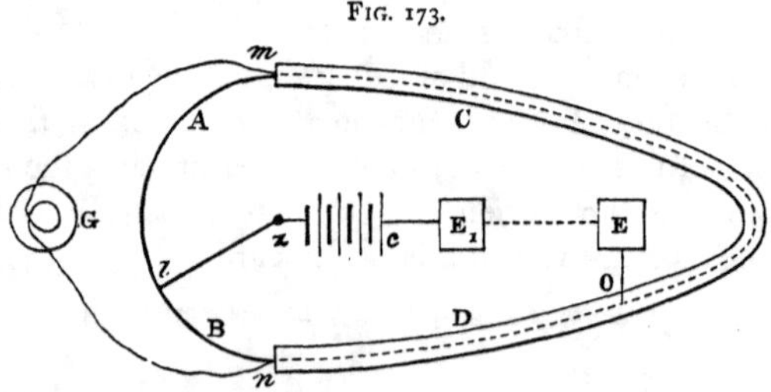

and the defective wire. The resistance *r* required to bring the galvanometer to zero is obviously equal to twice the resistance of the wire between the distant station and the fault.

Perhaps a still better method is given by arranging the Wheatstone balance as shown in Fig. 173, where the fault, supposed to be at *o*, forms part of the circuit connecting the pole *c* to the metallic conductor subdivided at *o*.

The variation of the resistance of the fault does not affect the result: it will indeed cause a greater or less deflection in the galvanometer until the desired balance is effected, but it will not alter the relative resistances of the several parts of the circuit required to reduce the deflection to zero. The test is made by adjusting the resistances A and B until no deflection is obtained; then, calling C and D the resistances of the conductors separating *m* and *n* respectively from the fault, we have $\frac{A}{B} = \frac{C}{D}$. Then the resistance of C + D being called L, the above equation gives the value of $C = \frac{A\,L}{A + B}$.

§ 5. The following is a plan for determining the position of a fault of high resistance in a submarine cable by a simultaneous test at each end. It takes into account the uniform leakage from each knot of the insulated cable, and can be carried out with much greater synchronism than is possible for the plans described in §§ 2 and 3, above. The connexions are shown in Fig. 174. G is a galvanometer;

FIG. 174.

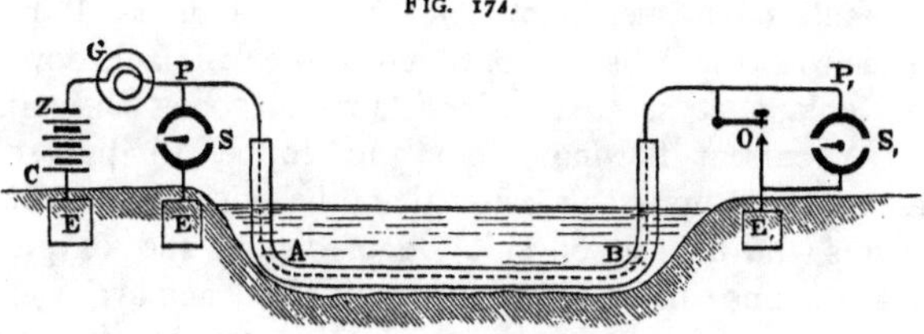

S an electrometer at the same station; S_1 an electrometer at the distant station, where the end of the submerged cable is insulated; the battery C Z has one pole connected with the galvanometer G, and the other pole to earth; let *k* be the resistance of the unit length of the conductor, and *i* the resistance of the unit length of insulated wire to conduction across the sheath; then let *l* be the length of the cable. Let λ be the distance of the fault from the galvanometer station; let P_1 be the potential at the distant station; let P be the potential at the near station, and C the current observed on the galvanometer.

$$\text{Let } a = \sqrt{\frac{k}{i}}$$

$$,, \quad F = P + \frac{k}{a} C - P_1 \varepsilon^{al}$$

$$,, \quad D = P_1 \varepsilon^{-al} + \frac{k}{a} C - P$$

$$\text{Then } \lambda = \frac{1}{2a} \log_e \frac{F}{D} \quad . \quad . \quad . \quad . \quad 3^b.$$

The measurements must be made in one consistent system of units. Absolute measurement in centimètres, grammes, and seconds may be used for the whole series.

The test requires two instruments by which P and P_1 can be measured in absolute measure.

§ 6. A fault of insulation in a submarine cable is generally due to a hole in the dielectric. This hole is gradually enlarged by the action of the current, although the polarization at the fault often seems to seal it up for a time. Rapid reversals with 100 cells or more tend to break a fault down, *i.e.* to enlarge it, so that its resistance becomes insignificant. A current flowing from the copper to the sea apparently seals up a fault better than the opposite current. It causes the deposit of chloride of copper and oxygen, whereas the zinc current causes a deposit of salt and hydrogen. The bubbles of gas formed under great pressure in time burst the film of deposited salts, and the fault temporarily breaks down. When this occurs with the negative current, no further damage occurs in general than a slight enlargement of the fault; but by the positive current a slow but certain erosion of the copper is produced, which always ends in producing a complete and sudden loss of continuity in the conductor. No warning is given of the impending fatal injury; for so long as the slenderest thread of copper remains no sensible diminution occurs in the resistance of the line. Signallers prefer to keep a cable positive to the sea, because they get better signals, the currents received being stronger, and less liable to the derangements produced by the sudden variations of a fault. The practice is, however, reprehensible. A faulty cable should always be kept negative relatively to the sea. It is possible to send very good signals through a cable or line in which there is a fault of such magnitude that its resistance is far less than that of the conductor between the stations. Nothing is absolutely fatal to communication except a want of continuity in the conductor.

Sometimes the fault is made by the presence of some foreign body in the insulator. When metal, such as a piece of broken wire, is driven through the dielectric connecting the conducting wire with the sea, or with the metal sheathing, a fault of no sensible resistance is produced, and this class of fault is easily recognised by the absence of polarization.

§ 7. A fault of the second class, *i.e.* involving want of continuity, may be combined with one of the first class: thus the cable or land-line may not only be broken, but may be in more or less perfect connexion with the earth at the fracture. In this case simultaneous tests at both ends are impracticable. We can only measure the resistance of each unbroken portion of the cable, and guess from the polarization what is likely to be the fraction of the whole resistance observed due to the fault. We can in any such case safely fix a maximum distance beyond which the fault cannot lie. With the minimum of polarization the bare copper end of a cable usually has a resistance equal to several miles of the conducting wire.

A fault of the second class not unfrequently occurs with perfect insulation. The conductor is broken, but insulated at the fracture. In a submarine cable the distance of the insulated fracture can then be measured very exactly by measuring the capacity of the cable between the fracture and the shore. The capacity per mile being known, this test gives the distance with great exactitude. On a land line the insulation is seldom good enough to allow this test to be rigorously applied.

§ 8. The position of a fault of the third kind—contact between neighbouring conductors—can easily be fixed if the contact is local, and of small resistance. We need only measure the resistance of the loop formed by the contact, and half this is evidently the resistance corresponding to the distance of the fault. When the contact is imperfect, its position can be very accurately determined by the aid of a third wire, if this be well insulated : to do this, treat one of the two wires in contact as an earth, leaving it uninsulated: and by the loop test described § 4 above, fix the position of the point of contact on the other wire, this contact being now in effect an ordinary fault of the first class.

The position of the contact can also be ascertained without a third wire, by a Wheatstone's balance test. To do this, the connections are arranged as follows, Fig. 124: r_{i} and r_{iii} are resistance coils, r_{ii} and r_{iv} are the two subdivisions of one of the two faulty line wires, subdivided at E by the contact; the point B_1 is the further end of the line, and is put to earth ; the branch r is made up of the galvanometer, and of the earth at B_1; the wire joining the battery with E is the second line wire in contact with the first at E; the further end of the second line is insulated.

Then, calling x and y the two subdivisions of the first line wire, we have $x = r_{ii}$, $y = r_{iv}$ and $r_{i} : r_{iii} = x : y$; whence, knowing r_{i} and r_{iii}, x and y can be found.

CHAPTER XXVI.

USEFUL APPLICATIONS OF ELECTRICITY, OTHER THAN TELEGRAPHIC.

§ **1.** ELECTRICITY has been applied in so many ways to the useful arts that a large separate treatise might be written on these applications. In this book a few only of these applications can be mentioned, and these must be very cursorily described, under the heads of Electro-Metallurgy, Electric Light, Medical Applications, the Firing of Mines, Clocks, governors and chronoscopes.

ELECTRO-METALLURGY.

§ **2.** In metallurgy electricity finds a threefold application. 1. To electro-plating, such as gilding or silvering objects. 2. To the reproduction by metallic casts of objects of any form. 3. To the reduction of metals from their ores. When our object is to coat a metal with a thin metallic film of some other metal, we immerse the object to be coated in a solution of some salt of the metal to be deposited. We pass a current from the bath to the object, so as to decompose the salt and deposit the metallic positive ion on the object, which is a negative electrode. By the choice of a proper salt, a proper strength of solution, and a proper strength of current, the film can be made adhesive. When copper objects are to be gilt, they are treated as follows :—They are first heated, to dispel any fatty matter from their surface; they are next plunged while still hot in very dilute nitric acid, which removes any coating of oxide or suboxide of copper; they are then rubbed with a hard brush, washed in distilled water, and dried in gently heated sawdust. They are still further cleaned by being rapidly immersed in ordinary nitric acid, and next in a mixture of nitric acid, bay salt, and soot. The objects thus prepared, so as to have a uniformly clean metallic surface, are immersed in a bath containing a solution of some salt of gold.

The objects are attached to the zinc pole of a battery consisting of three or four elements, the other pole of which is connected with an electrode of gold also plunged in the bath. The passage of the current decomposes the salt, deposits gold on the object, and causes the dissolution of an equal quantity of gold from the gold electrode. The time required for the operation depends on the thickness of coating required. One grain of gold and 10 grains of cyanide of potassium in every 200 grains of water form a suitable bath. Silver, bronze, brass, German silver, and some other metals can be directly gilt in this manner; but in order to gild iron, steel, zinc, tin, or lead, it is found necessary to electroplate them first with copper. The bath from which copper is deposited is a saturated solution of sulphate of copper. The positive electrode must then be a copper plate. A bath for the deposition of silver consists of two grains of cyanide of silver and two parts of cyanide of potassium in every two hundred grains of water; the positive electrode must be a silver plate.

§ **3**. The reproduction of objects in metal by electricity is effected by a thick deposit of the metal in a mould, the surface of which has been so treated as to be a good conductor. The deposit is obtained from a bath by the passage of a current, precisely as the deposit required for electro-plating is produced.

The mould, if made of metal, should be slightly coated with some fatty substance. A brush rapidly passed through a smoky flame, and then used as it were to dirty the mould, is said to be sufficient to prevent adhesion. Ganot mentions Street's fusible alloy, consisting of 5 parts of lead, 8 of bismuth, and 3 of tin, as suitable for moulds of metallic objects. Stearine is used to prepare moulds of plaster objects; these are first immersed in melted stearine and withdrawn quickly;

some of the stearine is absorbed by the pores of the plaster; the surface is next coated with graphite or with black lead rubbed on with a brush. The stearine mould can then be taken. The interior surface of the mould is covered with graphite to make it conduct.

Gutta percha moulds may be prepared by pressing gutta percha heated in warm water against the surface of the object to be copied, which should previously be covered with graphite to prevent adhesion. The mould must also be coated with graphite to make it conduct. Any of these moulds, used as a negative electrode in a bath of sulphate of copper, will become filled with a copper deposit, which reproduces the original object. This process is of great use to printers. Copper plates are beautifully reproduced by its means.

§ **4**. The reduction of ores has never been carried out on any large scale, but several of the rarer metals have only become known to us by the decomposition of their salts under the action of the electric current. Davy obtained potassium for the first time by decomposing a slightly moistened fragment of hydrate of potash by a current from 200 or 250 cells. Sodium can be obtained in a similar way; but other methods are now known, which are commercially preferable.

Barium, calcium, magnesium, aluminium, &c., can be obtained by electrolytic methods.

The ores of silver, lead, and copper have been treated by electric processes, many details of which will be found in the 'Traité d'Électricité et de Magnétisme,' by Messrs. Becquerel, vol. ii.

ELECTRIC LIGHT.

§ **5**. When the points of two pencils of charcoal or graphite, attached by thick wires to the two poles of a galvanic battery of forty or fifty Grove's elements, are placed for a moment in contact and then withdrawn, so as to remain about one eighth of an inch distant, a current will flow round the circuit crossing the arc from pencil to pencil, and at this spot emitting a most brilliant light.

The name *voltaic arc* is often used to designate that portion of a continuous current where there is a gaseous conductor. The voltaic arc is in most cases luminous. Its colour depends on the gas traversed, and its intensity is closely connected with the density of the gas. With rarefied gases, as in the Geissler tubes described above, a comparatively feeble glow is obtained; in air, the intensity of the electric light may be as great as $\frac{1}{3}$ that of sun-light, according to experiments of Fizeau and Foucault. The air is much heated at the point of passage, and its resistance thereby reduced; if the current be momentarily interrupted, the E. M. F. of the battery will be unable to re-establish the voltaic arc, unless the points are again brought very close or into contact, to be withdrawn as before when the current has been established; the reason being that the E. M. F. which is sufficient to send the current across hot air is insufficient when this air is cooled. The carbon of the pencils is consumed in the production of the light. The positive electrode is much more rapidly consumed than the negative electrode, and becomes hollow at the point. In order to render the light available for practical use, the graphite pencils must be held in a lamp, so constructed that the opening between the points remains sensibly in one place. In these lamps there must therefore be a feed supplying the pencils in the ratio in which they are found to be consumed. The lamp must also be furnished with some contrivance by which, if the voltaic arc is extinguished from any cause, the graphite points will instantly fall together, re-establish the arc, and again separate to the normal distance for the greatest intensity of light. Lamps fulfilling these conditions more or less perfectly by means of electro-magnetic gearing have been invented by Mr. F. H. Holmes, M. Serrin, M. Dubosc, and others.

Mr. Holmes's lamp has been used for lighthouse illumination with success.

An electromotive force of about eighty volts is apparently the least with which a good electric light can be produced, and the resistance of the circuit (exclusive of the voltaic arc) must not much exceed 12 or 15 ohms. Sir William Thomson has produced a good light with eighty Daniell's cells of the construction and dimensions described § 12, Chap. XV. These cells remained in good condition for several months, so that the light could be obtained at any moment by merely closing a circuit. Grove's cells will only act well for a few hours after being filled, and give out noxious fumes.

Mr. Waring produces an intense electric light by the incandescence of mercury vapour. The current is passed along a thin stream of mercury, which it volatilizes. The mercury is hermetically enclosed. This light has a greenish tinge. A rapid succession of sparks from a Ruhmkoff coil will also produce a somewhat feeble light.

The electric light may be made use of in photography, and the examination of its spectrum presents many points of great interest to the physicist.

FIRING OF MINES.

§ **6**. This is effected by passing a current through a film of semi-insulating substance, which becomes red hot, and fires a detonating mixture or gunpowder. A fuse is prepared to which two insulated wires are led. The ends of these wires are imbedded in a thin solid gutta percha rod :. they do not join, but end in a little layer of the priming composition, which is an intimate mixture of subsulphide of copper, subphosphide of copper, and chlorate of potassium. The whole is surrounded by gunpowder. A feeble current will not heat the priming composition to redness, but a powerful current, even if short, will develop enough heat by its passage to ignite the powder. The current is generally produced by the discharge of a condenser, and this condenser is often charged by a frictional electric machine. A vulcanite plate machine as designed by Ebner is much used with a condenser consisting of a sheet of india rubber with tinfoil armatures rolled up so as to form a cylinder. A magneto-electric current or a battery current may be used. When the mine or torpedo is to be fired by the discharge of a condenser, a fine wire is better than a thick one, in order that the capacity of the conductor may be small: with the same object the thickness of the dielectric should be considerable, and the very best insulation is necessary.

The detonating mixture may also be fired by heating to redness a fine platinum wire stretched between the two ends of the copper wires: the platinum wire should be coated with fulminate of mercury. A voltaic battery is required with this arrangement, which has the double advantage that the condition of the conductors can from time to time be tested by feeble currents which will not explode the charge, and that it allows several insulated conductors to be laid in one cable, which plan cannot be followed when the mine is fired by the discharge from a condenser, owing to the powerful current then induced in the neighbouring wires, which would fire all the mines whenever a current was passed along a single wire. The platinum fuse can be fired when the insulation of the conductors is very defective.

MEDICAL APPLICATIONS.

§ 7. Electricity in its passage through the body may produce very marked physiological effects. The simple passage of a current from one hundred cells produces a somewhat disagreeable disturbance or tingling at the point where it enters or leaves the body. This feeling is considerably more intense at the moment when the current begins and ceases than at any other time. When a powerful current of very short duration is sent through the body, as from a Leyden jar of moderate size charged to the potential of several hundred volts, the disturbance is felt throughout the frame, and is well known as an electric shock. The disturbance produced may be so great as to produce illness or death, and many persons who are killed by lightning are killed by the simple shock resulting from the sudden discharge of electricity from their bodies, which had been inductively electrified from the clouds; the lightning passing from cloud to cloud discharges these, and the escape of the electricity from the body previously charged produces the shock. The rapid succession of currents produced by rotating magneto-electric arrangements produce a singular numbness if passed through the body, so that a man holding two electrodes from which these short rapidly alternating currents flow cannot let them fall, but holds them convulsively. The very first discovery of the electric current by Galvani was due to the contraction of a muscle of a frog under the influence of the current. From all these facts it cannot be doubted that electricity may be of use as a curative agent; the medical man may find in it a means of producing important modifications in the condition of the body; but the author is unable to speak with any confidence of the applications as yet made of this agent. Rapidly alternating magneto currents are the most popular, but he is not aware that thoroughly scientific experiments have been made on the effects produced, or on the real magnitude of the currents employed. Valuable results may have been and may be attained, but it is for medical men to decide how far these have or have not been the results of some happy accident. The application of electricity, unhappily, can easily be made the subject of quackery without detection.

The actual cautery can be applied by platinum wire heated by an electric current in parts of the body which could not be reached in any other way.

CLOCKS, GOVERNORS AND CHRONOSCOPES.

§ 8. There are many other useful applications of electricity. Mr. Alexander Bain drives clocks by a small current acting on a propelment, the speed of which is regulated by a pendulum. The propelment acts like the propelment of the dial telegraph instruments. The same inventor followed by others controls distant clocks from one standard clock by electro-magnets set in action by currents. The pendulum of the distant clock oscillates freely if keeping perfect time, but is slightly retarded or accelerated by an electro-magnet if before or behind time. Time guns or other time signals are also given from observatories by the aid of electric currents.

Electricity is made use of in one form of governor to regulate the speed of machinery. When the speed is excessive, the governor balls by their divergence complete a contact which permits a current of electricity to produce friction by the action of an electro-magnet.

Electricity is made use of to light the gas in one species of motor gas engine, and electric sparks have been used to light gas lamps.

Electric chronoscopes measure time to thousandths of a second, and by their aid the speed of projectiles is ascertained: the plan in general being that the projectile at one part of its path interrupts one circuit, and at another part a second circuit, by cutting wires. The interruptions determine sparks which leave their record on prepared paper or a metallic surface, moving with known velocity: the distance between the records of the sparks serves therefore to measure the time occupied by the projectile in passing from one wire to the next. In this little treatise these and many other important applications can barely be enumerated. As the science becomes more familiarly known, the extent and number of useful applications will day by day increase.

CHAPTER XXVII.

ATMOSPHERIC AND TERRESTRIAL ELECTRICITY.

§ **1.** NOT much is known of the distribution of electricity on the surface of the earth. According to Sir William Thomson the most probable distribution is analogous to that which would be produced if the earth's surface generally were charged with negative electricity held as a charge on the inner armature of a condenser, the outer armature of which was in the upper regions of the atmosphere, the lower part of which acts as the dielectric. Electrified masses of air moving at no great distance from the earth's surface are continually altering the distribution of electricity, which is, however, generally found to be negative on the earth's surface. The modes of investigating the density of electrification and the sign of the electricity at the earth's surface are analogous to the method of the proof plane. Some conductor in contact with the earth is insulated, brought indoors, and the sign of its electrification ascertained by an electrometer. We here speak of the electrification of the surface, not of the potential, at points of the air which must be separately investigated. We cannot treat air as we can the earth, because it is an insulator, and will not part with its electricity to any conductor analogous to a proof plane.

§ **2.** The potential of the earth's surface is assumed as the zero or datum from which all other potentials are measured; nevertheless we know that the potentials of different places on and in the earth differ considerably, sometimes to the extent of several hundred volts, though this is rare. We obtain this information from the currents observed to flow through wires joining parts of the earth widely separated. These currents being known, and the resistance of the circuit being known, the E. M. F. due to differences of potential between the ends of the wire can be inferred with certainty. The difference of potential between the two sides of the Atlantic is often not more than one or two volts, and generally points joined by the sea are nearly at one potential. This condition is, however, liable to be disturbed from time to time, and these disturbances are called electric storms. Statistics of the distribution of potential over the earth's surface have not yet been compiled.

§ **3.** Any conductor at the end of which a flame is burning, or any small pipe from which water drops, will very soon acquire the potential of the air where the flame burns or the water is dropping; for if there is any difference of potential between the conductor and the air near the flame or tube end, it will cause an accumulation of electricity at the flame or tube end, and this electricity will then be conveyed away by the particles flying off in the flame or by the drops of water until there is no difference of potential between the conductor and the neighbouring air.

This fact enables us to measure the potential of the air at any point, or, in other words, to compare its potential with that of the earth. To do this, a conductor having a flame or water-dropping arrangement at one end is connected with one pair of quadrants of the reflecting electrometer; this pair of quadrants is thereby brought to the potential of the air at the spot to be tested. The other pair is connected with the earth, and the difference of potentials is then measured by the deflection of the electrometer in the usual way. Other forms of electrometer may be used. Sir William Thomson found that the potential of the air varied very rapidly near the surface of the earth. Thus he has observed a difference of potential between the earth and the air nine feet above it, equal to 430 volts in ordinary fair weather, and in breezes from the east and north-east as great a difference as this per foot of air. The potential is perpetually fluctuating, even in fair weather. Instruments have been in action for some time at Kew and elsewhere, recording continuously the differences of potential between the earth and one point in the air. The potential of the air appears to be generally positive in fine weather, and negative only during broken or rainy weather.

§ 4. The distribution of magnetic force on the surface of the earth has already been alluded to in Chap. VII. It is conceivable that this force may be wholly due to currents flowing round the earth, and maintained by the thermo-electric action due to the sun, or to some other cause connected with the rotation of the earth. Observation does not, however, as yet enable a decided opinion to be given on this point.

CHAPTER XXVIII.

THE MARINER'S COMPASS.

§ 1. The mariner's compass consists of a card pivotted on a vertical axis, and directed by having on its lower surface one, two, four, or more parallel magnets with similar poles pointing in similar directions. The magnets being free to turn in a horizontal plane, place themselves in the magnetic meridian. The object of using several magnets is to increase the magnetic moment for a given weight of steel. The upper surface of the card is divided into degrees and also into thirty-two parts, each containing $11^\circ\ 15'$; the thirty-two rays indicate the thirty-two points of the compass; the line joining the north and south points is parallel to the axes of the magnets. The north and south line indicates the magnetic meridian at each place. As was shown in Chapter VII., the declination varies at different times and at different places. The declination of the particular place at the particular time must be known by means of charts or otherwise before the true north or any other true course can be determined by the aid of the compass.

§ 2. The presence of any iron or steel in the neighbourhood of the compass alters the direction of the lines of force in the magnetic field, and causes what is termed a *deviation* of the north and south line from the magnetic meridian. In wooden ships, by a little care in placing the compass properly, deviation errors of any practical moment may be wholly avoided, but in iron ships they must be partly allowed for and partly compensated. The deviation in an iron ship is due to two causes—1st, the permanent magnetism of the ship; 2nd, the magnetism induced by the earth's magnetic force. We can compensate for the effect of the permanent magnetism by properly placing a permanent steel magnet in the neighbourhood of the compass, exerting an equal and opposite couple to that due to the ship.

We cannot compensate or can only very imperfectly compensate for the effect of induced magnetism, because it is impracticable to arrange a soft iron structure near the compass, such that its induced magnetism shall have an opposite and equal effect to that of the ship. The induced magnetism varies as the ship turns round horizontally. Thus when she bears north or south, her magnetic moment is much greater than when east or west. By testing experimentally in port the deviation on each course, a correction is obtained for that particular neighbourhood. The ship's induced magnetism also varies, however, as the direction and intensity of the earth's magnetic force varies; and no safe allowance can be made for errors resulting from this cause. Moreover the induced magnetism varies as the ship rolls, and (to a much less extent) as she pitches. The heeling error can be compensated, as was shown by the late Mr. Archibald Smith. The Admiralty Compass Manual, written by that gentleman in concert with Captain Evans, R.N., should be consulted by all who wish to understand the mariner's compass. The mathematical and practical investigations of Mr. Smith have been of the very highest utility in adding both to our scientific knowledge and to the practical utility of the mariner's compass.

The *prismatic compass* and *azimuth compass* are compasses fitted with contrivances by which the bearings of objects can be taken.

INDEX.

Spottiswoode & Co., Printers, New-street Square, London.

TEXT-BOOKS OF SCIENCE, MECHANICAL AND PHYSICAL, ADAPTED FOR THE USE OF ARTISANS AND OF STUDENTS IN PUBLIC AND SCIENCE SCHOOLS.

Now in course of publication, in small 8vo. each volume containing about Three Hundred pages,

A SERIES OF

ELEMENTARY WORKS ON MECHANICAL AND PHYSICAL SCIENCE,

FORMING A SERIES OF

TEXT-BOOKS OF SCIENCE

ADAPTED FOR THE USE OF ARTISANS AND OF STUDENTS IN PUBLIC AND OTHER SCHOOLS.

The First Thirteen of the Series, edited by T. M. GOODEVE, M.A. Barrister-at-Law, Lecturer on Applied Mechanics at the Royal School of Mines; and the remainder by C. W. MERRIFIELD, F.R.S. an Examiner in the Department of Public Education, and late Principal of the Royal School of Naval Architecture and Marine Engineering, South Kensington.

THE PUBLICATION of the Series of Books intitled Text-Books of Science was undertaken by Messrs. LONGMANS & Co. in consequence of a belief (which was based upon the Reports of the Public Schools and the Schools Inquiry Commissions, as well as the evidence taken before several Parliamentary Committees) that a want existed of a Series of Elementary Works in the various branches of Mechanical and Physical Science suitable for general use in Schools, Colleges, and Science Classes, and for the self-instruction of Working Men.

The cordial reception given to the various volumes of this Series, as they appeared, by all sections of the Public Press, and the fact that upwards of 50,000 Text-Books have been already sold, have shewn that Messrs. LONGMANS & Co. were fully justified in their belief in the existence of such a want, and have also warranted the supposition that, as regards the subjects hitherto treated, the want has now been supplied. Encouraged by this success, Messrs. LONGMANS & Co. have *determined to extend their series, and have accordingly entered into* negotiations for the production of works on *Astronomy*, *Agricultural Chemistry*, *Chemical Philosophy*, *Botany*, *Photography* and *Mineralogy*. The greatest care will continue to be taken to invite contributions from those men only who are acknowledged to be masters of the subjects entrusted to them, and whose names will be a sufficient guarantee for the excellence of their work.

The plan of the forthcoming volumes will be similar to that of those already issued; they are intended to serve for the use of practical men, as well as for exact instruction in the subjects of which they treat; and it is hoped that, while retaining the logical clearness and simple sequence of thought which are essential to the making of a good scientific treatise, the style and subject-matter will be found to be within the comprehension of working men, and suited to their wants. The books will not be mere manuals for immediate application, nor University text-books, in which mental training is the foremost object; they are meant to be *practical treatises, sound and exact in their logic, and with every theory and every process reduced to the stage of direct and useful application, and illustrated by well-selected examples from familiar processes and facts.*

TEXT-BOOKS PREPARING FOR PUBLICATION, TO BE EDITED BY T. M. GOODEVE, M.A.

ECONOMICAL APPLICATIONS OF HEAT.

Including Combustion, Evaporation, Furnaces, Flues, and Boilers.

By C. P. B. SHELLEY, Civil Engineer, and Professor of Manufacturing Art and Machinery at King's College, London.

With a Chapter on the Probable Future Development of the Science of Heat, by C. WILLIAM SIEMENS, F.R.S.

THE STEAM ENGINE.

By T. M. GOODEVE, M.A. Barrister-at-Law, Lecturer on Mechanics at the Royal School of Mines.

THE FOLLOWING TO BE EDITED BY C. W. MERRIFIELD, F.R.S.

SOUND.

By W. G. ADAMS, M.A. Professor of Natural Philosophy and Astronomy, King's College, London.

PHYSICAL OPTICS.

By G. G. STOKES, M.A. D.C.L. Fellow of Pembroke College, Cambridge; Lucasian Professor of Mathematics in the University of Cambridge; and Secretary to the Royal Society.

ELEMENTS OF MACHINE DESIGN.

With Rules and Tables for Designing and Drawing the Details of Machinery. Adapted to the use of Mechanical Draughtsmen and Teachers of Machine Drawing.

By W. CAWTHORNE UNWIN, B. Sc. Assoc. Inst. C.E. Professor of Hydraulic and Mechanical Engineering at Cooper's Hill College.

PHYSICAL GEOGRAPHY.

By the Rev. GEORGE BUTLER, M.A. Principal of Liverpool College; Editor of 'The Public Schools Atlas of Modern Geography.'

ASTRONOMY.

By ROBERT S. BALL, LL.D. F.R.S. Andrews Professor of Astronomy in the University of Dublin, and Royal Astronomer of Ireland.

AGRICULTURAL CHEMISTRY.

By ROBERT WARINGTON.

PHOTOGRAPHY.

By Captain ABNEY, Royal Engineers, F.R.A.S. & F.C.S.

STRUCTURAL AND PHYSIOLOGICAL BOTANY.

By Dr. OTTO WILHELM THOMÉ, Ordinary Professor of Botany at the School of Science and Art, Cologne. Translated and Edited by A. W. BENNETT, M.A. B.Sc. F.L.S. Lecturer on Botany at St. Thomas's Hospital.

MINERALOGY.

By H. BAUERMAN, Museum of Practical Geology.

CHEMICAL PHILOSOPHY.

By WILLIAM A. TILDEN, D.Sc. Lond. F.C.S. Lecturer on Chemistry in Clifton College.

TEXT-BOOKS, NOW PUBLISHED, EDITED BY T. M. GOODEVE, M.A.

THE ELEMENTS OF MECHANISM.

Designed for Students of Applied Mechanics. By T. M. GOODEVE, M.A. Barrister-at-Law. Lecturer on Mechanics at the Royal School of Mines. New Edition, revised; with 257 Figures on Wood. Price 3*s.* 6*d.*

METALS, THEIR PROPERTIES AND TREATMENT.

By CHARLES LOUDON BLOXAM, Professor of Chemistry in King's College, London; Professor of Chemistry in the Department of Artillery Studies, and in the Royal Military Academy, Woolwich. With 105 Figures on Wood. Price 3*s.* 6*d.*

INTRODUCTION TO THE STUDY OF INORGANIC CHEMISTRY.

By WILLIAM ALLEN MILLER, M.D. LL.D. F.R.S. late Professor of Chemistry in King's College, London; Author of 'Elements of Chemistry, Theoretical and Practical.' New Edition, revised; with 71 Figures on Wood. Price 3*s.* 6*d.*

ALGEBRA AND TRIGONOMETRY.

By the Rev. WILLIAM NATHANIEL GRIFFIN, B.D. sometime Fellow of St. John's College, Cambridge. Price 3*s.* 6*d.*

NOTES ON THE ELEMENTS OF ALGEBRA AND TRIGONOMETRY;

With SOLUTIONS of the more difficult QUESTIONS. By the Rev. WILLIAM NATHANIEL GRIFFIN, B.D. sometime Fellow of St. John's College, Cambridge. Price 3*s.* 6*d.*

PLANE AND SOLID GEOMETRY.

By the Rev. H. W. WATSON, formerly Fellow of Trinity College, Cambridge, and late Assistant-Master of Harrow School. Price 3*s.* 6*d.*

THEORY OF HEAT.

By J. CLERK MAXWELL, M.A. LL.D. Edin. F.R.SS. L. & E. Professor of Experimental Physics in the University of Cambridge. New Edition, revised; with 41 Woodcuts and Diagrams. Price 3*s.* 6*d.*

TECHNICAL ARITHMETIC AND MENSURATION.

By CHARLES W. MERRIFIELD, F.R.S. an Examiner in the Department of Public Education, and late Principal of the Royal School of Naval Architecture and Marine Engineering, South Kensington. Price 3*s.* 6*d.*

KEY TO MERRIFIELD'S TEXT-BOOK OF TECHNICAL ARITHMETIC AND MENSURATION.

By the Rev. JOHN HUNTER, M.A. one of the National Society's Examiners of Middle-class Schools; formerly Vice-Principal of the National Society's Training College, Battersea. Price 3*s.* 6*d.*

ON THE STRENGTH OF MATERIALS AND STRUCTURES.

The Strength of Materials as depending on their quality and as ascertained by Testing Apparatus; the Strength of Structures, as depending on their form and arrangement, and on the Materials of which they are composed. By JOHN ANDERSON, C.E. LL.D. F.R.S.E. Superintendent of Machinery to the War Department. Price 3*s.* 6*d.*

ELECTRICITY AND MAGNETISM.

By FLEEMING JENKIN, F.R.SS. L. & E. Professor of Engineering in the University of Edinburgh. New Edition, revised. Price 3*s.* 6*d.*

WORKSHOP APPLIANCES.

Including Descriptions of the Gauging and Measuring Instruments, the Hand Cutting Tools, Lathes, Drilling, Planing, and other Machine Tools used by Engineers. By C. P. B. SHELLEY, Civil Engineer, Hon. Fellow and Professor of Manufacturing Art and Machinery at King's College, London. With 209 Figures on Wood. Price 3*s.* 6*d.*

PRINCIPLES OF MECHANICS.

By T. M. GOODEVE, M.A. Barrister-at-Law, Lecturer on Applied Mechanics at the Royal School of Mines. Price 3*s.* 6*d.*

TEXT-BOOKS, NOW PUBLISHED, EDITED BY C. W. MERRIFIELD, F.R.S.

INTRODUCTION TO THE STUDY OF ORGANIC CHEMISTRY.

The CHEMISTRY of CARBON and its COMPOUNDS.

By HENRY E. ARMSTRONG, Ph.D. F.C.S. Professor of Chemistry in the London Institution. With 8 Figures on Wood. Price 3*s.* 6*d.*

MANUAL OF QUALITATIVE ANALYSIS AND LABORATORY PRACTICE.

By T. E. THORPE, Ph.D. F.R.S.E. Professor of Chemistry in the Andersonian University, Glasgow; and M. M. PATTISON MUIR. Price 3*s.* 6*d.*

QUANTITATIVE CHEMICAL ANALYSIS.

By T. E. THORPE, Ph.D. F.R.S.E. Professor of Chemistry in the Andersonian University, Glasgow. With 88 Figures on Wood. Price 4*s.* 6*d.*

TELEGRAPHY.

By W. H. PREECE, C.E. Divisional Engineer, Post Office Telegraphs; and J. SIVEWRIGHT, M.A. Superintendent (Engineering Department) Post Office Telegraphs. With 160 Figures and Diagrams. Price 3*s.* 6*d.*

RAILWAY APPLIANCES;

A Description of Details of Railway Construction subsequent to the completion of the Earthworks and Masonry, including a short Notice of Railway Rolling Stock. By JOHN WOLFE BARRY, Member of the Institution of Civil Engineers. With 207 Figures and Diagrams. Price 3*s.* 6*d.*

*** *To be followed by other Works on other Branches of Science.*

London, LONGMANS & CO.

www.ingramcontent.com/pod-product-compliance
Lightning Source LLC
LaVergne TN
LVHW020104110826
845151LV00001B/52